Innovative Methods and Materials in Structural Health Monitoring of Civil Infrastructures

Innovative Methods and Materials in Structural Health Monitoring of Civil Infrastructures

Editors

Raffaele Zinno
Serena Artese

MDPI • Basel • Beijing • Wuhan • Barcelona • Belgrade • Manchester • Tokyo • Cluj • Tianjin

Editors
Raffaele Zinno
University of Calabria
Italy

Serena Artese
University of Calabria
Italy

Editorial Office
MDPI
St. Alban-Anlage 66
4052 Basel, Switzerland

This is a reprint of articles from the Special Issue published online in the open access journal *Applied Sciences* (ISSN 2076-3417) (available at: https://www.mdpi.com/journal/applsci/special_issues/Structural_Infrastructures).

For citation purposes, cite each article independently as indicated on the article page online and as indicated below:

LastName, A.A.; LastName, B.B.; LastName, C.C. Article Title. *Journal Name* **Year**, *Volume Number*, Page Range.

ISBN 978-3-0365-0754-5 (Hbk)
ISBN 978-3-0365-0755-2 (PDF)

© 2021 by the authors. Articles in this book are Open Access and distributed under the Creative Commons Attribution (CC BY) license, which allows users to download, copy and build upon published articles, as long as the author and publisher are properly credited, which ensures maximum dissemination and a wider impact of our publications.

The book as a whole is distributed by MDPI under the terms and conditions of the Creative Commons license CC BY-NC-ND.

Contents

About the Editors vii

Raffaele Zinno and Serena Artese
Innovative Methods and Materials in Structural Health Monitoring of Civil Infrastructures
Reprinted from: *Appl. Sci.* **2021**, *11*, 1140, doi:10.3390/app11031140 1

Hanwei Zhao, Youliang Ding, Satish Nagarajaiah and Aiqun Li
Longitudinal Displacement Behavior and Girder End Reliability of a Jointless Steel-Truss Arch Railway Bridge during Operation
Reprinted from: *Appl. Sci.* **2019**, *9*, 2222, doi:10.3390/app9112222 5

Kang Yang, Youliang Ding, Peng Sun, Hanwei Zhao and Fangfang Geng
Modeling of Temperature Time-Lag Effect for Concrete Box-Girder Bridges
Reprinted from: *Appl. Sci.* **2019**, *9*, 3255, doi:10.3390/app9163255 25

Stefania Fortino, Petr Hradil, Keijo Koski, Antti Korkealaakso, Ludovic Fülöp, Hauke Burkart and Timo Tirkkonen
Health Monitoring of Stress-Laminated Timber Bridges Assisted by a Hygro-Thermal Model for Wood Material
Reprinted from: *Appl. Sci.* **2021**, *11*, 98, doi:10.3390/app11010098 41

Bernardino Chiaia and Valerio De Biagi
Archetypal Use of Artificial Intelligence for Bridge Structural Monitoring
Reprinted from: *Appl. Sci.* **2020**, *10*, 7157, doi:10.3390/app10207157 63

Ekin Ozer and Maria Q. Feng
Structural Reliability Estimation with Participatory Sensing and Mobile Cyber-Physical Structural Health Monitoring Systems
Reprinted from: *Appl. Sci.* **2019**, *9*, 2840, doi:10.3390/app9142840 75

Massimiliano Pepe, Domenica Costantino and Alfredo Restuccia Garofalo
An Efficient Pipeline to Obtain 3D Model for HBIM and Structural Analysis Purposes from 3D Point Clouds
Reprinted from: *Appl. Sci.* **2020**, *10*, 1235, doi:10.3390/app10041235 93

Meho Saša Kovačević, Mario Bačić, Irina Stipanović and Kenneth Gavin
Categorization of the Condition of Railway Embankments Using a Multi-Attribute Utility Theory
Reprinted from: *Appl. Sci.* **2019**, *9*, 5089, doi:10.3390/app9235089 113

Francesco Di Stefano, Miriam Cabrelles, Luis García-Asenjo, José Luis Lerma, Eva Savina Malinverni, Sergio Baselga, Pascual Garrigues and Roberto Pierdicca
Evaluation of Long-Range Mobile Mapping System (MMS) and Close-Range Photogrammetry for Deformation Monitoring. A Case Study of Cortes de Pallás in Valencia (Spain)
Reprinted from: *Appl. Sci.* **2020**, *10*, 6831, doi:10.3390/app10196831 131

Xi Chu, Zhixiang Zhou, Guojun Deng, Xin Duan and Xin Jiang
An Overall Deformation Monitoring Method of Structure Based on Tracking Deformation Contour
Reprinted from: *Appl. Sci.* **2019**, *9*, 4532, doi:10.3390/app9214532 153

Ivan Duvnjak, Suzana Ereiz, Domagoj Damjanović and Marko Bartolac
Determination of Axial Force in Tie Rods of Historical Buildings Using the
Model-Updating Technique
Reprinted from: *Appl. Sci.* **2020**, *10*, 6036, doi:10.3390/app10176036 . . . . **173**

Serena Artese and Giovanni Nico
TLS and GB-RAR Measurements of Vibration Frequencies and Oscillation Amplitudes of Tall
Structures: An Application to Wind Towers
Reprinted from: *Appl. Sci.* **2020**, *10*, 2237, doi:10.3390/app10072237 . . . . **189**

Serena Artese and Raffaele Zinno
TLS for Dynamic Measurement of the Elastic Line of Bridges
Reprinted from: *Appl. Sci.* **2020**, *10*, 1182, doi:10.3390/app10031182 . . . . **209**

Jing Yang, Peng Hou, Caiqian Yang and Yang Zhang
Study on the Method of Moving Load Identification Based on Strain Influence Line
Reprinted from: *Appl. Sci.* **2021**, *11*, 853, doi:10.3390/app11020853 . . . . **233**

**Hoofar Shokravi, Hooman Shokravi, Norhisham Bakhary, Seyed Saeid Rahimian Koloor and
Michal Petrů**
Health Monitoring of Civil Infrastructures by Subspace System Identification Method:
An Overview
Reprinted from: *Appl. Sci.* **2020**, *10*, 2786, doi:10.3390/app10082786 . . . . **249**

About the Editors

Raffaele Zinno holds a Ph.D. in Chemical Technologies and New Materials (1995), and a M.E. in Civil Engineering (1988) from the University of Calabria (Italy). In 1992–1994 he was a visiting doctoral student at the Department of Mechanical and Aerospace Engineering, West Virginia University (Morgantown, WV, USA). From 1997 to 1998, he was a researcher at the CNR, Central Institute for Industrialization and Building Technology, Rome. Since 1998, he has been Associate Professor of Strength of Materials and Structural Engineering at the University of Calabria, Department of Environmental Engineering. In 2018, he obtained the Italian National Scientific Qualification (ASN) as full professor in his scientific area. His main research interests are: structural health monitoring (SHM); structural uses of wood and related experiments; theoretical analysis of structures in composite material (problems of damage in composite material, bending and buckling of structures in ordinary and advanced materials, etc.); experiments on innovative materials (composites, fiber-reinforced, or high-performance concretes, self-diagnostics, smart, etc.), and on structures made with them; design of structures in innovative materials; and problems in the dynamic analyses of structures. He founded and directs the Structural Monitoring, structural Advanced materials, structural Rehabilitation, structural Testing Laboratory (SMART Lab) at the University of Calabria.

Serena Artese holds a Ph.D. and M.S. in Building and Architectural Engineering. In 2014, she was a visiting doctoral student at the Universitat Politècnica de València (UPV), Spain, Departamento de Ingeniería Cartográfica, Geodesia y Fotogrametría (DICGF). She is an Adjunct Professor of Geomatics at the University of Calabria, Italy. She holds the National Scientific Qualification (ASN) as an Associate Professor of Geomatics and is a founding member of Spring Research s.r.l. Spin-Off of the University of Calabria. Her research interests include the monitoring of buildings, infrastructure, and landslides using new methodologies and sensors; geomatics and surveying techniques; 3D modeling; and GIS applied to emergency and cultural heritage. She has participated as a speaker in several international conferences and is the author or coauthor of several articles and conference papers.

Editorial

Innovative Methods and Materials in Structural Health Monitoring of Civil Infrastructures

Raffaele Zinno * and Serena Artese

Department of Enviromental Engineering, University of Calabria, Ponte Bucci, Cubo 44/A, 87036 Arcavacata di Rende (CS), Italy; serena.artese@unical.it
* Correspondence: raffaele.zinno@unical.it

Citation: Zinno, R.; Artese, S. Innovative Methods and Materials in Structural Health Monitoring of Civil Infrastructures. *Appl. Sci.* **2021**, *11*, 1140. https://doi.org/10.3390/app11031140

Received: 13 January 2021
Accepted: 24 January 2021
Published: 26 January 2021

Publisher's Note: MDPI stays neutral with regard to jurisdictional claims in published maps and institutional affiliations.

Copyright: © 2021 by the authors. Licensee MDPI, Basel, Switzerland. This article is an open access article distributed under the terms and conditions of the Creative Commons Attribution (CC BY) license (https://creativecommons.org/licenses/by/4.0/).

Starting from the early years of the 21st century, the problem of monitoring the physical–chemical–mechanical conditions of structures and infrastructures for civil use began to be thought in a significantly different way than in the last century.

Since the appearance of the most common materials for civil use, in particular, reinforced concrete and steel, the structures made with them have been thought of almost as having an infinite life and subjected to checks only occasionally and, almost always, after traumatic events that could have strongly damaged them, such as earthquakes or landslides.

In the last twenty years, however, it has been realized that these materials have a long life but not indefinitely long. It has begun to be understood more and more that, as with many materials of the past, for example wood, carrying out correct and methodical maintenance is important and can greatly extend the useful life of a structure or infrastructure. This involves enormous economic gain for society, by also reducing environmental pollution, caused by demolished materials and/or materials needed for new buildings.

The constructions of the last century, however, are much more complex than the relatively simple masonry and wood structures of the years prior to the industrial revolution of the 19th century, although some exceptions were present at the beginning of the 19th century [1]. Skyscrapers, suspension and cable-stayed bridges, dams, particular industrial buildings, etc. they do not have easy maintenance and, often, the points where the degradation is concentrated are not immediately identifiable, also in consideration of the different and complex types of load to which they are subjected, often of a dynamic nature such as vehicular traffic, earthquakes, wind, etc., as well as environmental effects (particularly corrosive atmospheres, humid environments, etc.).

For large structures, especially publicly owned ones, an attempt was made to define a periodic control and monitoring plan, with maintenance interventions also scheduled. Despite this, both due to phenomena of scarce funding and bad use of funds, this programming has often not been correctly implemented and there have also been collapses that have produced numerous human victims (for those who know and frequent Italy, it is sadly famous the collapse of the Morandi bridge in Genoa in August 2018).

For smaller structures or for those of private property, then, the maintenance of this programming is even more complicated and difficult to perform, also because the owners do not always have the skills to perceive the danger that a damaged structure can entail.

All the arguments previously described have led to the modern concept of structural health and the need for its monitoring, similarly to what happens for human health. Indeed, considering the rapid and incisive diffusion of the concept, the acronym SHM has been introduced to briefly indicate Structural Health Monitoring, which is an interdisciplinary subject that incorporates synergistic knowledge and experience technologies in civil, mechanical, control and computer engineering.

In the first instance SHM foresees the detection of any structural damage, alerting the technicians in charge of maintenance to the need for inspection and repair and/or reinforcement. Very often, SHM systems stop at this level, but more and more frequently they begin to predict the localization and quantification of damage. The thing is not easy and it is almost always necessary to introduce artificial intelligence techniques, such as neural networks or genetic algorithms, whose "training" is obtained by simulating the behavior of the structure in the presence of various damages using computational mechanics. Finally, also for the purpose of planning maintenance and evaluating the compatibility of its economic cost, there is the prevision of the remaining life. In summary, SHM includes: (1) detection, (2) localization, (3) quantification and (4) prevision of remaining life.

Being clear that the SHM is multidisciplinary, the guest editors proposed to the scientific community to contribute to this Special Issue of *Applied Sciences* international journal, discussing the innovations relating to the methods and materials that allow to evaluate the health status of a structure or infrastructure, limited to those of civil use.

Several topics are related to the Special Issue, such as intrinsically diagnostic materials [2], or composites [3], modeling of structures using finite element methods [4], and recently SHM connected with IoT [5]. Problems of surveying structures, methods useful for remote monitoring of infrastructures, in particular, bridges [6] or works of historical-monumental importance were also addressed [7]. Recently, the guest editors have also started to evaluate the economic effects of the SHM, in particular on the estimated value of residential buildings [8].

Since SHM is interdisciplinary, several research fields are involved. Among them we can emphasize the following ones:

(a) The Computational Mechanics. In fact, it is important to simulate the intact and damaged structure in order to identify the presence of the damage and evaluate its position and entity. Without it, the cases that an Artificial Intelligence method uses to train oneself to interpret the data coming from the sensors to arrive at the correct definition of the damage, could hardly be populated. Fundamental concepts in Elasticity, Mechanics and Wave Propagation, Signal Processing, Application of the Finite Element Method in SHM, Spectral Finite Element Method, are some of the important aspects for SHM of competence of this discipline [9].

(b) The Theoretical Analysis of Structures, especially the dynamic one. It is necessary to develop static and dynamic analysis methods, adapting well-known theories to be able to simulate the behavior of a structure and to interpret the data coming from the sensors, which measure static quantities (translations and rotations, temperatures, inclinations, etc.) or dynamic ones (accelerations, vibrations, etc.) [10,11].

(c) The New Materials. There are self-diagnostic materials that can become sensors themselves and make it possible to identify the damage. Furthermore, materials capable of self-repairing and materials that allow a structure to change shape or to react to stresses are appearing (i.e., smart, adaptative and intelligent materials). In the near future, these materials will be of great importance in structural monitoring [12].

(d) The Electronics, Sensors, often in connection with the Internet of Things (IoT), in the measurement of mechanical and structural characteristics. It is important to develop new accelerometers, less intrusive and expensive, possibly connected wirelessly. However, also classic sensors such as strain gauges, inclinometers, LVDTs, etc., must be rethought as a function of modern monitoring, hopefully to be carried out remotely and for a large number of structures [13].

(e) The Artificial Intelligence (Neural Networks, Genetic Algorithms, Machine Learning, etc.) that now pervades most of the technological applications and that allows to make the structure as well as smart even cognitive, that is, able to predict future damage, thus significantly helping scheduled maintenance [14].

(f) Finally, Geomatics and surveying techniques in the continuous and discontinuous monitoring of structures. Today, monitoring can make use of modern techniques for measuring displacements and other geometric quantities, such as detection and measurement by using radio waves (RADAR), or light (LiDAR), photogrammetry, multispectral satellite images, radar synthetic aperture (SAR) [15], infrared thermography, and image analysis methods, including digital image correlation (DIC), ground penetration radar (GPR) and remote acoustics [16]. Another important issue is the classification of buildings and neighborhoods for carrying out emergency plans [17,18].

About the Present Special Issue

Several interesting contributions are proposed in this special issue. The influence of temperature on the displacements of bridge and their structural elements is analyzed by Zhao et al. [19] and by Yang et al. [20]. The effect of humidity is taken into account for timber bridges by Fortino et al. [21].

The Structural Health Monitoring Systems is performed with the Artificial Intelligence by Chiaia and De Biagi [22], while Ozer and Feng propose the Participatory Sensing and Mobile Cyber-Physical [23].

Pepe et al. present a pipeline to obtain 3D Model for HBIM from 3D Point Clouds [24].

An application concerning the GPR survey for infrastructure maintenance is developed by Kovačević et al. [25].

Applications in the field of geomatics are also proposed. The assessment of the structural health status can be performed with different geomatic techniques by means of static measurements [26–28] and dynamic ones [29–31].

The latest contribution is a review of the Subspace System Identification (SSI) method for health monitoring [32].

Author Contributions: Conceptualization, R.Z. and S.A.; writing—original draft preparation, R.Z. and S.A.; writing—review and editing, R.Z. and S.A.; supervision, R.Z. All authors have read and agreed to the published version of the manuscript.

Funding: This research received no external funding.

Acknowledgments: The guest editors of this Special Issue thank all the Authors of the published notes who have certainly offered a contribution to establish the state of the art on the subject and to identify new challenges for the near future, relating to all the aspects listed above, for the common purpose of constantly assessing the structural health of civil infrastructures. The guest editors express their gratitude for the kind assistance of the editorial office, especially for Luca Shao (Managing Editor) for his valuable contribution.

Conflicts of Interest: The authors declare no conflict of interest.

References

1. Ruggieri, N.; Tampone, G.; Zinno, R. Typical failures, seismic behavior and safety of the "Bourbon system" with timber framing. *Adv. Mater. Res.* **2013**, *778*, 58–65. [CrossRef]
2. Chiarello, M.; Zinno, R. Electrical conductivity of self-monitoring CFRC. *Cem. Concr. Compos.* **2005**, *27*, 463–469. [CrossRef]
3. Bruno, D.; Lato, S.; Zinno, R. Nonlinear analysis of doubly curved composite shells of bimodular material. *Compos. Eng.* **1993**, *3*, 419–435. [CrossRef]
4. Zinno, R.; Barbero, E.J. A three-dimensional layer-wise constant shear element for general anisotropic shell-type structures. *Int. J. Numer. Methods Eng.* **1994**, *37*, 2445–2470. [CrossRef]
5. Zinno, R.; Artese, S.; Clausi, G.; Magarò, F.; Meduri, S.; Miceli, A.; Venneri, A. Structural Health Monitoring (SHM). In *The Internet of Things for SmartUrban Ecosystems*; Springer International Publishing: Cham, Switzerland, 2019; pp. 225–249. [CrossRef]
6. Artese, S.; Achilli, V.; Zinno, R. Monitoring of Bridges by a Laser Pointer: Dynamic Measurement of Support Rotations and Elastic Line Displacements: Methodology and First Test. *Sensors* **2018**, *18*, 338. [CrossRef]
7. Artese, S.; Lerma, J.L.; Aznar Molla, J.; Sánchez, R.M.; Zinno, R. Integration of surveying techniques to detect the ideal shape of a dome: The case of the Escuelas Pías Church in Valencia. *Int. Arch. Photogramm. Remote Sens. Spat. Inf. Sci.* **2019**, *XLII-2/W9*, 39–43. [CrossRef]
8. Arcuri, N.; De Ruggiero, M.; Salvo, F.; Zinno, R. Automated valuation methods through the cost approach in a BIM and GIS integration framework for smart city appraisals. *Sustainability* **2020**, *12*, 7546. [CrossRef]
9. Gopalakrishnan, S.; Ruzzene, M.; Hanagud, S. *Computational Techniques for Structural Health Monitoring*; Springer Series in Reliability Engineering; Springer: Cham, Switzerland, 2011; ISBN 978-0-85729-284-1.
10. Sarah, J.; Hejazi, F.; Rashid, R.S.M.; Ostovar, N. A review of dynamic analysis in frequency domain for structural health monitoring. *IOP Conf. Ser. Earth Environ. Sci.* **2019**, *357*, 012007. [CrossRef]
11. Zagari, G.; Zucco, G.; Madeo, A.; Ungureanu, V.; Zinno, R.; Dubina, D. Evaluation of the erosion of critical buckling load of cold-formed steel members in compression based on Koiter asymptotic analysis. *Thin-Walled Struct.* **2016**, *108*, 193–204. [CrossRef]

12. Han, B.; Ding, S.; Yu, X. Intrinsic self-sensing concrete and structures: A review. *Meas. J. Int. Meas. Confed.* **2015**, *59*, 110–128. [CrossRef]
13. Tokognon, C.A., Jr.; Gao, B.; Yun Tian, G.; Yan, Y. Structural Health Monitoring Framework Based on Internet of Things: A Survey. *IEEE Internet Things J.* **2017**, *4*, 619–635. [CrossRef]
14. Bao, Y.; Chen, Z.; Wei, S.; Xu, Y.; Tang, Z.; Li, H. The State of the Art of Data Science and Engineering in Structural Health Monitoring. *Engineering* **2019**, *5*, 234–242. [CrossRef]
15. Artese, G.; Fiaschi, S.; Di Martire, D.; Tessitore, S.; Fabri, M.; Achilli, V.; Ahmed, A.; Borgstrom, S.; Calcaterra, D.; Ramondini, M.; et al. Monitoring of land subsidence in ravenna municipality using integrated SAR—GPS techniques: Description and first results. *Int. Arch. Photogramm. Remote Sens. Spat. Inf. Sci. ISPRS Arch.* **2016**, *41*, 23–28. [CrossRef]
16. Barrile, V.; Fotia, A.; Leonardi, G.; Pucinotti, R. Geomatics and Soft Computing Techniques for Infrastructural Monitoring. *Sustainability* **2020**, *12*, 1606. [CrossRef]
17. Francini, M.; Artese, S.; Gaudio, S.; Palermo, A.; Viapiana, M.F. To support urban emergency planning: A GIS instrument for the choice of optimal routes based on seismic hazards. *Int. J. Disaster Risk Reduct.* **2018**, *31*, 121–134. [CrossRef]
18. Artese, S.; Achilli, V. A gis tool for the management of seismic emergencies in historical centers: How to choose the optimal routes for civil protection interventions. *Int. Arch. Photogramm. Remote Sens. Spat. Inf. Sci.* **2019**, *XLII-2/W11*, 99–106. [CrossRef]
19. Zhao, H.; Ding, Y.; Nagarajaiah, S.; Li, A. Longitudinal Displacement Behavior and Girder End Reliability of a Jointless Steel-Truss Arch Railway Bridge during Operation. *Appl. Sci.* **2019**, *9*, 2222. [CrossRef]
20. Yang, K.; Ding, Y.; Sun, P.; Zhao, H.; Geng, F. Modeling of Temperature Time-Lag Effect for Concrete Box-Girder Bridges. *Appl. Sci.* **2019**, *9*, 3255. [CrossRef]
21. Fortino, S.; Hradil, P.; Koski, K.; Korkealaakso, A.; Fülöp, L.; Burkart, H.; Tirkkonen, T. Health Monitoring of Stress-Laminated Timber Bridges Assisted by a Hygro-Thermal Model for Wood Material. *Appl. Sci.* **2021**, *11*, 98. [CrossRef]
22. Chiaia, B.; De Biagi, V. Archetypal Use of Artificial Intelligence for Bridge Structural Monitoring. *Appl. Sci.* **2020**, *10*, 7157. [CrossRef]
23. Ozer, E.; Feng, M.Q. Structural Reliability Estimation with Participatory Sensing and Mobile Cyber-Physical Structural Health Monitoring Systems. *Appl. Sci.* **2019**, *9*, 2840. [CrossRef]
24. Pepe, M.; Costantino, D.; Restuccia Garofalo, A. An Efficient Pipeline to Obtain 3D Model for HBIM and Structural Anaysis Purposes from 3D Point Clouds. *Appl. Sci.* **2020**, *10*, 1235. [CrossRef]
25. Kovačević, M.S.; Bačić, M.; Stipanović, I.; Gavin, K. Categorization of the Condition of Railway Embankments Using a Multi-Attribute Utility Theory. *Appl. Sci.* **2019**, *9*, 5089. [CrossRef]
26. Di Stefano, F.; Cabrelles, M.; García-Asenjo, L.; Lerma, J.L.; Malinverni, E.S.; Baselga, S.; Garrigues, P.; Pierdicca, R. Evauation of Long-Range Mobile Mapping System (MMS) and Close-Range Photogrammetry for Deformation Monitoing. A Case Study of Cortes de Pallás in Valencia (Spain). *Appl. Sci.* **2020**, *10*, 6831. [CrossRef]
27. Chu, X.; Zhou, Z.; Deng, G.; Duan, X.; Jiang, X. An Overall Deformation Monitoring Method of Structure Based on Tracking Deformation Contour. *Appl. Sci.* **2019**, *9*, 4532. [CrossRef]
28. Duvnjak, I.; Ereiz, S.; Damjanović, D.; Bartolac, M. Determination of Axial Force in Tie Rods of Historical Buildings Using the Model-Updating Technique. *Appl. Sci.* **2020**, *10*, 6036. [CrossRef]
29. Artese, S.; Nico, G. TLS and GB-RAR Measurements of Vibration Frequencies and Oscillation Amplitudes of Tall Structures: An Application to Wind Towers. *Appl. Sci.* **2020**, *10*, 2237. [CrossRef]
30. Artese, S.; Zinno, R. TLS for Dynamic Measurement of the Elastic Line of Bridges. *Appl. Sci.* **2020**, *10*, 1182. [CrossRef]
31. Yang, J.; Hou, P.; Yang, C.; Zhang, Y. Study on the Method of Moving Load Identification based on Strain Influence Line. *Appl. Sci.* **2021**, *11*, 853. [CrossRef]
32. Shokravi, H.; Shokravi, H.; Bakhary, N.; Rahimian Koloor, S.S.; Petrů, M. Health Monitoring of Civil Infrastructures by Subspace System Identification Method: An Overview. *Appl. Sci.* **2020**, *10*, 2786. [CrossRef]

Article

Longitudinal Displacement Behavior and Girder End Reliability of a Jointless Steel-Truss Arch Railway Bridge during Operation

Hanwei Zhao [1], Youliang Ding [1,*], Satish Nagarajaiah [2] and Aiqun Li [3]

1. Key Laboratory of C&PC Structures of the Ministry of Education, Southeast University, Nanjing 210096, China; wudizhw_0@126.com
2. Department of Civil, Environmental Engineering and Mechanical Engineering, Rice University, Houston, TX 77005, USA; nagaraja@rice.edu
3. Beijing Advanced Innovation Center for Future Urban Design, Beijing University of Civil Engineering and Architecture, Beijing 100044, China; liaiqun@bucea.edu.cn
* Correspondence: civilding@seu.edu.cn

Received: 4 May 2019; Accepted: 28 May 2019; Published: 30 May 2019

Abstract: The long length and complex service load form conflicts with the low limits of longitudinal and transverse displacements of jointless bridge design. The longitudinal displacements of the Nanjing Dashengguan Yangtze River Bridge, a jointless steel-truss arch railway bridge, and its girder end reliability are investigated in this article. The time–frequency characteristics of the longitudinal displacements of bearings and expansion joints are analyzed using the empirical wavelet transform. The long-term characteristics of the longitudinal displacements of bearings and expansion joints in the operation period are explored. Furthermore, the relative transverse displacements of the bridge girder end are calculated using longitudinal displacement monitoring data. The mechanical behaviors of the expansion device under relative transverse displacements are studied. The reliability of expansion devices and crossing trains under the effects of relative transverse displacements is studied using kernel density estimation. The main results demonstrate that: (1) The longitudinal displacements of bearings and expansion joints are mainly influenced by environmental temperature. (2) The maximum relative transverse displacement of the expansion joint is close to 1 mm in long-term bridge operation, with the transverse rail deflection at the expansion device approaching 1 mm, which reduces the stability of cross high-speed trains.

Keywords: structural health monitoring; jointless bridge; high-speed railway; bearing; expansion device; displacement analysis

1. Introduction

High-speed railways have provided remarkable mobility in densely populated areas, with bridges playing a vital role in their design. Long-span bridges help high-speed railway lines cross large rivers, deep valleys, bays, and other natural barriers [1]. To avoid the adverse impacts of expansion joints on crossing trains, high-speed bridges can only have expansion joints at the girder ends of the main bridge structures. The long length and complex temperature field of the jointless bridge are in conflict with the low limits required for temperature-induced displacements. Hence, the temperature effects on bridge displacements—especially at the girder end—must be investigated, which is the focus of this article.

The study of the temperature effects on long-span bridges is an important focus of civil engineers, who wish to master the behavior of bridge structures under both day–night and seasonal temperature changes. Some researchers have investigated the temperature behavior of the bridge structure using finite element (FE) simulations [2,3]. Considering the complicated nature of the actual temperature field in a bridge during operation, additional studies have attempted to analyze the thermal effects on bridge structures based on measured data and long-term monitoring data [4–9]. Furthermore, scholars have explored the behavior and performance of bridge structures under diurnal temperature loads via the validation of FE results using monitoring data [10–12]. Analysis of the temperature-induced behavior of bridge structures has solved a variety of engineering problems for many major infrastructure systems by employing such continuous studies.

However, such studies have typically focused on the mechanical behavior of the girder end under service loads, especially under temperature loads, which are typically minimal [13,14]. With the advancement of displacement-measurement technology [15,16] and time–frequency analytical techniques [17–20], the time–frequency and long-term characteristics of bridges can now be analyzed based on structural health monitoring (SHM) data. Due to the redundancy and complexity of train-bridge systems [21–23], girder expansion on the longitudinal direction must be unimpeded, and the transverse dislocation between two girder ends (the main bridge and ramp bridge) is not allowed.

Based on long-term SHM data, this article investigates the longitudinal behaviors and girder end reliability of a jointless steel-truss arch railway bridge in operation. The time–frequency and long-term characteristics of the longitudinal displacements of the bearings and expansion joints are analyzed. The influence of the bridge's longitudinal displacements due to the environmental temperature is demonstrated based on the monitoring data. The wear lifetimes of the bearings caused by cumulative longitudinal sliding displacements are predicted. The relative transverse displacements of the bridge girder end are calculated using the monitoring data from the longitudinal displacements. The mechanical behaviors of the expansion device under the relative transverse displacements are simulated via FE modeling. The fatigue reliability of the expansion devices and the reliability of crossing trains under the in-service relative transverse displacements are studied, which can help bridge engineers to maintain this high-speed railway bridge during diurnal operation.

2. Bridge Description and Longitudinal Displacement Monitoring

The Nanjing Dashengguan Yangtze River Bridge is the longest six-track high-speed railway bridge in the world. As shown in Figure 1, the bridge consists of two continuous steel-truss arches and four approach spans, without expansion joints over the total length of 1272 m. The heights of the truss ribs in the main spans vary from 12 m at the crown to 96 m at the spring line. The bridge has six tracks, including two tracks on the downstream side for the Beijing–Shanghai (B–S) high-speed railway, two tracks on the upstream side for the Shanghai–Wuhan–Chengdu (S–W–C) railway and two tracks on the outer sides of the bridge deck for the Nanjing Metro. The design load of the six tracks is greater than 600 kN/m along the longitudinal direction of the bridge. As of 2018, the train speed on the two tracks of the B–S high-speed railway have been as great as 350 km/h, with all six tracks now in operation. The bridge structure features three main trusses spaced 15 m apart in the transverse direction. Three specially designed ball-steel bearings are used on each pier. At Pier 4, the center truss rests on a fixed bearing, and the bearings for the two side trusses allow for transverse motion; on the other piers, the bearings for the center truss allow for longitudinal motion, while the bearings for the side trusses allow for both longitudinal and transverse motion. A total of eight specially designed expansion devices are installed at the expansion joint of the girder ends at Piers 1 and 7 (Figure 2), where each railway track (B–S and S–W–C) is installed and with one device at the Beijing/Shanghai girder end. The stock rail and the fixed side of the expansion device are located at the main bridge, while the switch rail and the free side of the expansion device are located at the ramp bridge, with the stock rail spanning the expansion joint. On the free side of the expansion device, the expansion box and sliding support, the sleeper and rail, the sleeper and guide-rail can slide relative to each other in the longitudinal direction of the bridge. The ramp bridge on the Beijing side uses a 30-meter-wide steel-truss bridge (and a 30-meter-wide steel girder), while the ramp bridge on the Shanghai side uses two independent 15-meter-wide concrete girders.

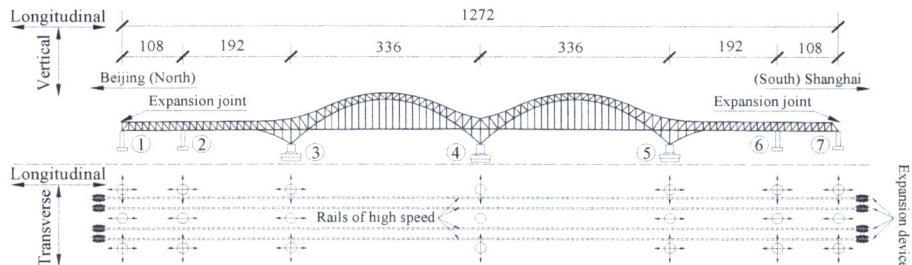

Figure 1. Bridge and its bearings and expansion devices (units: meters).

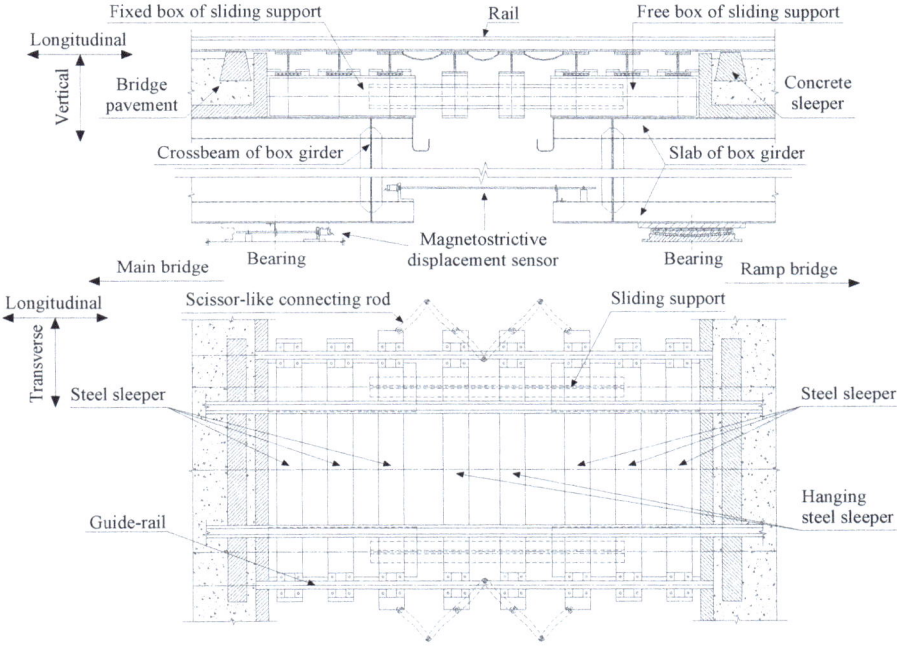

Figure 2. Expansion devices and bearings at Pier 7.

Because of the unusual characteristics of the Nanjing Dashengguan Yangtze River Bridge, including the long length of the girder without any expansion joints, and the fact that this bridge is used by high-speed trains, a long-term SHM system was installed in 2012. Considering the significant need for safety in high-speed train systems, contact sensors cannot be used on the rails and rail-related devices. To monitor the service state of the bridge, magnetostrictive displacement sensors are employed to measure the longitudinal displacements of the bearings and expansion joints. The locations of longitudinal displacement measurement are shown in Figure 3. The downstream and upstream bearings on Piers 1–3 and 5–7 are measured relative to the longitudinal displacements between the upper and lower plates of the bearing. The expansion joints on Piers 1 and 7 are measured relative to the longitudinal displacements between the main bridge and the ramp bridge. The sampling rate of the magnetostrictive displacement sensors is 1 Hz. The magnetostrictive displacement sensors were installed on the expansion joints in 2017. As shown in Figure 3, all of the sensors are 1 m from the center line of Truss 1, 2, or 3. Moreover, an atmospheric thermo-hygrometer is employed at the arch foot between the southern middle span and the southern side span to measure the atmospheric temperature and humidity, with the sampling rate of the atmospheric thermo-hygrometer being 1 Hz.

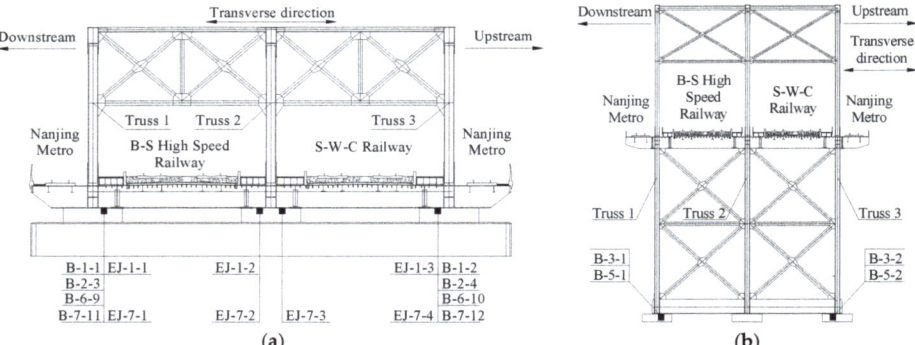

Figure 3. Locations of the magnetostrictive displacement sensors: (**a**) Pier 1 (2, 6, 7); (**b**) Pier 3 (5).

3. Behavior Analysis of the Bridge Longitudinal Displacements

Long-span jointless railway bridges are affected by temperature, trains, and any other service loads. The analysis of a bridge's longitudinal displacements under service loads can help scholars and engineers master the normal load-response behaviors of the bridge for diurnal monitoring, which is the premise for ensuring unimpeded longitudinal expansion of the bridge.

3.1. Time–Frequency Characteristics of the Longitudinal Displacements of the Bridge

A bridge structure is subject to multiple loads during diurnal operation. The time–frequency analysis of bridge responses can help engineers capture the input–output mechanism of the bridge and then determine the main factors of influence in terms of the bridge responses.

Empirical wavelet transform (EWT) is a new self-adaptive time–frequency decomposition method that was first proposed by Gilles [24]. In contrast to the empirical mode decomposition (EMD) which lacks strict mathematical derivations and requires a relatively long calculation time [25], the EWT inherits the adaptivity of the EMD and the mathematical theory of wavelet transformation. The EWT adaptively segments the Fourier spectrum by extracting the maximum point in the frequency domain to separate the signal into the different modes and then constructs adaptive bandpass filters in the frequency domain to establish orthogonal wavelet functions and extract amplitude modulation–frequency modulation (AM–FM) components that have a compact support Fourier spectrum. The principle of EWT for a signal is as follows:

Assume that the signal consists of N AM–FM components as

$$f(t) = \sum_{i=0}^{N-1} f_i(t) \qquad (1)$$

Standardizing the Fourier spectral range of the signal to $[0, \pi]$, $N + 1$ boundaries are needed to divide the Fourier spectral range $[0, \pi]$ into N intervals for all of the components. Excluding the spectral boundaries 0 and π, $N - 1$ boundaries must be determined. (ω_{n-1}, ω_n) is the boundary of the nth interval ($\omega_0 = 0$, $\omega_N = \pi$), $1 \leq n \leq N - 1$. The first nth maxima are in the corresponding nth interval. Define a transition with a width $T_n = 2\tau_n$ centring on each ω_n, with the order $\tau_n = \gamma \omega_n$, $\gamma = \min((\omega_{n+1} - \omega_n)/(\omega_{n+1} + \omega_n))_n$. After determining the interval (ω_{n-1}, ω_n), the empirical wavelet is defined as the bandpass filter on each interval. Based on the *Meyer* wavelet, the empirical wavelet function $\hat{\psi}_n(\omega)$ and the empirical scale function $\hat{\phi}_n(\omega)$ are defined as

$$\hat{\psi}_n(\omega) = \begin{cases} 1, & (1+\gamma)\omega_n \leq |\omega| \leq (1-\gamma)\omega_{n+1} \\ \cos\left[\frac{\pi}{2}\beta\left(\frac{1}{2\gamma\omega_{n+1}}(|\omega| - (1-\gamma)\omega_{n+1})\right)\right], & (1-\gamma)\omega_{n+1} \leq |\omega| \leq (1+\gamma)\omega_{n+1} \\ \sin\left[\frac{\pi}{2}\beta\left(\frac{1}{2\gamma\omega_n}(|\omega| - (1-\gamma)\omega_n)\right)\right], & (1-\gamma)\omega_n \leq |\omega| \leq (1+\gamma)\omega_n \\ 0, & \text{other} \end{cases} \quad (2)$$

$$\hat{\phi}_n(\omega) = \begin{cases} 1, |\omega| \leq (1-\gamma)\omega_n \\ \cos\left[\frac{\pi}{2}\beta\left(\frac{1}{2\gamma\omega_n}(|\omega| - (1-\gamma)\omega_n)\right)\right], (1-\gamma)\omega_n \leq |\omega| \leq (1+\gamma)\omega_n \\ 0, \text{other} \end{cases} \quad (3)$$

where $\hat{\bullet}$ denotes a Fourier transform, and $\beta(x)$ can be an arbitrary function such that

$$\beta(x) = \begin{cases} 0, & \text{if } x \leq 0 \\ & \text{and } \beta(x) + \beta(1-x) = 1, \forall x \in [0,1] \\ 1, & \text{if } x \geq 1 \end{cases} \quad (4)$$

In accordance with the work of Gilles (2013), this article uses $\beta(x) = x^4 (35 - 84x + 70x^2 - 20x^3)$.

Referring to the construction method of the classic wavelet transform, the detail coefficients $W\varepsilon f(n,t)$ and the approximation coefficients $W\varepsilon f(0,t)$ of the empirical wavelet transform can be calculated from the inner product as

$$W_f^\varepsilon(n,t) = \langle f, \psi_n \rangle = \int f(\tau)\overline{\psi_n(\tau - t)}d\tau = \left(f(\omega)\overline{\hat{\psi}_n(\omega)}\right)^\vee \quad (5)$$

$$W_f^\varepsilon(0,t) = \langle f, \phi_1 \rangle = \int f(\tau)\overline{\phi_1(\tau - t)}d\tau = \left(f(\omega)\overline{\hat{\phi}_1(\omega)}\right)^\vee \quad (6)$$

where $\overline{\bullet}$ denotes the complex conjugate; $\bullet^\vee$ denotes the inverse Fourier transform; and $\langle \bullet, \bullet \rangle$ denotes the inner product.

Then, the reconstruction of signal $f(t)$ can be expressed as

$$f(t) = W_f^\varepsilon(0,t)^*\phi_1(t) + \sum_{n=1}^N W_f^\varepsilon(n,t)^*\psi_n(t) = \left(\hat{W}_f^\varepsilon(0,\omega)\phi_1(\omega) + \sum_{n=1}^N W_f^\varepsilon(n,\omega)\psi_n(\omega)\right)^\vee \quad (7)$$

where $\bullet^*$ denotes a convolution.

As a result, the AM–FM components $f_i(t)$ (in Equation (1)) decomposition by EWT can be expressed as

$$f_0(t) = W_f^\varepsilon(0,t)^*\phi_1(t) \quad (8)$$

$$f_k(t) = W_f^\varepsilon(k,t)^*\psi_k(t), k = 1,2,3,\cdots,N-1 \quad (9)$$

Using EWT (Equation (1) to Equation (9)), a monitoring sequence can be smart decomposed to several sequences in different frequency bands. Then, the bridge response can be self-adaptively decomposed to different sequences which correspond to different load effects. For example, the temperature-induced response is in the low frequency band and the train-induced response is in the high frequency band.

Figure 4 shows the atmospheric temperature trend of the bridge and its power density over the course of one day in 2017, where the main frequency of atmospheric temperature data in that day is less than 0.0076 Hz according to power density analysis. Figure 5 shows the longitudinal displacement data and the low-frequency portion of the data for the bearings and expansion joints at the south girder end (Pier 7) for the bridge in the same day. The low-frequency portion is obtained using the frequency domain filter, the low pass Finite Impulse Response (FIR) filter and the EWT, respectively. The low-frequency domain of the frequency domain filter and the low-pass FIR filter is set to 0~0.0076 Hz.

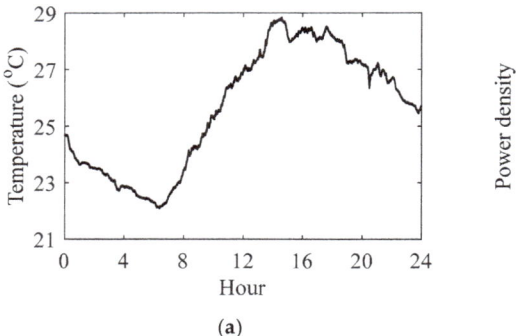
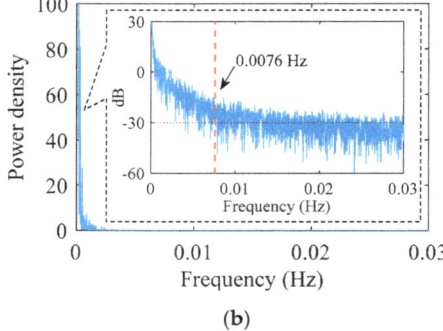

Figure 4. Atmospheric temperature and its frequency spectrum over the course of one day: (**a**) time history; (**b**) power density.

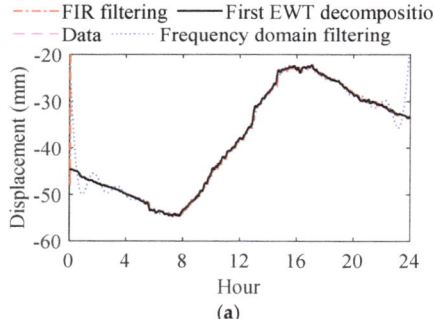
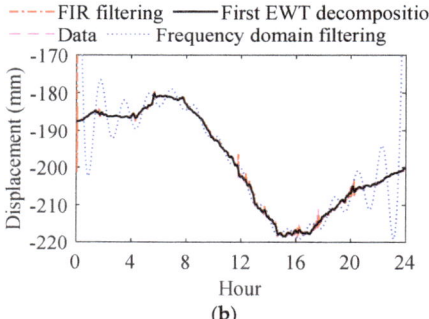

Figure 5. Longitudinal displacements and the associated low-frequency portion at the south girder end over the course of one day: (**a**) bearing (B-7-11); (**b**) expansion joint (EJ-7-1).

From Figure 5, the EWT can obtain the low-frequency portion of the signal more efficiently than either the frequency domain filter or the low pass FIR filter. Specifically, the frequency domain filter distorts the beginning and end of the signal, and the low-pass FIR filter distorts the beginning of the signal if the signal does not begin at zero. According to the atmospheric temperature trend in Figure 4 and the EWT results in Figure 5, the longitudinal displacements of both the bearings and the expansion joints are correlated with environmental temperature, although the correlation is not absolutely consistent.

Figure 6 shows the results of the EWT decomposition of the longitudinal displacements of the bearings and expansion joints at the south girder end (Pier 7) of the bridge for the same day. The adaptive frequency boundaries of the EWT decomposition in Figure 6a are (0, 0.0076), (0.0076, 1.0844), (1.0844, 0.5); the adaptive frequency boundaries of the EWT decomposition in Figure 6b are (0, 0.0077), (0.0077, 0.0795), (0.0795, 0.1722), (0.1722, 0.4680), (0.4680, 0.5). The EWT is a self-adaptive signal decomposition method that depends on the power density results of a given data set. The different frequency boundaries of the bearing and the expansion joint means the two signals have different frequency spectrum characteristics. The frequency spectrum of the bearing signal has three main crests, while the frequency spectrum of the expansion joint signal has five main crests.

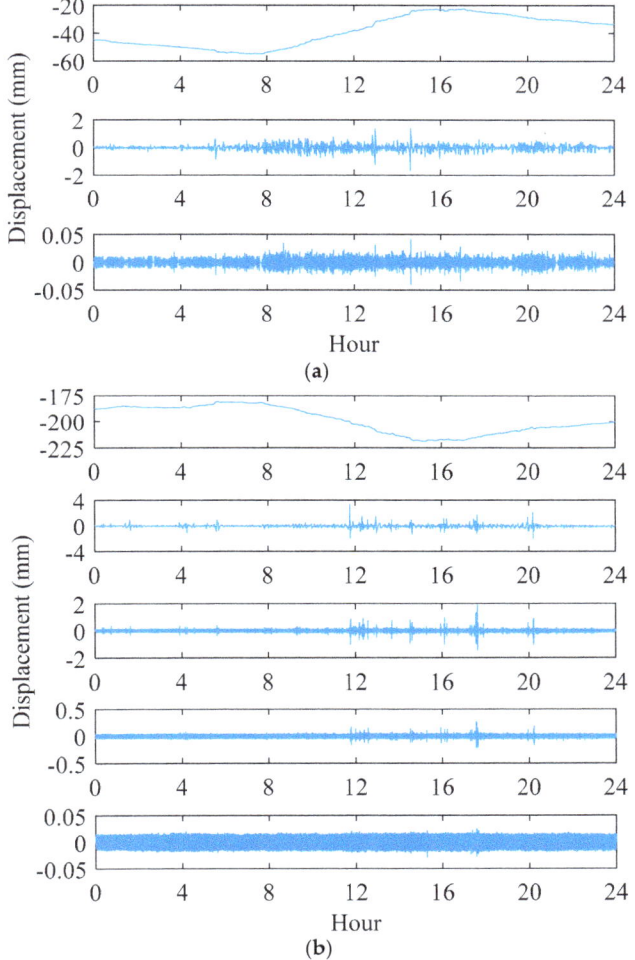

Figure 6. Results of the empirical wavelet transform (EWT) decomposition of the longitudinal displacements at the south girder end over the course of one day: (**a**) bearing (B-7-11); (**b**) expansion joint (EJ-7-1).

As shown in Figure 6a, the initial decomposition part of the longitudinal displacements of the bearings is due to the environmental temperature, whereas the second portion exhibits a nonstationary signal, predominantly between 06:00–24:00 h during that same day, which is consistent with the times during which trains cross the bridge. The third decomposition part is mainly the free responses of the structure due to environmental excitation. As shown in Figure 6b, the first decomposition part of the longitudinal displacements of the expansion joints is mainly associated with the effects of temperature on the main bridge structure; the second, third, and fourth decomposition parts exhibit unstable fluctuations at the times when the temperature changes, which may be due to the superposition of the temperature-induced responses of the main bridge and the ramp bridge, where differences in the temperature variations and temperature-induced behaviors of the main bridge and the ramp bridge lead to higher frequency behavior when the environmental temperature changes significantly because the longitudinal displacements of the expansion joints are more sensitive to temperature variation than the bearings. The fifth decomposition part mainly represents the free responses of the structure under environmental excitation; train crossings have no significant influence on the longitudinal displacements of the expansion joints because there is no nonstationary fluctuation throughout the duration of the train crossing time (06:00–24:00 h) in Parts 1~5.

3.2. Long-Term Characteristics of the Longitudinal Displacements of the Bridge

The long-term analysis of monitoring data can indicate the importance of understanding bridge responses and show the spatio-temporal characteristics of the structural behaviors of in-service bridges.

Figure 7 shows the trend in the longitudinal displacements at the south girder end (Pier 7) of the bridge from June to December in 2017 and includes the trend of atmospheric temperature for comparison. The dashed line (in color) denotes the longitudinal displacement data, and the solid line (in black) represents the temperature data. The temperature data exhibits missing time windows, with only six periods of temperature data measured from June to December. The time series cross-correlation of the displacement and temperature data in these six documented periods is analyzed, and the time delay ratios between the displacement and temperature sequences in the six periods are (0, 0, 0, 0, less than 0.5%, less than 0.5%). The time lag effect between the longitudinal displacement and environmental temperature is small. This phenomenon may be because bearings and expansion devices can directly release temperature-induced longitudinal displacements (the secondary effect and nonlinear behavior are not obvious).

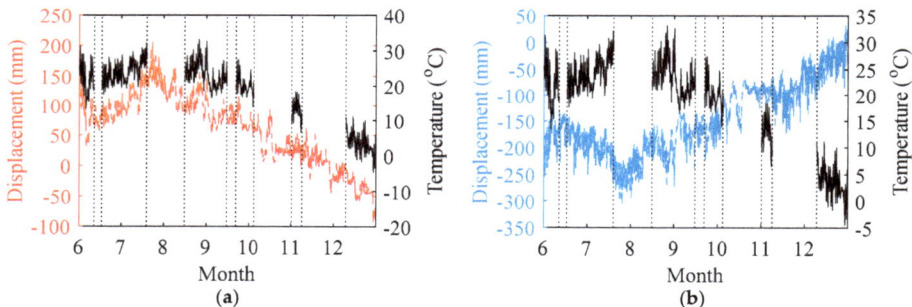

Figure 7. Trend of longitudinal displacements at the south girder end from June to December in 2017 (compared with the trend of atmospheric temperature): (**a**) bearing (B-7-11); (**b**) expansion joint (EJ-7-1).

Figure 8 shows the long-term correlation between longitudinal displacement of the bearings and expansion joints at the south girder end and the atmospheric temperature over the seven-month study period. The R in Figure 8 indicates the correlation coefficient of the data points. The first decomposition part of the EWT results slightly increases the temperature correlation when compared with original longitudinal displacement data, because the original data include some nonlinear effects of the operation loads. However, the correlation between the original longitudinal displacement data and the atmospheric temperature is already quite high. As shown in Figures 7 and 8, the longitudinal displacements of the bearing and the expansion joint both have a high correlation with environmental temperature. This result is due to the use of modified ultra-high molecular weight polyethylene (Modified UHMWPE) in the sliding plate structure of the bearings and expansion devices, which effectively releases the longitudinal displacements of the bridge during operation.

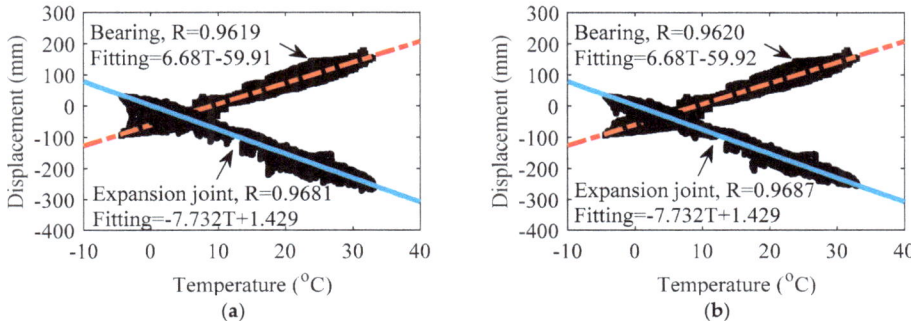

Figure 8. Correlation between longitudinal displacement and atmospheric temperature at the south girder end (bearing (B-7-11) and expansion joint (EJ-7-1)) from June to December in 2017: (**a**) original longitudinal displacement data; (**b**) first decomposition part of the EWT results.

Figure 9 shows the longitudinal displacement of the bearings on Piers 1–3 and 5–7 versus the distance of each pier from Pier 4. In this figure, each value for a pier is the mean of the two side bearings on the same pier. Figure 9a is the average value for 2015, and Figure 9b is the linear fitting line for all of the data at 1600 s sampling intervals in 2015. As shown in Figure 9a, the longitudinal displacement is linearly related to the distance from Pier 4. As shown in Figure 9b, the longitudinal displacements have high values (absolute) for 8–10 months of the one year, and the bearings at the girder ends (Piers 1 and 7) have the highest values (greater than 150 mm) compared with the bearings on the other piers. It is worth noting that the bearings at the girder ends have the lowest capacity (20 MN) and the highest longitudinal displacement. The sliding limit of a bearing is positively correlated with the capacity of the bearing: the higher the capacity of the bearing, the larger the size of the bearing, and the higher its sliding limit. Hence, the low-capacity bearing at the girder end must be checked in the bridge design and maintenance. The general limit of the longitudinal sliding displacement of bearings with 20 MN capacity is ±200 mm, and in this case, the longitudinal displacement of the bearings in the girder ends is close to this limit. The limit of the longitudinal sliding displacement of the expansion device with two hanging steel sleepers is ±600 mm, which means that the longitudinal displacement of the expansion joints is at the low level (Figure 7b).

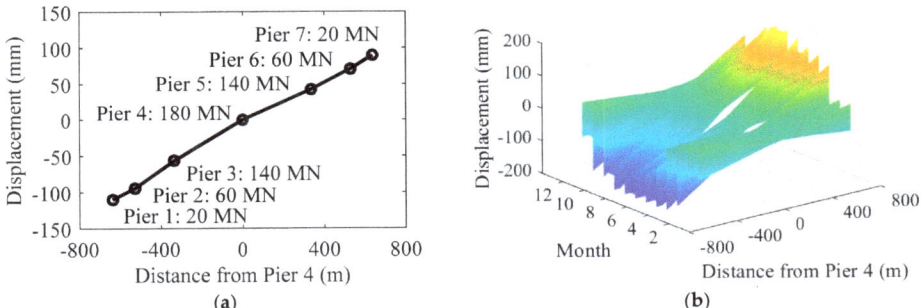

Figure 9. Longitudinal displacements of the bearings on Piers 1~3 and 5~7 versus their distances from Pier 4 in 2015: (**a**) annual average value with the capacity of bearings; (**b**) linear fitting line by each 1600 s of data.

The Ministry of Railways of the People's Republic of China has specified that the Modified UHMWPE used in bearings and expansion devices must be restricted to a low level of wear in the cumulative displacement of 50 km for major projects [26]. With regards to the longitudinal displacement data recorded in 2015, the cumulative longitudinal displacement of each bridge bearing in this one year are shown in Table 1. The lifetime for each bearing to achieve 50 km of displacement is determined from this one year of data, where the value for each pier is the average of its two side bearings (upstream and downstream). As seen in Table 1, for a major bridge such as the Nanjing Dashengguan Yangtze River Bridge with a design life of 100 years, all bearings will surpass their ideal situation (more than 50 km of displacement within the bridge's lifetime), especially the bearings on Piers 2, 3, and 5, which will surpass the constraint within 50 years. The Modified UHMWPE will wear approximately 5 microns of thickness for every km of cumulative sliding displacement, such that the wear thickness of the bearings of Pier 3 is approximately 550.9 microns in 100 years. For every km of displacement, it is recommended to check the volume of the silicone grease (used as the lubricant for the sliding plate) on the bearings and expansion devices, and at every 50 km, it is recommended to inspect the wear state of the sliding plate on the bearings and expansion devices. It is worth noting that the bridge bearings at Piers 3 and 5 (the arch foot bearings) have the shortest lifetime (45.38 years), while the bridge bearings at Piers 1 and 7 (the girder end bearings) have the longest lifetime (81.24 years). The arch foot bearings experience a greater cumulative longitudinal displacement in the same amount of time due to the short-distance reciprocating motion of the bridge.

Table 1. Cumulative longitudinal displacement and wear details of the bridge bearings.

Location of Bearings	Pier 1	Pier 2	Pier 3	Pier 5	Pier 6	Pier 7
Cumulation Displacement in One Year (unit: km)	0.6165	1.0910	1.1018	1.0489	0.9456	0.6154
Wear Rate in One Year (Compared with 50 km)	0.0123	0.0218	0.0220	0.0210	0.0189	0.0123
Wear Time Every 50 km (unit: Years)	81.10	45.83	45.38	47.67	52.88	81.24

4. Girder End Reliability under the Relative Transverse Displacements

The sliding motion of the bearings and expansion joints in the longitudinal direction releases deformation due to temperature change and ensures the normal operation of the long-span jointless bridge. However, the unequal longitudinal displacement of the upstream part relative to the downstream part of the bridge results in the cross-sectional rotation of the girder, as well as the relative transverse displacement between the main bridge and ramp bridge at the expansion joint. Furthermore, these unequal displacements generate the secondary stresses of expansion devices and rail transverse deflections. Because the maximum longitudinal displacement of the bridge always occurs at the girder end, the transverse behavior (caused by unequal longitudinal displacement) of the girder end should be considered with greater scrutiny.

4.1. Transverse Behavior of the Girder End Calculated by the Longitudinal Displacement Data

Based on the longitudinal displacement data of the bearings and expansion joints on the bridge girder end, the cross-section rotation of the girder and the relative transverse displacement of the expansion joint can be calculated using the same principle (as shown in Figure 10). L can be the transverse length of a girder or an expansion joint, and H can be the distance between two sensors. The $L/\cos\theta$ in Step 4 is the length of the triangular hypotenuse in Step 3. It should be noted that because the calculated θ is small (less than 6.5×10^{-5} rad), $L/\cos\theta - L$ (the relative transverse displacement between two measured points, i.e., the transverse deformation of the girder cross-section) must be chosen, and not θ (cross-section rotation), to characterize the transverse behavior of the girder.

According to the calculation presented in Figure 10, Figure 11 shows the transverse deformation of the girders and the relative transverse displacement of the expansion joints at the bridge girder ends in 2017. Figure 11a is the transverse deformation of the girder (30 m apart, calculated from B-7-11 and B-7-12) at Pier 7 from June to December; Figure 11b is the relative transverse displacement of the downstream expansion joint (15 m apart, calculated from EJ-7-1 and EJ-7-2) at Pier 7 from June to December; Figure 11c is the relative transverse displacement of the expansion joint (30 m apart, calculated from EJ-7-1 and EJ-7-4) at Pier 7 from June to December; and Figure 11d is the relative transverse displacement of the expansion joint (30 m apart, calculated from EJ-1-1 and EJ-1-3) at Pier 1 from June to July.

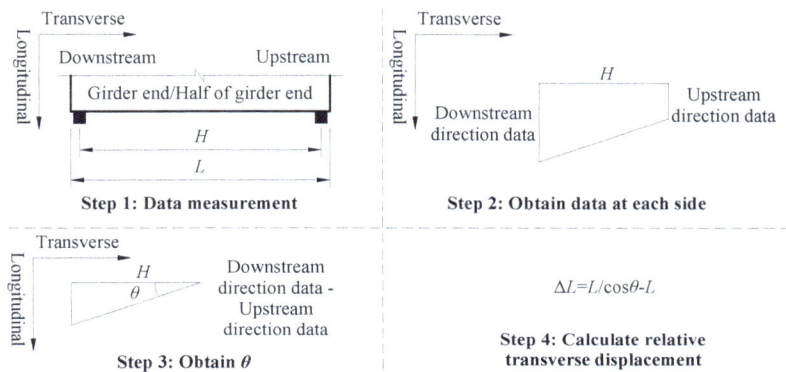

Figure 10. Process used to calculate the relative transverse displacement based on longitudinal displacement data.

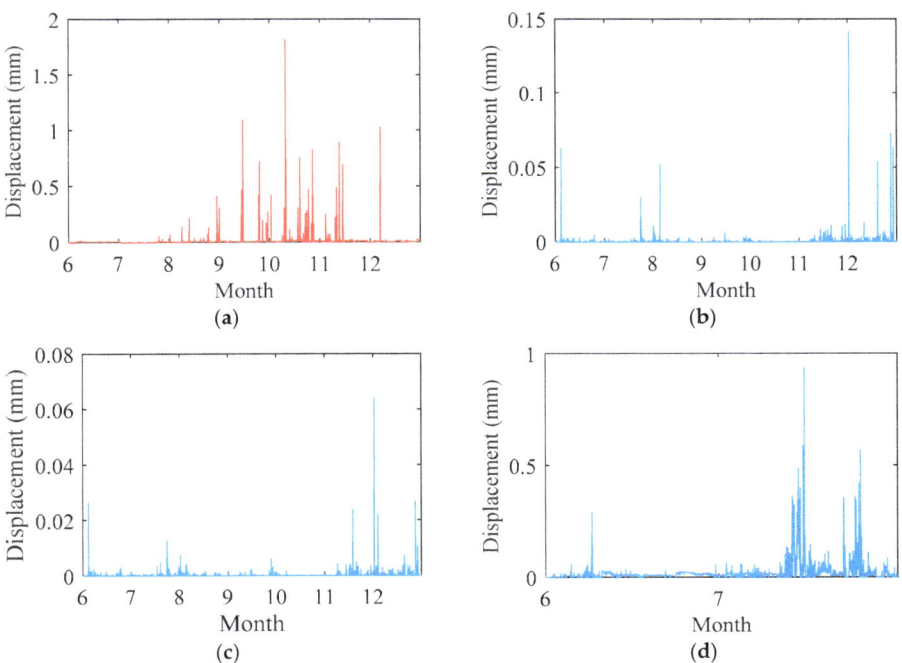

Figure 11. Relative transverse deformation/displacement at the girder ends: (a) main bridge girder (from B-7-11 and B-7-12); (b) expansion joint (from EJ-7-1 and EJ-7-2); (c) expansion joint (from EJ-7-1 and EJ-7-4); (d) expansion joint (from EJ-1-1 and EJ-1-3).

As shown in Figure 11, the relative transverse displacement of the downstream expansion joint (15 m, including two expansion devices) is not significantly smaller than that of the whole expansion joint (30 m, including four expansion devices) due to the relative transverse displacement of the expansion joint, which depends not only on the main bridge deformation but also the ramp bridge deformation. Specifically, the 30-m-wide ramp bridge on Pier 7 includes two independent 15-m-wide girders. The maximum transverse deformation of the girder is nearly 2 mm, and the maximum relative transverse displacement of the expansion joint is nearly 1 mm.

The transverse deformation (2 mm) is a low level of deformation for a steel girder with a transverse length of 30 m. However, for the expansion joint (and the expansion devices in the expansion joint), which does not have any transverse stiffness, the effect of 1 mm of relative transverse displacement between the main bridge and the ramp bridge must be investigated further. Furthermore, according to the study of Li et al. [27] and the design documents [28–30], the stability of crossing high-speed trains declines significantly when 1 mm of relative transverse displacement is present at the expansion joint (including the expansion devices within the joint). The maximum value of the relative transverse displacement of the expansion joint in Figure 11d is consistent with the daily inspection results of the bridge, which aids in the rapid resetting of the transverse rail deflection at the expansion joint. However, the mechanical behavior of the expansion device (including rail deflection) under the relative transverse displacements must be evaluated more precisely.

4.2. Mechanical Characteristics of the Expansion Device under Relative Transverse Displacements

A three-dimensional FE model of the expansion device is established using ANSYS to analyze the mechanical behavior of the expansion device under relative transverse displacements. The rails, the steel sleepers, the expansion boxes and the sliding support in the expansion boxes, the guide-rails, and the scissor-like connecting rod are established using the SOLID187 element with the corresponding parameters for steel, while the concrete sleepers are established using the SOLID187 element with the parameters for concrete. According to the design documents and the engineering experience, the tensile yield strength and tensile strength of the steel used on the expansion device are 370 and 510 MPa (Chinese Q370qE steel), and the tensile yield strength and tensile strength of the steel rail are 460 and 880 MPa (Chinese U71Mn steel); the elastic modulus of all steel materials is set to 2.06×10^5 N/mm^2 before yielding, and is set to 0 N/mm^2 after yielding; the elastic modulus of concrete sleeper is set to 3.6×10^4 N/mm^2; the damping ratio of the steel material is set to 0.02, while the damping ratio of the concrete material is set to 0.05. The dynamic effects of the operating loads are small for the expansion device, such that the stiffness and damping of all of the fasteners and the sliding plates are neglected. All of the fasteners are modeled by coupling the degrees of freedom (DOF) in the corresponding directions, while all of the sliding plates are modeled by releasing the DOF in the corresponding directions. The expansion box and sliding support, the sleeper and rail, and the sleeper and guide-rail at the fixed side are coupled in the longitudinal (X), transverse (Y), and vertical (Z) directions. The expansion box and sliding support, the sleeper and rail, and the sleeper and guide-rail at the free side are coupled in the Y and Z directions. The scissor-like connecting rods are coupled with the steel sleepers in the X, Y, and Z directions at the connected nodes, allowing the connecting rods to rotate around the Y axis.

Figure 12 shows the scalar displacement contours and the von Mises stress contours of the expansion device under 1 mm relative transverse displacement using three types of loading methods. Figure 12a presents the results under −0.5 mm of displacement on the fixed side and +0.5 mm of displacement on the free side; Figure 12b displays the results under 1 mm of displacement on the free side; and Figure 12c,d show the results under −1 mm of displacement on the fixed side. In the FE model of the expansion device (Figure 12), the left side is fixed (main bridge side), and the right side is free (ramp bridge side). The displacement loads are applied at the nodes of the expansion box bottom and the concrete sleeper bottom. The FE model of the expansion device is symmetric along the upstream and downstream directions (the transverse direction is the Y direction), and all three loading methods can represent the typical situation of the expansion device under relative transverse displacements. The three loading methods used to generate the 1-mm relative transverse displacement generate different displacement contours but nearly the same stress contours.

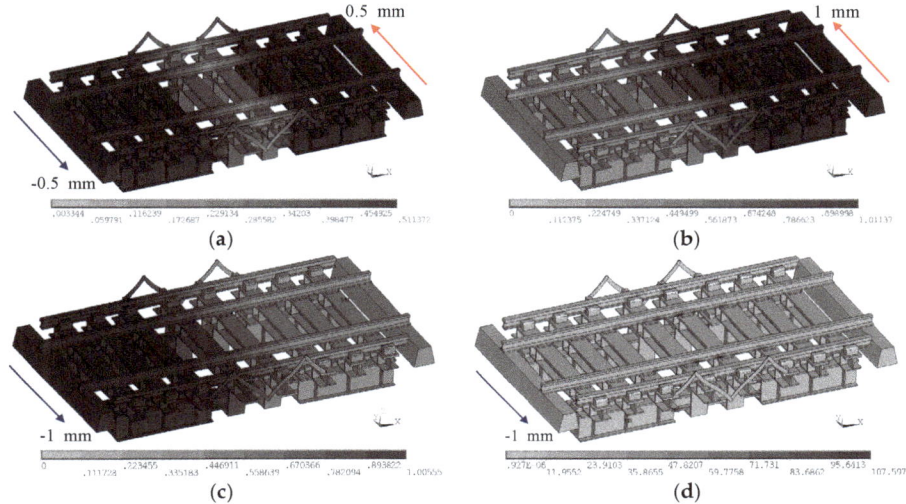

Figure 12. Mechanical contours of the expansion device under 1 mm relative transverse displacement: (**a**) displacement contour under −0.5 mm on the fixed side and +0.5 mm on the free side (unit: mm); (**b**) displacement contour under +1 mm on the free side (unit: mm); (**c**) displacement contour under −1 mm on the fixed side (unit: mm); (**d**) stress contour under −1 mm on the fixed side (unit: MPa).

As shown in Figure 12, the displacement maximum under 1 mm of displacement on the free side is 0.58% larger than that under −1 mm of displacement on the fixed side; when subjected to 1 mm of relative transverse displacement, the rails, guide-rails, and scissor-like connecting rods, which are connected to the steel sleepers, exhibit relatively higher displacements, and the contact areas of the hanging steel sleepers with the rails, guide rails, sliding supports, and scissor-like connecting rods will have relatively higher stresses. However, the stress of the entire structure remains at a low level comparing with the yield strength of steel materials. It is worth noting that, according to the time–frequency analysis results of the monitoring data, the displacement load of the expansion device is mainly due to the main bridge (fixed side), where Figure 12c,d represent the most common situation encountered by the expansion device in bridge operation.

For the expansion device at the girder end, two types of important factors can be used to indicate the reliability of the bridge structure and train crossings: first, the transverse deflection of the rail indicates the stability of train crossing; second, the stress that develops within the expansion device members indicates the degree of fatigue sustained by the expansion device. Table 2 shows the maximum mechanical responses of important factors and their location under −1 mm of transverse displacement on the fixed side (as in Figure 12c,d). The two other types of loading methods give rise to almost the same maximum values and locations in displacement.

Based on the monitoring data, the change in the relative transverse displacements is a gradual process. Hence, the dynamic effects of the expansion device under relative transverse displacement is neglected. What should be explained here is that if the dynamic effects are considered, then the top point of the scissor-like connecting rods will reach 1.88 mm under the transient loading of −1 mm of displacement on the fixed side (the static result is 0.825 mm), which is the dynamic displacement maximum point of the whole device. However, the maximum von Mises stress of the scissor-like connecting rod is still at a low level (less than 48 MPa), which occurs at the top of the rotating shaft of the scissor-like connecting rods due to the limited amount of stress that can be generated by the transverse rigid swing of the scissor-like connecting rods.

Table 2. Device mechanical responses under −1 mm of transverse displacement on the fixed side.

Mechanical Responses	Maximum Value	Maximum Value Location
Transverse deflection of rails	−1.00069 mm	Rails fastened on the steel sleepers at the fixed side.
von Mises stress of sliding supports	107.60 MPa	Contact areas of the sliding support and the fixed box mouth.
von Mises stress of guide rails	48.05 MPa	Contact areas of the guide rails and the hanging steel sleeper (including the fasteners).
von Mises stress of rails	88.44 MPa	Contact areas of the rails and the hanging steel sleeper (including the fasteners).
von Mises stress of scissor-like connecting rods	42.72 MPa	Middle rotating shafts of the scissor-like connecting rods connected with the steel sleeper.

4.3. Reliability and Early Warning of the Girder End of the Train-Bridge System

Adding the calculated relative transverse displacements to the fixed side (the main bridge side) of the expansion device FE model, the mechanical behaviors of the expansion device during the operating period can be calculated. Then, the reliability analysis of the girder end train-bridge system can be conducted based on the monitoring data. Considering the missing data of sensor EJ-1-3 (valid only in June and July, 2017), and the maximum relative transverse displacements that occur at EJ-1-1 to EJ-1-3 in July 2017, the relative transverse displacements used for reliability analysis come from the data of EJ-1-1 to EJ-1-3 for June through July 2017.

Most of the long-term responses will not obey a single common distribution (e.g., normal distribution). The relative transverse displacement data-driven calculated responses of expansion device FE model do not obey a simple probability distribution too. Hence, a non-parametric estimation should be used in the description of their statistical features [31–33]. The kernel density estimation is used to calculate the statistical features of the mechanical behaviors of the expansion device. A kernel distribution is defined by a smoothing function and a bandwidth value, which control the smoothness of the resulting density curve. The kernel density estimator is the estimated probability density function (PDF) of a random variable. For any discrete variables of x, the kernel density estimator's formula is given by

$$f_h(x) = \frac{1}{nh}\sum_{i=1}^{n} K\left(\frac{x-x_i}{h}\right) \quad (10)$$

where $x_1, x_2, \ldots, x_n$ are samples from an unknown distribution; n is the sample size; $K(\cdot)$ is the kernel smoothing function, which uses a normal distribution function in this article; and h is the bandwidth.

The kernel estimator for the cumulative distribution function (CDF), for any real values of x, is given by

$$F_h(x) = \int_{-\infty}^{x} f_h(t)dt = \frac{1}{n}\sum_{i=1}^{n} G\left(\frac{x-x_i}{h}\right) \quad (11)$$

where $G(x) = \int_{-\infty}^{x} K(t)dt$. In this article, the smoothing function uses *normal*, h equals (data maximum − data minimum)/1000.

For the fatigue reliability of the girder end (expansion device), the kernel PDF of the data-driven calculated von Mises stresses and their rainflow equivalent amplitudes belonging to the calculated maximum points of the sliding supports, guide rails, rails, and scissor-like connecting rods are shown in Figure 13. Under the diurnal relative transverse displacements, the von Mises stress amplitudes are mainly in the range of 0–3 MPa. The maximum von Mises stress of the sliding support is 100.71 MPa, but only 214 amplitudes exceed 10 MPa in the 2-month study period. According to the fatigue design codes of Europe and United States [34,35], the expansion device will not reach the cut-off limit of fatigue (the lowest stress amplitude that will lead to fatigue in the steel; the lowest cut-off limit of Europe code [34] is 15 MPa, and the lowest cut-off limit of US code [35] is 10 MPa) in most cases. The fatigue reliability of the expansion device under the diurnal relative transverse displacements is sufficient. The fatigue of the girder end may depend more on the vertical train loads during bridge operation.

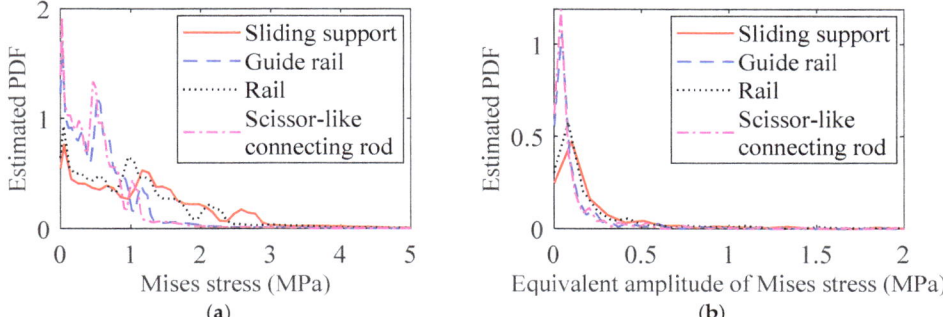

Figure 13. Kernel probability density function (PDF) of the von Mises stresses belonging to the calculated maximum points of the expansion device members: (**a**) data-driven calculated results; (**b**) rainflow equivalent amplitude.

For the train crossings, the transverse deflection of the rail is used to analyze the girder end reliability. Figure 14 gives the transverse rail deflections under the load necessary to achieve −1 mm of relative transverse displacement on the fixed side. The rail deflects at approximately 1 mm in the longitudinal distance of 2000 mm. For a China Railway High-speed 3 (CRH3) type electric multiple unit (EMU) train with cars that are 25 m in length, the rail transverse deflection will give an impulse to each crossing train wheel. The transverse stability of the crossing high-speed trains will be significantly weakened. Considering the cumulative effect of some other factors (e.g., the transverse track irregularity), the transverse instability of the train running will be further weakened. The transverse deflections of the rail should be monitored carefully during diurnal bridge operation.

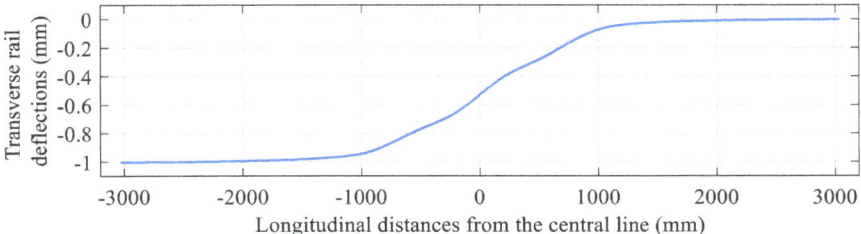

Figure 14. Transverse deflections of the rail under −1 mm relative transverse displacement of the fixed side.

Next, assuming that **Z** is the discrete data of rail transverse deflections, the probability (CDF value) that the element of **Z** is less than ar can be expressed as

$$P = P(z \leq ar) = \int_{-\infty}^{ar} f_{\mathbf{Z}}(z)\mathrm{d}z \qquad (12)$$

where $f_Z(z)$ is the PDF of Z; r is the resistance of the transverse rail deflections at the expansion device, which is determined by the code limit or engineering experience; and a is the statistical redundancy of the early warning.

Figure 15 shows the kernel PDF and kernel CDF of transverse rail deflections based on the data of EJ-1-1 to EJ-1-3 from June through July 2017. As shown in Figure 15b, the value of ar is approximately 0.6 mm at the probability of 99.995%. The train running stability limit of transverse rail deflections for the expansion device (r) is 1 mm, and the statistical redundancy of the early warning (a) can be set as 0.6. If the calculated transverse rail deflections based on the real-time monitoring data exceed 0.6 mm, a deceleration instruction should be sent to the crossing high-speed trains until the rails at the girder end can be inspected and repaired. The determination of the reliable probability in the diurnal early warning depends on the maintenance experience of the bridge engineers.

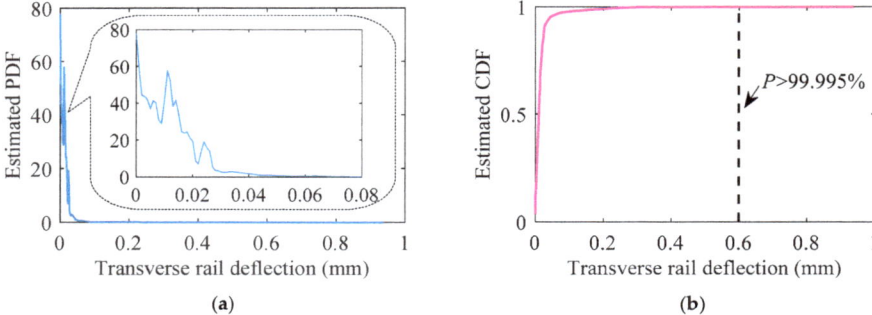

Figure 15. Statistical features of transverse rail deflections based on the kernel density estimation: (**a**) PDF; (**b**) cumulative distribution function (CDF).

The original intention of bridge health monitoring is to detect bridge damage as early as possible, to facilitate the rapid repair of a bridge subject to diurnal operation. With the accumulation of monitoring data, the reliable probability and the statistical redundancy of the early warning system can be updated for more informed long-term bridge operation. This approach promotes smart maintenance of the girder end of the long-span high-speed railway bridge.

5. Conclusions

With regards to the Nanjing Dashengguan Yangtze River Bridge, the analysis in this study mainly consists of two parts. The first part is an explanation of the behavior of the bridge based on its measured longitudinal displacement data, while the second part is an estimation of the girder end reliability based on the calculated relative transverse displacements and expansion device in an FE model. Four main conclusions are drawn, with the first and second conclusions corresponding to the first part of the study, and the third and fourth conclusions corresponding to the second part.

(1) The empirical wavelet transform can adaptively decompose the data of bridge longitudinal displacements, and the time–frequency characteristics can be captured to investigate the input–output mechanism of the bridge structure. The bridge longitudinal displacements are mainly influenced by temperature change. The longitudinal displacements of the expansion joints are more sensitive to the temperature variations than those of the bearings, because the longitudinal displacements of the expansion joints develop not only from the main bridge but also due to the ramp bridge.

(2) The time lag effect between the longitudinal displacement and environmental temperature is small for the bearings and expansion joints. The longitudinal displacements of the bearings and the expansion joints both exhibiting a high degree of correlation with environmental temperature. The bridge bearings at the girder ends generally have the largest longitudinal displacements and may possess insufficient reserve redundancy at the sliding limit. The wear lifetimes of the bridge bearings

are predicted based on the cumulative longitudinal sliding displacement (50 km) of the design. Ideally, the bridge bearings will exhibit lifetimes in the range of 45–82 years, which are less than the bridge design lifetime (100 year).

(3) Based on the longitudinal displacement data of the bearings and expansion joints at the bridge ends, the transverse deformation of the main-bridge girder ends, and the relative transverse displacement of the expansion joints between the main-bridge girder-ends and the ramp-bridge girder-ends are calculated. The maximum transverse deformation of each girder end is close to 2 mm, while the maximum relative transverse displacement of the expansion joint is close to 1 mm during long-term bridge operation. Under 1 mm of relative transverse displacement at the main-bridge side of the expansion device, the transverse deflection of the rails is 1.00069 mm, while the stress of the entire structure of expansion device remains at a low level.

(4) The long-term von Mises stress of the expansion device is at a low level, which should not result in fatigue under the anticipated transverse displacements (due to temperature effects) during bridge operation. However, the transverse rail deflections have approached 1 mm, which will reduce the stability of crossing high-speed trains at the location of the expansion device. Close attention should be paid to the stability of the train crossings. Using kernel density estimation, a non-parametric estimation of long-term transverse rail deflections is conducted. The probability of the transverse rail deflections being under 0.6 mm is estimated to be 99.995%. The reliability threshold of the transverse rail deflection can be used to indicate the deceleration instructions given to high-speed trains and the rapid inspection of expansion devices.

Author Contributions: Conceptualization, Y.D.; Methodology & Analysis, Writing—original draft, H.Z.; Writing—review and editing, S.N. and A.L.

Funding: This research was funded by the National Key R&D Program of China (grant no. 2017YFC0840200), the National Natural Science Foundation of China (grant nos. 51438002 and 51578138).

Conflicts of Interest: The authors declare no conflict of interest.

References

1. Zhao, H.W.; Ding, Y.L.; Geng, F.F.; Li, A.Q. RAMS evaluation for a steel-truss arch high-speed railway bridge based on SHM system. *Struct. Monit. Maintenance* **2018**, *5*, 79–92.
2. Treyssède, F. Finite element modeling of temperature load effects on the vibration of local modes in multi-cable structures. *J. Sound Vib.* **2018**, *413*, 191–204. [CrossRef]
3. Liu, Y.; Qian, Z.; Hu, J.; Jin, L. Temperature Behavior and Stability Analysis of Orthotropic Steel Bridge Deck during Gussasphalt Pavement Paving. *J. Bridge Eng.* **2018**, *23*, 04017117. [CrossRef]
4. Cao, Y.; Yim, J.; Zhao, Y.; Wang, M. Temperature effects on cable stayed bridge using health monitoring system: A case study. *Struct. Health Monit.* **2010**, *10*, 523–537.
5. Van Le, H.; Nishio, M. Time-series analysis of GPS monitoring data from a long-span bridge considering the global deformation due to air temperature changes. *J. Civ. Struct. Health Monit.* **2015**, *5*, 415–425. [CrossRef]
6. Sun, Z.; Zou, Z.; Zhang, Y. Utilization of structural health monitoring in long-span bridges: Case studies. *Struct. Control Health Monit.* **2017**, *24*, e1979. [CrossRef]
7. Xia, Q.; Zhang, J.; Tian, Y.; Zhang, Y. Experimental Study of Thermal Effects on a Long-Span Suspension Bridge. *J. Bridge Eng.* **2017**, *22*, 04017034. [CrossRef]
8. Maguire, M.; Roberts-Wollmann, C.; Cousins, T. Live-Load Testing and Long-Term Monitoring of the Varina-Enon Bridge: Investigating Thermal Distress. *J. Bridge Eng.* **2018**, *23*, 04018003. [CrossRef]
9. Erazo, K.; Sen, D.; Nagarajaiah, S.; Sun, L. Vibration-based structural health monitoring under changing environmental conditions using Kalman filtering. *Mech. Syst. Sig. Process.* **2019**, *117*, 1–15. [CrossRef]
10. Liu, Y.; Zhang, H.; Liu, Y.; Deng, Y.; Jiang, N.; Lu, N. Fatigue reliability assessment for orthotropic steel deck details under traffic flow and temperature loading. *Eng. Fail. Anal.* **2017**, *71*, 179–194. [CrossRef]
11. Liu, Y.; Zhang, S. Probabilistic Baseline of Finite Element Model of Bridges under Environmental Temperature Changes. *Comput.-Aided Civ. Infrastruct. Eng.* **2017**, *32*, 581–598. [CrossRef]

12. Zhao, H.W.; Ding, Y.L.; Nagarajaiah, S.; Li, A.Q. Behavior Analysis and Early Warning of Girder Deflections of a Steel-Truss Arch Railway Bridge under the Effects of Temperature and Trains: Case Study. *J. Bridge Eng.* **2019**, *24*, 05018013. [CrossRef]
13. Guo, T.; Liu, J.; Huang, L. Investigation and control of large cumulative girder movements of long-span steel suspension bridges. *Eng. Struct.* **2016**, *125*, 217–226. [CrossRef]
14. Liu, Z.; Phares, B.M.; Greimann, L.F. Use of Longitudinal Expansion Joints in Wide-Bridge Applications to Reduce Deck Cracking. *J. Bridge Eng.* **2016**, *21*, 04016068. [CrossRef]
15. Sekiya, H.; Maruyama, O.; Miki, C. Visualization system for bridge deformations under live load based on multipoint simultaneous measurements of displacement and rotational response using MEMS sensors. *Eng. Struct.* **2017**, *146*, 43–53. [CrossRef]
16. Luo, L.; Feng, M.Q.; Wu, Z.Y. Robust vision sensor for multi-point displacement monitoring of bridges in the field. *Eng. Struct.* **2018**, *163*, 255–266. [CrossRef]
17. Nagarajaiah, S. Adaptive passive, semiactive, smart tuned mass dampers: identification and control using empirical mode decomposition, hilbert transform, and short-term fourier transform. *Struct. Control Health Monit.* **2009**, *16*, 800–841. [CrossRef]
18. Liu, W.; Cao, S.; Chen, Y. Seismic Time–frequency Analysis via Empirical Wavelet Transform. *IEEE Geosci. Remote Sens. Lett.* **2016**, *13*, 28–32. [CrossRef]
19. Nagarajaiah, S.; Erazo, K. Structural monitoring and identification of civil infrastructure in the United States. *Struct. Monit. Maintenance* **2016**, *3*, 51–69. [CrossRef]
20. Wang, D.; Zhao, Y.; Yi, C.; Tsui, K.L.; Lin, J. Sparsity guided empirical wavelet transform for fault diagnosis of rolling element bearings. *Mech. Syst. Sig. Process.* **2018**, *101*, 292–308. [CrossRef]
21. Wang, G.X.; Ding, Y.L.; Song, Y.S.; Wu, L.Y.; Yue, Q.; Mao, G.H. Detection and Location of the Degraded Bearings Based on Monitoring the Longitudinal Expansion Performance of the Main Girder of the Dashengguan Yangtze Bridge. *J. Perform. Constr. Facil.* **2016**, *30*, 04015074. [CrossRef]
22. Zhao, H.W.; Ding, Y.L.; An, Y.H.; Li, A.Q. Transverse Dynamic Mechanical Behavior of Hangers in the Rigid Tied-Arch Bridge under Train Loads. *J. Perform. Constr. Facil.* **2017**, *31*, 04016072. [CrossRef]
23. Ding, Y.L.; Zhao, H.W.; Deng, L.; Li, A.Q.; Wang, M.Y. Early Warning of Abnormal Train-Induced Vibrations for a Steel-Truss Arch Railway Bridge: Case Study. *J. Bridge Eng.* **2017**, *22*, 05017011. [CrossRef]
24. Gilles, J. Empirical wavelet transform. *IEEE Trans. Signal Process.* **2013**, *61*, 3999–4010. [CrossRef]
25. Huang, N.E.; Shen, Z.; Long, S.R.; Wu, M.C.; Shih, H.H.; Zheng, Q.; Yen, N.-C.; Tung, C.C.; Liu, H.H. The empirical mode decomposition and the Hilbert spectrum for nonlinear and non-stationary time series analysis. In Proceedings of the Royal Society of London. Series A: Mathematical, Physical and Engineering Sciences, London, UK, 8 March 1998; Royal Society: London, UK; pp. 903–995.
26. Ministry of Railways of the People's Republic of China. *TB/T 2331-2013: Pot Bearing for Railway Bridge*; China Railway Publishing House: Beijing, China, 2014. (In Chinese)
27. Li, Y.L.; Xiang, H.Y.; Wan, T.B.; Ren, H.Q. Performance of Train Running over Expansion Joints at Beam Ends of Long-span Railway Bridge. *J. Chin. Railway Soc.* **2012**, *34*, 94–99. (In Chinese)
28. Railway Engineering Research Institute of CARS. *Temporary Technical Regulations for the Expansion Devices in the Girder End of the Dashengguan Yangtze River Bridge and Jinan Yellow River Bridge on the Beijing–Shanghai High-Speed Railway*; Chinese Academy of Railway Sciences: Beijing, China, 2010. (In Chinese)
29. National Railway Administration of the People's Republic of China. *TB 10621-2014: Code for Design of High Speed Railway*; China Railway Publishing House: Beijing, China, 2016. (In Chinese)
30. National Railway Administration of the People's Republic of China. *TB 10082-2017: Code for Design of Railway Track*; China Railway Publishing House: Beijing, China, 2018. (In Chinese)
31. Botev, Z.I.; Grotowski, J.F.; Kroese, D.P. Kernel Density Estimation via Diffusion. *Ann. Stat.* **2010**, *38*, 2916–2957. [CrossRef]
32. Balomenos, G.P.; Padgett, J.E. Fragility Analysis of Pile-Supported Wharves and Piers Exposed to Storm Surge and Waves. *J. Waterw. Port Coastal Ocean Eng.* **2018**, *144*, 04017046. [CrossRef]
33. Memmolo, V.; Pasquino, N.; Ricci, F. Experimental characterization of a damage detection and localization system for composite structures. *Measurement* **2018**, *129*, 381–388. [CrossRef]

34. CEN. *EN1993-1-9: Eurocode 3: Design of Steel Structures: Part 1–9: Fatigue*; European Committee for Standardization: Brussels, Belgium, 2005.
35. AASHTO. *Guide Specifications for Fatigue Evaluation of Existing Steel Bridges*; American Association of State Highway and Transportation Officials: Washington, DC, USA, 1990.

 © 2019 by the authors. Licensee MDPI, Basel, Switzerland. This article is an open access article distributed under the terms and conditions of the Creative Commons Attribution (CC BY) license (http://creativecommons.org/licenses/by/4.0/).

Article

Modeling of Temperature Time-Lag Effect for Concrete Box-Girder Bridges

Kang Yang [1], Youliang Ding [1,*], Peng Sun [2], Hanwei Zhao [1] and Fangfang Geng [3]

1. School of Civil Engineering, Key Laboratory of C&PC Structures of the Ministry of Education, Southeast University, Nanjing 210096, China
2. Department of Civil and Environmental Engineering, University of Michigan, Ann Arbor, MI 48109, USA
3. School of Architecture Engineering, Nanjing Institute of Technology, Nanjing 211167, China
* Correspondence: civilchina@hotmail.com

Received: 1 July 2019; Accepted: 7 August 2019; Published: 9 August 2019

Featured Application: According to field research results, a time-lag between structural response and temperature load is commonly encountered in practice. Due to the fact that it cannot be neglected for accurate structure health monitoring, a phase-shifting method is proposed; with the method, the time-lag effect can be effectively reduced, leading to a sound understanding of temperature load and its effect.

Abstract: It is common to assume the relationship between temperature and temperature response is instantaneous in bridge health monitoring systems. However, a time-lag effect between temperature and thermal strain response has been documented by the analysis of monitored field data of concrete box-girder s. This effect is clearly reflected by the ring feature in the temperature-strain correlation curve. Inevitably, the time-lag effect has an adverse impact on the accuracy and reliability of state assessment and real-time warning for structural health monitoring (SHM) systems. To mitigate the influence of the time-lag effect, a phase-shifting method is proposed based on the Fourier series expansion fitting method. The time-domain signal is firstly converted into the frequency domain signal to compute the phase difference between temperature data and response strain data at each decomposed order. Subsequently, the total phase difference can be obtained by weighted summation. The signal processing effectively reduces the hysteresis loop area and enhances the correlation between the structural response data and the temperature data. When processing the daily data in different seasons, it is found that after subtraction by the proposed method, the linear feature becomes dominant in the relationship between temperature and the strain during long-term observation.

Keywords: structural health monitoring; temperature effects; time-lag effect; Fourier series expansion; box-girder bridges

1. Introduction

For concrete box-girder bridges, integral parts of the structural health monitoring (SHM) system include the structural state assessment system and the real-time warning system. A necessary prerequisite to achieve a reliable state assessment and real-time warning is the clear understanding of environmental load effects on the structure, especially the temperature related effects. In fact, the temperature effects have been investigated by many researchers. It is considered that the stress generated by the nonlinear temperature distribution is usually equivalent to the live load, and the temperature-induced stress is significant in the concrete structure [1]. Taysi et al. [2] studied the thermal characteristics of concrete under the influence of temperature variation on box-girder utilizing experiments and finite element simulation. Their work highlighted the distribution of thermal difference and its influencing factors. Besides, Catbas et al. [3] discovered that temperature effects possess an important impact on the reliability of full bridges by analyzing a vast amount of monitored data.

In order to analyze the influence of temperature effects on structure assessment [4,5], Huang et al. [6] applied Kalman filtering and Kalman cointegration to identify the damage and recognized that external effects (e.g., temperature) may mask the changes induced by structural damage. Moreover, Liang et al. and Li et al. [7,8] carried a sensitivity study based on the structural fundamental frequencies in which it has been found that variations in ambient temperature might lead to the misjudgment of structural health conditions. Their work emphasized on temperature effects, but the observation and analysis of temperature distributions need to be further investigated.

The temperature distribution is the state of temperature at various positions inside and outside a structure at a certain time [9]. Preliminary studies of the temperature effects only considered the general temperature effects. Xu et al. [10] examined more than 8 years' temperature displacements using mean values of different temperatures sensors and obtained the statistical law of temperature change. Xu et al. [11] studied the change of structural dynamic response using uniform temperature rise and fall models. Nevertheless, they all ignored the difference in temperature distribution.

To further study the temperature stress, the nonlinear effect should be taken into consideration, when horizontal and vertical temperature gradients exist [12,13]. The external factors affecting the temperature distribution of concrete structures are solar radiation, nighttime cooling, cold air flow, wind, rain, snow, and other meteorological factors [1]. The internal factors that affect the temperature distribution of the concrete are mainly determined by the thermophysical properties of the concrete and the geometrical dimensions of the components. The spatial heterogeneity and time-dependent nature of temperature distribution make it difficult to determine the exact relationship between temperature and structural response [14]. Considering the monitored response data of a real bridge inevitably includes the influences of live load and environmental factors, making the problem more complicated.

To explore the temperature influences, separation of the temperature-induced part in data should be performed beforehand. For instance, Chen et al. [15] used the linear fitting method to determine the relationship between ambient temperature and temperature-induced strain while Hedegaard et al. [16] separated the time-dependent deformations from the temperature-related deformations by means of linear regression. They all achieved the purpose of separating the temperature-induced parts in the raw structural response data.

Previously, some scholars have noticed discrepancies in different temperature history data of steel bridges. For example, Zhou et al. [17] analyzed the lateral temperature distribution and temperature time history of the steel box-girder, and found a difference existed in the lateral temperature distribution of the box-girder at the same time. Brownjohn et al. [18] revealed that temporal and spatial temperature variations dominate displacement in long-span bridges. Meanwhile, the difference can reach 5 and 12 °C in winter and summer seasons, respectively. Furthermore, Brownjohn et al. [19] noticed there exists a time-lag effect due to thermal inertia effects, giving us a clear insight into the temperature time-lag effect.

To account for this temperature time-lag phenomenon, Zhao et al. [20] selected the first five principal components as the main components to determine the overall response of the -structure. Moreover, Guo et al. [21] noted that displacement data and temperature data for a steel box-girder cable-stayed bridge represented a lag time of approximately 45 min. After directly shifting the temperature data by 45 min, they found the correlation between the two was significantly improved. However, the temperature distribution and the temperature-induced effects of concrete small box-girder bridges still need to be further investigated.

Based on the above analysis, this study establishes that there is a significant time-lag effect in concrete box-girder bridges. This effect can be handled by a phase shift method, which is illustrated in a case study using field monitoring (strain and temperature) data. As a result, the aims of this paper include: (1) Investigate the time-lag phenomenon and its basic characteristics of concrete box-girder bridges; (2) cope with the issue of insufficient correlation between temperature data and strain data.

2. Time-Lag Effect

In this section, the time-lag effect is demonstrated through the field measured data of a real bridge. Firstly, the general situation of the Lieshihe Bridge and its SHM systems will be introduced. Then the collected data is illustrated and the time-lag effect is put forward. Finally, the character of the time-lag effect in different seasons is discussed.

2.1. Introduction of Lieshihe Bridge

Concrete small box-girder are commonly used in medium and small span bridges. Thanks to the stability and the structural characteristics suitable for both positive and negative bending moments, the cross-section form (concrete small box-girder) is becoming more and more popular in designing small and medium span bridges. This study focuses on the temperature effects on small concrete box-girders.

Lieshihe Bridge is a typical small box-girder bridge, located in Jiangsu province, China. This case study is used as an example to illustrate the thermal load on a bridge and the time-lag effect of temperature-induced strains on the bridge. The superstructure of Lieshihe Bridge is a 5-span continuous box-girder, as shown in Figure 1a. Each of the spans has a length of 25 m, with the total length of the bridge reaching 125 m. The main beams are small box-girder with a height of 1.5 m, as shown in Figure 1b. On the longitudinal axis, the locations for temperature and strain sensors are arranged in the middle section of the second span, while their positions on the lateral axis are illustrated in Figure 1b.

(a) Photo of Lieshihe River Bridge

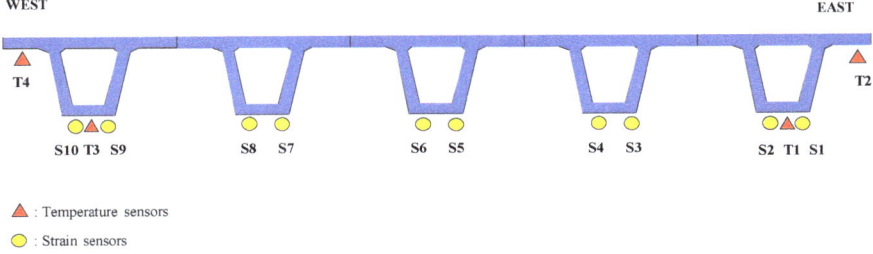

(b) Deployment locations of the sensors

Figure 1. Photo of Lieshihe River Bridge and deployment locations of sensors.

2.2. Time-Lag Effect

Concrete is a non-homogeneous material that consists of multiple different phases with complex thermal properties; heat transfer through complex geometries common in structural applications often results in nonuniform temperature distributions [22]. Temporal, spatial, and structural characteristics exist in temperature distributions on the bridge structure. Apart from the difference in temperature distributions (between different seasons and different days), a notable feature of the time delay between temperature and temperature-induced response indeed exists. The phenomenon that the temperature-induced structural response lags following the temperature is referred to as the time-lag phenomenon of the temperature-induced response. A considerable time-lag effect between temperature and temperature-induced strain can be found in concrete box-girder bridges.

The measured data of temperature and strain history at location S10 on 4 April 2017 is shown in Figure 2a,b, respectively. Furthermore, the correspondence of temperature and strain is plotted in Figure 2c.

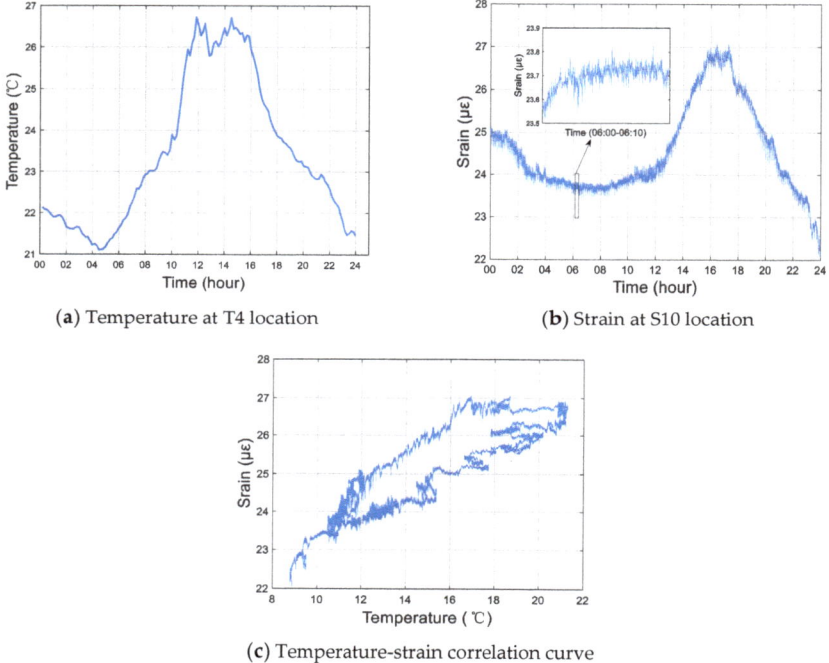

Figure 2. Time history plots of (a) temperature, (b) strain, and (c) temperature-strain correlation curve on 4 April 2017.

In Figure 2a, from the 24 h time history data, starting from 00:00 AM to 23:59 PM, the temperature data shows a trend of decreasing between 00:00 AM and 04:15 AM, then an increasing between 04:15 AM and 15:00 PM, and a downward trend between 15:00 PM and 23:59 PM at last. By comparing Figure 2a,b, the strain is found to follow the same overall trend as the temperature. The curve of temperature vs. corresponding strain at the same time are plotted in Figure 2c. The relationship between the two represents a fusiform annular shape, showing a significant nonlinear correlation. As a structural response, strain changes cyclically due to the process of heating and cooling from sunshine, which lags the temperature change, as a result showing the annular feature. The strain data is inevitably contaminated by some live load, such as traffic load from moving vehicles, which determine a high number of local small-period fluctuations, as is shown in Figure 2b. The strain obtained after the removal of the live load is the temperature-induced strain. The separation methods will be introduced in Section 4.1.

The primary cause of the temperature time-lag effect may be the hysteresis of temperature transfer, as transfer of heat throughout the concrete cross section takes time. This hysteresis manifests as uneven temperature distributions, called thermal gradients or thermal inertia effects [19]. As strain represents an overall response of the structure measured at a single point, considering only one temperature measurement point may cause some inconsistencies. This paper investigates the temperature time-lag effect in concrete box-girder bridges, but further research is required to reveal the ultimate cause of this phenomenon. It's worth noting that the time-lag effects mentioned in this paper all refer to the time scale of a single day. Considering that the daily trend of temperature is most distinct and has a direct influence on the structure, this paper mainly deals with the daily time-lag effect.

2.3. Seasonal Characteristics

Firstly, the data of a typical day in spring and summer was selected, respectively. Then, the time-lag curve of temperature and temperature-induced strain was drawn, see Figure 3. It is obvious that the temperature and response strain curve possess typical seasonal characteristics. By comparing Figure 3a,b, it is observed that the hysteresis loop in spring, as shown in Figure 3a, is relatively more compact than that in summer, as shown in Figure 3b. It reflects that in spring the change in temperature is comparatively slower. Hence, the degree of strain response lagging temperature is also less than that in summer. It also gives us a new way to measure the extent of time-lag effect through the hysteresis loop area.

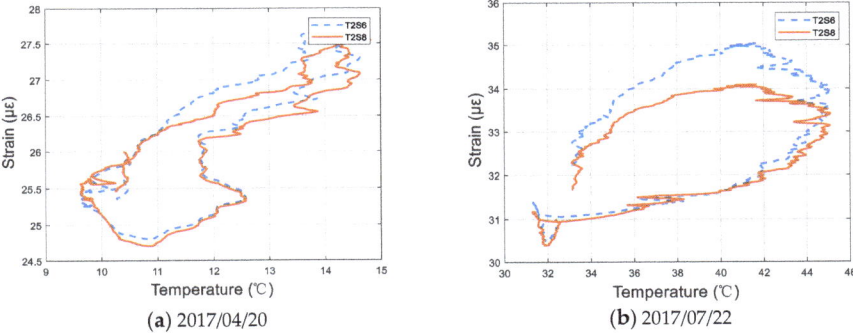

Figure 3. Plots of strain vs. temperature.

Furthermore, to illustrate the time-lag effect as a common phenomenon of concrete box-girder bridges, this paper investigated three months of data. Additionally, the hysteresis loop area of temperature and strain data of every single day are plotted in Figure 4. From Figure 4, a time-lag phenomenon of different extent in almost every day indeed exists. Moreover, the hysteresis loop area in summer (July) is larger than that in winter (November).

In summary, this part gives us a clear understanding of the time-lag phenomenon of temperature effects, a very common phenomenon in concrete box-girder bridges. Moreover, this phenomenon in the form curve annular feature can be directly reflected by the temperature vs. strain graph. Based on the analysis of data in different seasons, a seasonal characteristic was found.

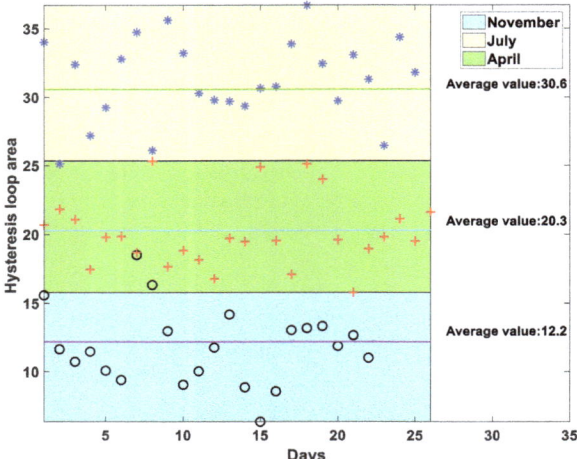

Figure 4. Hysteresis loop area in three months of 2017.

3. Methods

For eliminating the time-lag effect, the premise is obtaining the phase difference between the temperature and the temperature-induced strain. To scientifically and accurately acquire the phase difference, a method based on Fourier fitting [23] is proposed.

Since any continuous periodic signal can be composed of a set of appropriate sinusoids, the Fourier transform was firstly performed on the temperature and the temperature-induced strain [24]. Specifically, the original data was fitted to obtain the phase difference of various orders [25]. Then, the phase difference between the two signals was weighted and summed, so the total phase difference was obtained. Furthermore, the flow chart of the Fourier series expansion fitting method is shown in Figure 4.

According to Figure 5, the specific processes of the method are as follows.

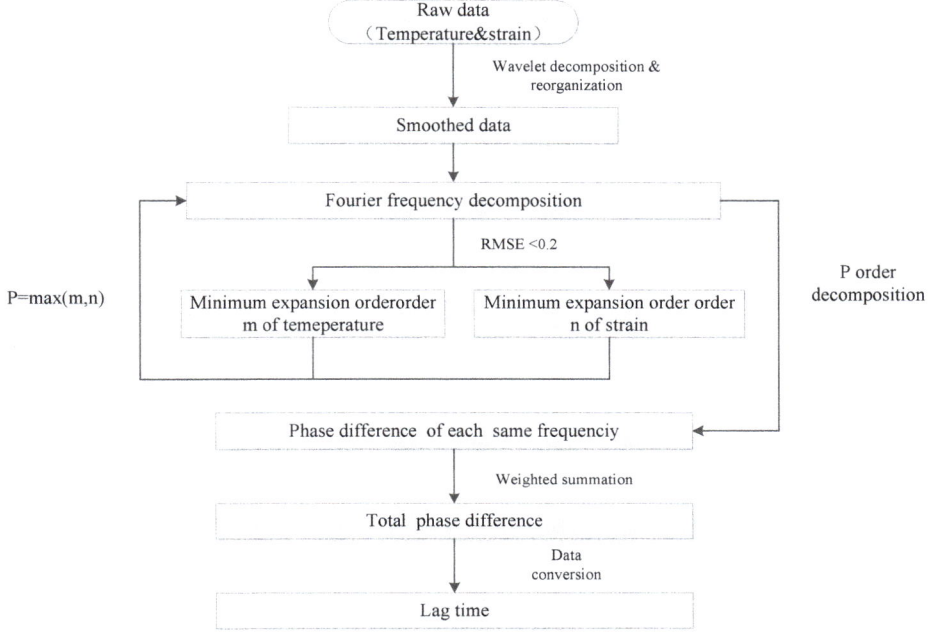

Figure 5. The flow chart of the Fourier series expansion.

3.1. Temperature-Induced Response Separation

The temperature history data and structural strain data of the same day are represented as f_{temp} and f_{sr}, respectively. The wavelet decomposition and reconstruction method is used to separate the measured strain data. Consequently, the temperature-induced strain $f_{sr,tem}$ and live-load strain $f_{sr,load}$ are extracted. The principle and process of the method are described by Zhao et al. [20].

3.2. Fourier Frequency Decomposition

The Fourier frequency decomposition in the time domain of $f_{sr,tem}$ and f_{temp} is used to obtain the frequency components of the data signal. The specific steps are as follows.

3.2.1. The Fourier Expansion

According to the Fourier expansion [26], the phase corresponding to each frequency component can be obtained as

$$f_{fourier}(x) = \frac{a_0}{2} + \sum_{k=1}^{\infty} a_k \cos(k\lambda x) + \sum_{k=1}^{\infty} b_k \sin(k\lambda x) = \frac{a_0}{2} + \sum_{k=1}^{\infty} c_k \sin(k\lambda x + \phi_k) \quad (1)$$

where $a_0, a_1, a_2, \ldots, a_k$ are the cosine coefficients of the Fourier factor and $b_1, b_2, \ldots, b_k$ are the sine coefficients of the Fourier factor. And x is a discrete time variable, as to the temperature and strain data in Lieshihe bridge, $x = 1/1440 \times [1, 2, 3, \ldots, 1440]$, owing to the sampling frequency is 1/1 min. λ is the reference frequency of the raw data, and ϕ_k is the phase of k-th order. Moreover, the amplitude of each order data signal is

$$c_k = \sqrt{a_k^2 + b_k^2} \quad (2)$$

The phase of each order of the signal is

$$\phi_k = \arctan(a_k/b_k), k = 1, 2, 3, \ldots \tag{3}$$

3.2.2. Determining the Minimum Order of the Fourier Series Expansion P

The expansion order is determined by trial-by-level trials until the residuals root mean square error (RMSE) meets the requirements. The S order RMSE(S) between the Fourier expansion value $f_{fourier}$ and the temperature data f_{temp} can be calculated by $RMSE(S) = \sqrt{\frac{\sum_{i=1}^{S}(f_{fourier} - f_{temp})}{S}}$, where S is the expansion order. The minimum expansion order M of temperature is determined by $RMSE(m) < 0.2$. Then, the expansion order N of strain data is also obtained in the same way. At last, the larger value of M and N is taken as P as the final expanded order.

The value of RMSE is related to the absolute value of the data value, the degree of dispersion, and so on. As a result, RMSE has no certain criterion for different kinds of data. The criterion of RMSE is determined by the correlation coefficient of fitted data and raw data in this paper.

3.3. Calculating the Phase Difference

The phase difference $\Delta\phi_i$ between the separated structural response data $f_{sr,tem}$ and the temperature data f_{temp} are solved at the same frequency. The specific steps are as follows.

3.3.1. Obtain the Phase of Temperature and Strain Respectively

Calculate the phase $\phi_{sr,k}$ of the structural response data $f_{sr,tem}$ and the phase $\phi_{temp,k}$ of the temperature data f_{temp} according to the Fourier series approximation expression [27]. The Fourier series approximation expression for the structural response data $f_{sr,fourier}$ and temperature data $f_{temp,fourier}$ are

$$f_{sr,fourier}(x) = \frac{a_{sr,0}}{2} + \sum_{k=1}^{P} c_{sr,k} \sin(k\lambda_{sr}x + \phi_{sr,k}) \tag{4}$$

$$f_{temp,fourier}(x) = \frac{a_{temp,0}}{2} + \sum_{k=1}^{P} c_{temp,k} \sin(k\lambda_{temp}x + \phi_{temp,k}) \tag{5}$$

where λ_{sr} and λ_{temp} represent the reference frequency of strain data and temperature data, respectively. They can be computed by $2\pi/L$. L is the length of the normalized cycle, which is closely related to the baseline period of the raw data and can be calculated automatically by the MATLAB Fourier series fitting program. Lambda varies with data of different days. It is mainly dependent on the shape feature of the data. As the shape of data in different single days is roughly similar, so the value of Lambda is approximate $5\pi/2$.

3.3.2. Calculate the Phase Difference

The phase difference $\Delta\phi_i$ can be calculated according to the following formula:

$$\Delta\phi_i = \phi_{temp,i} - \phi_{sr,i}, i = 1, 2, 3, \ldots, P \tag{6}$$

where $\phi_{temp,i}$ is the i-th order temperature data phase and $\phi_{sr,i}$ is the i-th order structure response data phase.

3.4. Determining the Total Phase Difference and Lag Time

Through a mass data research, the total phase difference can be obtained by the weighted summation of phase differences in each order. Moreover, the weight is proportional to the square of the frequency amplitude of each order.

$$w_j(x_1, x_2, \cdots x_P) = c_{temp,j}^2 / \sum_{j=1}^{P} c_{temp,j}^2 \qquad (7)$$

where w_j is the phase weight of the j-th order.

The delay effect on the correlation is eliminated or reduced by translating phase difference φ, which can be calculated by the following equation:

$$\varphi = \sum_{i=1}^{P} w_i \times \Delta \phi_i \qquad (8)$$

where φ is the total delay phase to be eliminated.

Since the temperature data changes in cycles of days and the overall trend is a half-sine function, the lag time can be determined from the ratio of the lag phase difference to the half cycle of the sine function:

$$T_{lag} = \varphi \times 1440 / \lambda_{sr} \qquad (9)$$

where T_{lag} is the lag time in minutes.

4. Case Study

This section includes two parts. First, the processing of separating temperature-induced strain from the field measured data is given. Then, according to the above data, the phase subtraction method based on the Fourier fitting method is described in detail and its effectiveness is verified.

4.1. Separation of Temperature-Induced Strain

The measured strain data includes the interaction of live load and temperature effects. Therefore, the separation of temperature strain is a prerequisite for thermal strain response studies. The measured data of the S6 strain measuring point of the bridge on 9 June 2017 is shown in Figure 6.

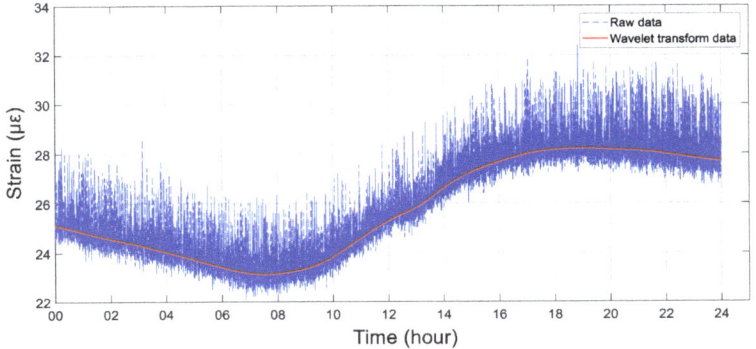

(**a**) Raw strain and temperature-induced strain.

Figure 6. *Cont.*

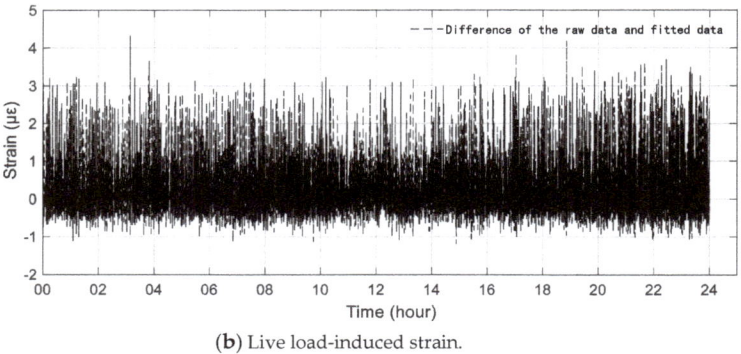

(**b**) Live load-induced strain.

Figure 6. Strain time history of the raw data and the processed data.

In Figure 6, the separation of temperature effects is achieved by the wavelet decomposition and reconstruction method [20]. As shown in Figure 6, this method can separate temperature strain effectively. The raw strain data is finally decomposed into temperature-induced strain (Figure 6a, the red line) and live load strain (Figure 6b). It is worth noting that the live load-induced strain in Figure 6b is larger than that in Figure 2b. The reason is the location of sensor S6 is in the center of the bridge where the girder will withstand more vehicle loads.

4.2. Phase Subtraction

The separated temperature-induced strain and temperature data was fitted subject to the Fourier decomposition. Furthermore, the temperature fitting curves of the 1st order, 4th order, and 8th order decomposition on 20 April 2017 are shown in Figure 7a. With the increase of the Fourier expansion order, the consistency was better, as shown in Figure 7b. When the data was expanded to the 6th order, the accuracy requirement was met.

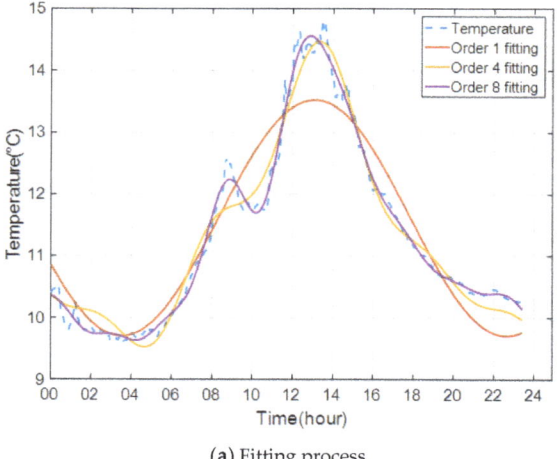

(**a**) Fitting process.

Figure 7. *Cont.*

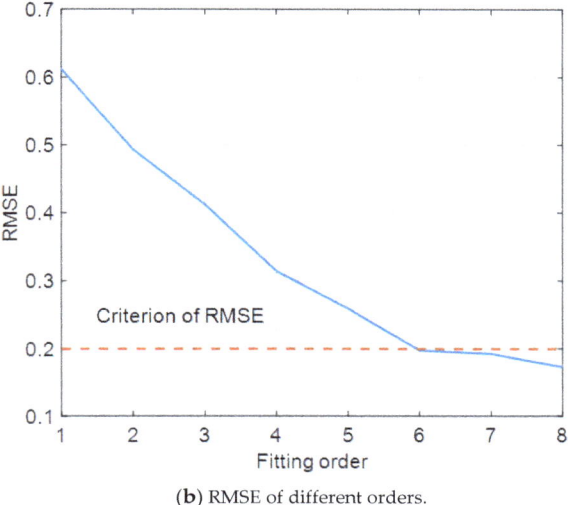

(**b**) RMSE of different orders.

Figure 7. The Fourier curve fitting results (20 April 2017).

The direct consequence of the time-lag phenomenon is the decline in correlation between the temperature and the temperature response. The hysteresis loop area directly reflects the degree of the time-lag phenomenon. Specifically, the larger the area, the more significant the lag effect. Therefore, the hysteresis loop area and the correlation coefficient can be used as indicators to verify the effectiveness of subtracting the time-lag phenomena by the Fourier series expansion method.

With the Fourier fitting method, the phase difference of the measured temperature and strain data on 22 July 2017 (summer) was calculated. The lag time of the S8 strain measuring point and T4 temperature measuring point can be calculated through Equation (9). As a result, the lag time was approximately 176 min in the summer season. Then, a translation of the temperature data by the lag time achieves the goal of subtracting the time-lag phenomenon. The temperature and strain correlation before and after the subtraction is shown in Figure 8. Using the same method to subtract the time-lag phenomenon in winter, Figure 9 is obtained. Meanwhile, through Equation (9), the lag time was calculated to be around 129 min in the winter season. As mentioned before, the displacement and temperature data for steel bridges possess a lag time of approximately 45 min in the research by Guo et al. [21]. As a result, the lag time in concrete structures is longer than that in steel structures.

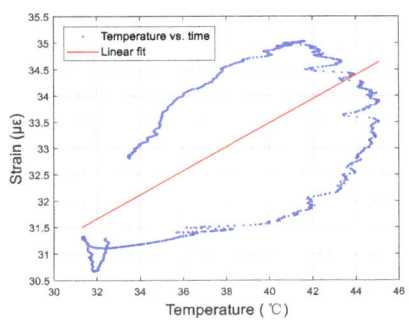

(**a**) Before the phase difference is translated

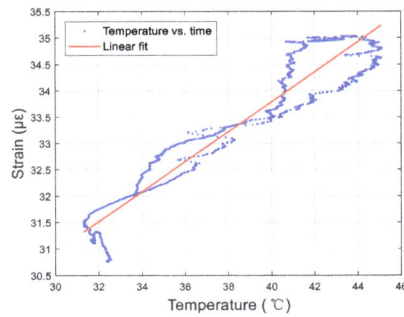

(**b**) After the phase difference is translated

Figure 8. Temperature vs. strain plots in summer (22 July 2017).

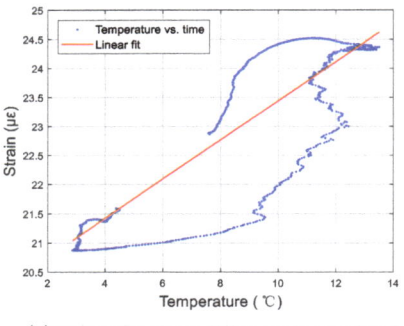 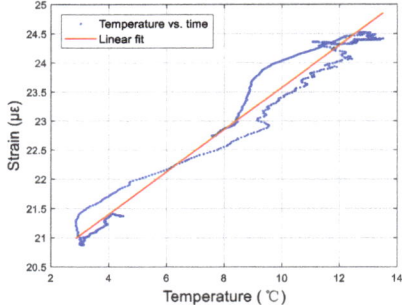

(a) Before the phase difference is translated (b) After the phase difference is translated

Figure 9. Temperature vs. strain plots in winter (27 November 2017).

By calculating the hysteresis areas and lag time of raw data and phase difference eliminated data, Table 1 was obtained. From Table 1, the hysteresis curve and lag times in winter with those in summer, respectively, were compared. Consequently, after subtracting, the hysteresis loop was notably reduced and the correlation coefficient was significantly improved. Therefore, the Fourier fitting method was verified to be effective.

Table 1. Comparison of indicators before and after phase difference elimination.

Date		Correlation Coefficient	Hysteresis Loop Area
22 July 2017	Before elimination	0.698	29.67
	After elimination	0.957	4.41
27 November 2017	Before elimination	0.885	13.65
	After elimination	0.974	3.27

With the proposed method, the time-lag phenomenon can be reduced. Furthermore, the long-term stable relationship between temperature load and the structural corresponding response can be more clearly reflected. Specifically, the same temperature and strain measurement points of four days in the four seasons of 2017 were selected to draw the temperature–strain curve. Then, the subtracted data curve was plotted in Figure 10. It is clear to see, although the specific distribution difference of strain varying with temperature exists in different seasons, the hysteresis loops in four seasons possess similar slopes overall. In particular, by eliminating the phase difference, the reduced curve can more clearly reflect the corresponding linear relationship characteristics of temperature-induced effects and temperature in different seasons. This feature reflects the specific mapping relationship between temperature and structure response and it is related to structure physical properties, such as structural geometry and material properties.

In summary, the wavelet decomposition and reconstruction method is useful and handy to extract temperature-induced strain. The lag time in a concrete structure is longer than that in a steel structure. Moreover, according to the comparison of hysteresis loop areas and correlation coefficient before and after subtracting the time-lag effect, the effectiveness was verified.

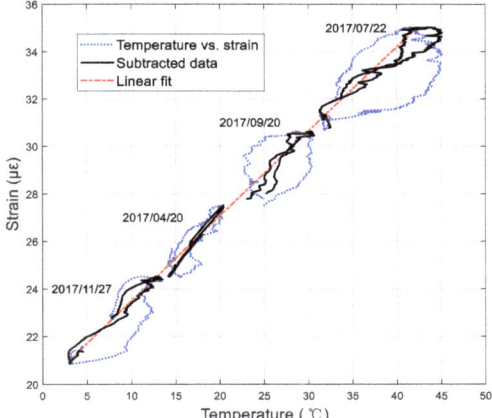

Figure 10. Temperature–strain curve and the subtracted data curve in different seasons.

5. Conclusions

Temperature is the most common and critical environmental load acting on bridge structures. The non-uniformity distribution and time-dependent character of the temperature leads to the complexity of temperature response. The existence of a structural response lag behind a reference temperature was investigated by analyzing the monitored data of a small concrete box-girder bridge. This time-lag effect causes interference between temperature load and structural response, leading to difficulties in the real-time warning of the SHM system. To address this problem, this study proposed a phase difference subtraction method based on the Fourier series fitting algorithm. Then, the method was verified to be effective through a case study. The main conclusions are as follows:

1. The temperature-induced effect of concrete small box-girder determine a time-lag phenomenon under the nonuniformed temperature load. Specifically, the time-lag curve in summer is fuller than that in winter, indicating that the time-lag phenomenon in summer is more notable.
2. The hysteresis loop areas and correlation coefficients can be used as two indicators to describe the extent of the time-lag effect.
3. The effectiveness of the Fourier series expansion and least-squares fitting method to subtract the time-lag effect was confirmed through a case study. Furthermore, using this method, the time-lag effect is reduced and the corresponding linear relationship characteristics between temperature load and temperature-induced strain in different seasons can be more clearly reflected.
4. The lag time in a concrete structure is longer than that in a steel structure.

Author Contributions: Original ideas K.Y. and Y.D.; data analysis, K.Y.; funding acquisition, Y.D.; methodology, K.Y. and H.Z.; writing—original draft preparation P.S.; writing—review and editing, P.S. and F.G.

Funding: This research was funded by the National Natural Science Foundation of China (51578138 and 51608258) and the Fundamental Research Fund for the Central Universities.

Conflicts of Interest: The authors declare no conflict of interest.

References

1. Xia, Q.; Cheng, Y.Y.; Zhang, J.; Zhu, F.Q. In-Service condition assessment of a long-span suspension bridge using temperature-induced strain data. *J. Bridge Eng.* **2017**, *22*, 04016124. [CrossRef]
2. Taysi, N.; Abid, S. Temperature distributions and variations in concrete box-girder bridges: Experimental and finite element parametric studies. *Adv. Struct. Eng.* **2015**, *18*, 469–486. [CrossRef]

3. Catbas, F.N.; Susoy, M.; Frangopol, D.M. Structural health monitoring and reliability estimation: Long span truss bridge application with environmental monitoring data. *Eng. Struct.* **2008**, *30*, 2347–2359. [CrossRef]
4. Kromanis, R.; Kripakaran, P. Predicting thermal response of bridges using regression models derived from measurement histories. *Comput. Struct.* **2014**, *136*, 64–77. [CrossRef]
5. Xiao, F.; Hulsey, J.L.; Balasubramanian, R. Fiber optic health monitoring and temperature behavior of bridge in cold region. *Struct. Control Health* **2017**, *24*, e2020. [CrossRef]
6. Huang, J.Z.; Li, D.S.; Li, H.N.; Song, G.B.; Liang, Y.B. Damage identification of a largecable-stayed bridge with novel cointegrated Kalman filter method under changing environments. *Struct. Control Health Monit.* **2018**, *25*, e2152. [CrossRef]
7. Liang, Y.B.; Li, D.S.; Song, G.B.; Feng, Q. Frequency Co-integration-based damage detection for bridges under the influence of environmental temperature variation. *Measurement* **2018**, *125*, 163–175. [CrossRef]
8. Xiao, F.; Chen, G.S.; Hulsey, J.L. Monitoring bridge dynamic responses using fiber Bragg grating tiltmeters. *Sensors* **2017**, *17*, 2390. [CrossRef]
9. Li, J.; Shang, M.; Liu, G.; Yang, T.; Pan, Y.; Zhou, J.; Zhao, Y. Two-step improvements of volumetric design method based on multi-point supported skeleton for asphalt mixtures. *Constr. Build. Mater.* **2019**, *217*, 456–472. [CrossRef]
10. Xu, Y.L.; Chen, B.; Ng, C.L.; Wong, K.Y.; Chan, W.Y. Monitoring temperature effect on a long suspension bridge. *Struct. Control Health Monit.* **2010**, *17*, 632–653. [CrossRef]
11. Xu, Z.D.; Wu, Z. Simulation of the effect of temperature variation on damage detection in a long-spancable-stayedbridge. *Struct. Health Monit.* **2007**, *6*, 177–189. [CrossRef]
12. Hedegaard, B.D.; French, C.E.W.; Shield, C.K. Effects of cyclic temperature on the time-dependent behavior of posttensioned concrete bridges. *J. Struct. Eng.* **2016**, *142*, 04016062. [CrossRef]
13. Ding, Y.L.; Zhou, G.D.; Li, A.Q.; Wang, G.X. Thermal field characteristic analysis of steel box girder based on long-term measurement data. *Int. J. Steel Struct.* **2012**, *12*, 219–232. [CrossRef]
14. Liu, H.; Wang, X.; Jiao, Y. Effect of temperature variation on modal frequency of reinforced concrete slab and beam in cold regions. *Shock Vib.* **2016**, *206*, 4792786. [CrossRef]
15. Chen, C.; Wang, Z.L.; Wang, Y.H.; Wang, T.; Luo, Z. Reliability assessment for PSC box-girder bridges based on SHM strain measurements. *J. Sens.* **2017**, *2017*, 8613659. [CrossRef]
16. Hedegaard, B.D.; French, C.E.W.; Shield, C.K. Time-dependent monitoring and modeling of I-35W St. Anthony falls bridge. I: Analysis of monitoring data. *J. Bridge Eng.* **2017**, *22*, 04017025. [CrossRef]
17. Zhou, L.R.; Xia, Y.; Brownjohn, J.M.W.; Koo, K.Y. Temperature analysis of a long-span suspension bridge based on field monitoring and numerical simulation. *J. Bridge Eng.* **2016**, *21*, 04015027. [CrossRef]
18. Brownjohn, J.M.W.; Koo, K.Y.; Scullion, A.; List, D. Operational deformations in long-span bridges. *Struct. Infrastruct.* **2015**, *11*, 556–574. [CrossRef]
19. Brownjohn, J.M.; Kripakaran, P.; Harvey, B.; Kromanis, R.; Jones, P.; Huseynov, F. Structural health monitoring of short to medium span bridges in the United Kingdom. *Struct. Monit. Maint.* **2016**, *3*, 259–276. [CrossRef]
20. Zhao, H.W.; Ding, Y.L.; Nagarajaiah, S.; Li, A.Q. Behavior analysis and early warning of girder deflections of a steel-truss arch railway bridge under the effects of temperature and trains: Case study. *J. Bridge Eng.* **2019**, *24*, 05018013. [CrossRef]
21. Guo, T.; Liu, J.; Zhang, Y.F.; Pan, S.J. Displacement monitoring and analysis of expansion joints of long-span steel bridges with viscous dampers. *J. Bridge Eng.* **2015**, *20*, 04014099. [CrossRef]
22. Zhang, S.; Li, Y.; Shen, B.; Sun, X.; Gao, L. Effective evaluation of pressure relief drilling for reducing rock bursts and its application in underground coal mines. *Int. J. Rock Mech. Min.* **2019**, *114*, 7–16. [CrossRef]
23. Zhou, P.; Zhang, W.; Wang, J.; Liu, J.; Su, R.; Xuemin, W. Multimode optical fiber surface plasmon resonance signal processing based on the Fourier series fitting. *Plasmonics* **2015**, *11*, 721–727. [CrossRef]
24. Ding, Y.L.; Li, A.Q. Temperature-induced variations of measured modal frequencies of steel box girder for a long-span suspension bridge. *Int. J. Steel Struct.* **2011**, *11*, 145–155. [CrossRef]
25. Yang, Z.-C. The least-square Fourier-series model-based evaluation and forecasting of monthly average water-levels. *Environ. Earth Sci.* **2018**, *77*, 328. [CrossRef]

26. Kvernadze, G. Approximation of the singularities of a bounded function by the partial sums of its differentiated Fourier series. *Appl. Comput. Harmon. Anal.* **2001**, *11*, 439–454. [CrossRef]
27. Li, W.L. Comparison of Fourier sine and cosine series expansions for beams with arbitrary boundary conditions. *J. Sound Vib.* **2002**, *255*, 185–194. [CrossRef]

© 2019 by the authors. Licensee MDPI, Basel, Switzerland. This article is an open access article distributed under the terms and conditions of the Creative Commons Attribution (CC BY) license (http://creativecommons.org/licenses/by/4.0/).

Article

Health Monitoring of Stress-Laminated Timber Bridges Assisted by a Hygro-Thermal Model for Wood Material

Stefania Fortino [1,*], Petr Hradil [1], Keijo Koski [1], Antti Korkealaakso [1], Ludovic Fülöp [1], Hauke Burkart [2] and Timo Tirkkonen [3]

1 VTT Technical Research Centre of Finland Ltd., P.O. Box 1000, VTT, 02044 Espoo, Finland; petr.hradil@vtt.fi (P.H.); keijo.koski@vtt.fi (K.K.); antti.korkealaakso@vtt.fi (A.K.); ludovic.fulop@vtt.fi (L.F.)
2 Standards Norway, P.O. Box 242, NO-1326 Lysaker, Norway; hbu@standard.no
3 Väylävirasto, Opastinsilta 12 A, 00520 Helsinki, Finland; timo.tirkkonen@vayla.fi
* Correspondence: stefania.fortino@vtt.fi; Tel.: +358-40-579-3891

Abstract: Timber bridges are economical, easy to construct, use renewable material and can have a long service life, especially in Nordic climates. Nevertheless, durability of timber bridges has been a concern of designers and structural engineers because most of their load-carrying members are exposed to the external climate. In combination with certain temperatures, the moisture content (*MC*) accumulated in wood for long periods may cause conditions suitable for timber biodegradation. In addition, moisture induced cracks and deformations are often found in timber decks. This study shows how the long term monitoring of stress-laminated timber decks can be assisted by a recent multi-phase finite element model predicting the distribution of *MC*, relative humidity (*RH*) and temperature (*T*) in wood. The hygro-thermal monitoring data are collected from an earlier study of the Sørliveien Bridge in Norway and from a research on the new Tapiola Bridge in Finland. In both cases, the monitoring uses integrated humidity-temperature sensors which provide the *RH* and *T* in given locations of the deck. The numerical results show a good agreement with the measurements and allow analysing the *MCs* at the bottom of the decks that could be responsible of cracks and cupping deformations.

Keywords: timber bridges; stress-laminated timber decks; monitoring; humidity-temperature sensors; wood moisture content; multi-phase models; finite element method

1. Introduction

Timber and engineered wood have increased their popularity as structural materials thank to their outstanding environmental performance, competitive price, mechanical properties, and relatively easy handling. However, the use of wood in unsheltered bridges is rather limited because of the exposure to the harsh climate conditions. Designers and structural engineers are mostly worried about the service life of the load-carrying structures which is recommended to be one hundred years in Europe [1].

Although evidence exists that structural wood can retain its strength through many centuries [2], it is very sensitive to the variable temperature (*T*) and moisture content (*MC*) which may lead to the material degradation and loss of its structural performance [3]. In some cases, the biotic damage can grow from inside out, and therefore the proper monitoring of internal material condition is essential in wooden bridges.

Stress-laminated timber decks (SLTDs) are composed of wood lamellas placed longitudinally between the supports of the bridge and compressed together with preloaded steel bars in the transverse direction (see [4] and the related references). This technology was developed in Canada in 1976 to replace nail-laminated wooden decks, which delaminated under cyclic loading and moisture variation. The first stress-laminated bridges were built in North America in 1980. The technology was adapted in Europe in mid 1980s and it was introduced in Australia, Japan and other countries since 1990. The greatest advantage of

laminated decks is that they form a stiff and solid base for the pavement, and therefore can redistribute the external loads to their supports. This effect is due to the prestressing action of the high-strength steel bars that squeeze the wooden lamellas together. The bar force, measured by using load cells, is typically from 89 to 356 kN [4].

Even though many of the originally built stress-laminated decks are performing well over three decades, it is essential to avoid errors during the construction and maintenance of the bridge. For instance, Scharmacher et al. [5] reported that blistering between wood and asphalt surface may occur, because of high MC of the deck and elevated asphalt temperature. This will affect the performance of the shear connection, but may also create conditions for water accumulation or ice formation under the asphalt surface.

Since the stress-laminating technology was developed in Canada and the northern parts of the United States, the effect of freezing temperatures has been thoroughly examined [4]. Laboratory tests revealed significant decrease of the bar forces of deck sections placed from a temperature of 21.1 °C to temperatures below zero ranging between −12.2 °C and −34.4 °C, strongly depending on the MC of the wood. Therefore, Wacker [4] recommends thermal design considerations in cold climates such as Alaska and Canada. This recommendation should also be applicable to the Nordic countries with similar weather conditions. Apparently, the simplest thermal design consideration is to keep the MC low in winter months to prevent the loss of pre-loading forces in the high-strength steel bars.

During the last decades, the development of timber bridges in European Nordic countries has been promoted by the joint effort of road authorities, timber industries and research organizations. A result of this cooperation was the Nordic Timber Bridges Programme [6]. Part of the activities under this programme was monitoring the long-term behaviour of wooden bridges in Norway financed by the Norwegian Public Roads Administration. Five of the monitored bridges in Norway have SLTDs and are located in Evenstad, Daleråsen, Flisa, Sørliveien and Måsør [7]. The bridges are built between 1996 and 2005 and are typically multi-span structures with glue laminated arches or trusses as the main load-carrying system. All of them have similar deck composed of 48 × 233 mm lamellas treated with creosote excepting the footbridge in Sørliveien (Figure 1), which has a deck of untreated spruce and a deck height of 333 mm, made of vertically sawn glulam beams. The SLT deck protecting the whole bridge structure is shown for Sørliveien Bridge (Norway) in Figure 1a. In addition, Figure 1b shows a detail of the SLT deck with the view of the wood lamellas and the steel bars for the same bridge.

Figure 1. Sørliveien Bridge. (**a**) Side view of the whole bridge structure. (**b**) Detail of the stress-laminated timber (SLT) deck protecting the bridge.

The efforts to promote timber bridges continued also after the end of Nordic Timber Bridges Programme. For instance, the Wood Building Programme (2016–2021) was launched in Finland as a government undertaking to increase the use of wood in urban development, public buildings, bridges and halls [8]. The programme is also seen as an efficient way of attaining the energy and climate targets to reduce Finland's carbon footprint by 2030. However, the number of laminated wooden deck bridges for vehicle traffic in Finland is still relatively small. One such bridge, carrying significant vehicle traffic, is the highway crossing recently erected in the Tapiola district of the city of Espoo. The Tapiola Bridge is now being permanently monitored under the supervision of the Finnish Transport Infrastructure Agency.

In addition to the durability problems, a common effect of moisture variation in SLTDs is the cupping deformation, which is usually measured as the uplift at the corner in the bottom surface of the deck [9]. A sharp increase of cupping is usually observed during wetting and only a partial decrease during drying. For the details, the reader is referred to Section 4.3 of the Durable Timber Bridges report [9].

The above review about performance of bridges shows that control of the *MC* in wooden parts is not only essential for the durability of the material, but for the whole superstructure as well. Variation of the *MC* directly affects structural integrity, serviceability and loading capacity of the bridge. Therefore, the monitoring techniques have a fundamental role in controlling the health of large structures exposed to outdoor climates, such as timber bridges. However, measurements obtained by the usual monitoring techniques based, e.g., on integrated humidity-temperature sensors, provide hygro-thermal measurements only in specific locations of the wood components.

As shown in the recent literature [10–12], advanced multi-phase models are an effective tool to assist the hygro-thermal monitoring of timber bridge components such as glulam beams. Compared to the single-phase (or single-Fickian) models for transient moisture transport in wood [13–15], where the *MC* is the only variable of a Fick's second law equation, the multi-phase models below the fibre saturation point (FSP) analyse two different water phases, i.e., the water vapour in lumens and the bound water in wood-cell walls. Starting from the seminal works of Krabbenhøft [16] and Frandsen [17], there was a strong effort to develop a multi-phase theory (often called multi-Fickian) for moisture transport in wood that includes the conversion rates between the different water phases. The multi-Fickian theory below the FSP is based on the identification of three phenomena occurring in cellular wood during moisture transfer, i.e., the diffusion of water vapour in the lumens, the sorption of bound water and the diffusion of bound water in the cell walls. In the multi-phase models available in the current literature, the two water phases are separated and the coupling between them is defined through a sorption rate [10–12,17–20]. Recently, Autengruber et al. [21], developed a whole multi-Fickian model including also the transport of free water in the lumens above the FSP. Therefore, in addition to the sorption rate between the two phases of water vapour and bound water, also the sorption rate between the free water and bound water phases, as well as the evaporation/condensation rate between the free water and the water vapour phases, need to be defined. These phenomena are schematized in Figure 2. For a complete description of the moisture transfer in wood, a sorption hysteresis characterized by two isotherms of adsorption and desorption was originally introduced in the multi-Fickian model by Frandsen [17]. In the present work, only case-studies with moisture states below the FSP are studied.

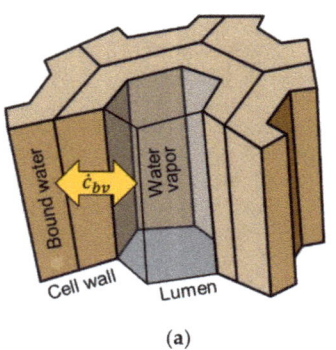 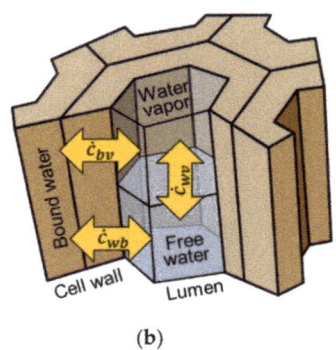

Figure 2. Scheme of the water phases and sorption phenomena in wood. (**a**) Below the fibre saturation point (FSP): bound water in the wood cell walls, water vapour in the lumens and sorption rate between bound water and water vapour phases ($\dot{c}_{bv}$). (**b**) Above the FSP: Bound water in the wood cell walls, water vapour and free water in the lumens, sorption rates between bound water and water vapour ($\dot{c}_{bv}$) and between water vapour and bound water ($\dot{c}_{wb}$), and evaporation/condensation rate between free water and water vapour ($\dot{c}_{wv}$).

As discussed by Svensson et al. [22] and Fragiacomo et al. [13], high values of moisture gradients due to high yearly and daily variations of relative humidities are the main causes of moisture induced stresses (*MIS*) perpendicular to the grains in wooden members. Larger yearly variations of *RH* (average values above 80%) and larger moisture gradients and *MIS* in timber cross sections were found under Northern European climates when compared to Southern European climates [13]. In [10,11] it was observed that, under Norther climates, high gradients in the vicinity of surfaces of bridge glulam beams during drying periods of the year are caused by high peaks of *RH* (above 85%) in conjunction with high daily variations of *RH* (above 50%). Knowledge on moisture gradients is therefore important to identify the zones prone to crack risk in wooden components, as shown in [12] for the case of a bridge glulam beam where the *MIS* were also calculated and discussed in relation to the moisture gradients. In [12] it was found that the most critical *MIS* are the tensile stresses perpendicular to the grain that can be also greater than the limits prescribed by the Eurocodes. Due to this, the uncoated bridge wooden beams may be exposed to the formation of moisture induced cracks and delamination. The structural significance of cracks in timber bridges under outdoor environments is discussed also in [23] where the asymmetric damage (longitudinal splitting cracks) is especially investigated.

Models for moisture transport, coupled with mechanical models, can be used to calculate the moisture induced cupping in stress laminated timber decks. The models can also allow the evaluation of the bar force losses during time, as shown in Section 4 of [9], were a single-phase model for moisture transport was used.

The novelty of the present paper is the use of a recent multi-phase model, proposed by some of the authors in [11], to assist the monitoring of SLTDs of bridges under Nordic European climates carried out by integrated humidity-temperature sensors. In particular, the hygro-thermal monitored data are collected from a previous study of the untreated deck of Sørliveien Bridge in Norway [24,25] and from the on-going monitoring of the painted and thick deck of Tapiola Bridge in the city of Espoo, Finland. Untreated and painted bridges are interesting cases to study in terms of their hygro-thermal performance. In this paper, the monitoring systems of the Sørliveien and Tapiola bridges are presented, and selected measurements are used for simulation by the finite element method (FEM). While the monitoring provides the *RH* and *T* in some locations of the analysed decks, the numerical model completes the health monitoring providing the overall hygro-thermal response of a representative volume of the deck in terms of distribution of *MC*, vapour pressure and T. In particular, the hygro-thermal response of the bottom deck, which is more affected by the external climate, is investigated.

2. Materials and Methods
2.1. Description of the Multi-Phase Model

Summarizing the multi-Fickian model presented in [11] for wood below the *FSP*, the variables for transient moisture transport are the concentration of bound water in the cell walls c_b, the concentration of water vapour in the cell lumens c_v, and the temperature T. Denoting by $\mathbf{D}_b$ and $\mathbf{D}_v$ the diffusion tensors for bound water and water vapour phases and by $\mathbf{K}$ the thermal conductivity tensor, the governing equations of the problem are:

$$\frac{\partial c_b}{\partial t} = -\nabla \cdot \mathbf{J}_b + \dot{c}_{bv} \tag{1}$$

$$\frac{\partial c_v}{\partial t} = -\nabla \cdot \mathbf{J}_v - \dot{c}_{bv} \tag{2}$$

$$c_w \varrho \frac{\partial T}{\partial t} = -\nabla \cdot \mathbf{J}_H - \nabla \cdot \mathbf{J}_b h_b - \nabla \cdot \mathbf{J}_v h_v + \dot{c}_{bv} h_{bv} \tag{3}$$

where ∇ is the nabla operator, $\mathbf{J}_b$ and $\mathbf{J}_v$ are the fluxes of bound water and water vapour, and $\mathbf{J}_H$ represents the thermal flux:

$$\mathbf{J}_b = -\mathbf{D}_b \nabla c_b, \; \mathbf{J}_v = -\mathbf{D}_v \nabla c_v, \; \mathbf{J}_H = -\mathbf{K} \nabla T \tag{4}$$

In Equation (3), c_w represents the specific heat and ϱ the wood density, the coupling term $\dot{c}_{bv}$ is the sorption rate between the two water phases (see Figure 1), h_b and h_v are the specific enthalpies and $h_{bv} = h_b - h_v$ is the specific enthalpy of the transition from the bound water to the water vapour. The moisture content *MC* is defined as c_b/ϱ_0 where ϱ_0 is the dry wood density. The sorption rate in Equations (1)–(3) is defined as:

$$\dot{c}_{bv} = H_c(\varrho_0 MC_{bl} - c_b) \tag{5}$$

where H_c represents the moisture dependent reaction rate and MC_{bl} is the moisture content in equilibrium with the relative humidity. In Equation (5), the MC_{bl} has the meaning of temperature-dependent sorption isotherms. These are defined by using the Anderson–McCarthy model (see Appendix A). In the present work, according to [11], an average between the temperature dependent adsorption and desorption isotherms is used, while a model for sorption hysteresis is not included.

Since the bound water cannot pass the external surfaces and it is restricted in the cell walls, the model includes only exchanges of vapour and heat with the ambient air. Therefore, the first boundary condition of Equation (6) holds on all the external surfaces in relation to variable c_b. For the other variables, the second and third boundary conditions in Equation (6) apply for the external surfaces exposed to the variable *RH* and T:

$$\mathbf{n} \cdot \mathbf{J}_b = 0, \; \mathbf{n} \cdot \mathbf{J}_v = k_v^w c_v' - k_v^a c_v^a, \; \mathbf{n} \cdot \mathbf{J}_v = k_T(T - T^a) \tag{6}$$

where $\mathbf{n}$ represents the outward normal direction to the surface, c_v^a and T^a are the water vapour concentration and temperature of the air, k_v^w and k_v^a the surface permeances corresponding to wood temperature and air temperature, and k_T is the thermal emission coefficient. The expressions of the permeances are reported in Appendix A. In Equation (6), $c_v' = c_v/\varphi$ represents the concentration of water vapour divided by the wood porosity φ. The concentration c_v is related to the partial vapour pressure p_v through the ideal gas law:

$$c_v = \varphi \, p_v M_{H2O} / RT \tag{7}$$

where R is the gas constant and M_{H2O} the molecular mass of water. The vapour pressure can be expressed as a function of the relative humidity RH:

$$p_v = RH \cdot p_{vs} \tag{8}$$

where p_{vs} is the saturated vapour pressure given by the semi-empirical Kirchhoff expression for the thermal ranges above the freezing point and by Teten's fitting for ice in the subfreezing temperature range [26]:

$$p_{vs} = \begin{cases} exp\left(53.421 - \frac{6516.3}{T} - 4.125 \ln(T)\right) & \text{for } T \geq 0°C \\ 100 \times 10^{\frac{9.5(T-273.15)}{T-7.65} + 0.7858} & \text{for } T < 0°C \end{cases} \tag{9}$$

All material parameters of the model are summarized in Table A1 of Appendix A. The model is suitable for wooden members sheltered from rain and without water traps or other contacts with water. It does not allow the modelling of liquid water in pores and can simulate only moisture states below the FSP.

2.2. Implementation of the Hygro-Thermal Model for Stress-Laminated Timber Deck in Abaqus Code

The selected commercial finite element software Abaqus provides a comfortable environment for the 3D model construction and the evaluation of results. The finite element to be used for the hygro-thermal analysis was defined in the user subroutine UEL to accommodate the three differential equations that describe the material model. The subroutine is reading the weather data from the database of measured temperatures and air relative humidities at every time increment and applies them as external loads on the exposed model surfaces. The shape functions for 8-nodes isoparametric brick elements are used and a weak form of the governing equations and their boundary conditions with three variables per node (bound water concentration, water vapour concentration and temperature) is implemented in the UEL.

The time integration is carried out using the fully implicit Euler scheme and the nonlinear system is solved using the Newton method at each time step. The subroutine allows to implement the FEM contributions to the residual vector and to the Jacobian iteration matrix.

The general scheme for the hygro-thermal modelling of the timber deck is shown in Figure 3 and the simplifications used are the following:

- The model is a 3D slice of the deck far from the ends. The bottom face is exposed to the humidity and temperature of the air. The top surface is exposed only to temperature, because the top of the deck is protected from moisture by the asphalt layer.
- The asphalt layer is not modelled.
- The lateral, back and front faces are internal surfaces and therefore are not exposed to the air temperature or moisture fluxes.
- The bottom surface is sheltered from rain and without water traps.
- The model does not include the effect of solar radiation.
- The height of the model represents the thickness of the timber deck and its width varies depending on the width of the lamella, with the mesh size of the FEM typically between 5 and 10 mm.
- The effect of glue between lamellas is not considered.

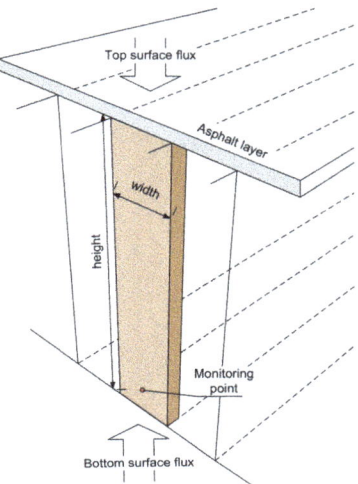

Figure 3. Scheme of a 3D vertical slice of the timber deck for the hygro-thermal analysis. The asphalt layer is not modelled.

The initial values for the variables of the differential problem are the following:

- The temperature T_0 is chosen equal to the air temperature at the beginning of the analysis.
- The concentration of bound water is calculated as $c_{b0} = \varrho_0 MC_0$, where MC_0 is the moisture content in equilibrium with the initial air relative humidity RH_0 at the beginning of the monitoring. This is obtained from the temperature dependent sorption isotherm listed in Table A2 of Appendix A.
- The concentration of water vapour c_{v0} is calculated by using Equations (7)–(9) and the RH_0 and T_0.

The fluxes acting on the 3D slice of the deck are as follows:

- The first boundary condition of Equation (6) applies on the top and bottom surfaces.
- The heat flux and thermal flux act on the bottom surface exposed to the air temperature and relative humidity.
- Only the heat flux acts on the top protected by the asphalt.
- There are no fluxes on the lateral (internal) surfaces.

The input material data used for both case-studies of the paper are the dry wood density $\varrho_0 = 450$ kg/m^3, the porosity $\varphi = 0.65$ and the coefficients of the diffusion tensors that are listed in Appendix A. The permeances for the uncoated wood used in the first case-study (k_w) and for the weak paint used in the second case-study (k_p) are listed in Table A3, and the thermal emission coefficient are listed in Table A1 of Appendix A.

The outputs are the moisture content MC, the vapour pressure p_v (obtained from the water vapour concentration c_v), and the temperature T in each element of the 3D model.

2.3. Case-Study: Sørliveien Bridge

Sørliveien Bridge (Figures 4 and 5) is a pedestrian bridge built in summer 2005 in Akershus County, Norway, crossing a local road [24,25]. The owner was the Norwegian Public Road Administration. It is a slab bridge with eight spans and a total length of 87 m. The longest span is 17 m. The stress-laminated timber deck (48 × 333 mm) is composed of spruce glulam planks, which are untreated except for the edge planks of creosote-impregnated pine wood. The top layer consists of 60 mm asphalt with a moisture membrane of polymer modified bitumen (Topeka 4S) underneath. The bridge has been instrumented in August 2005 and monitored since then. The instrumentation is situated at the northern end and is logged every fourth hours.

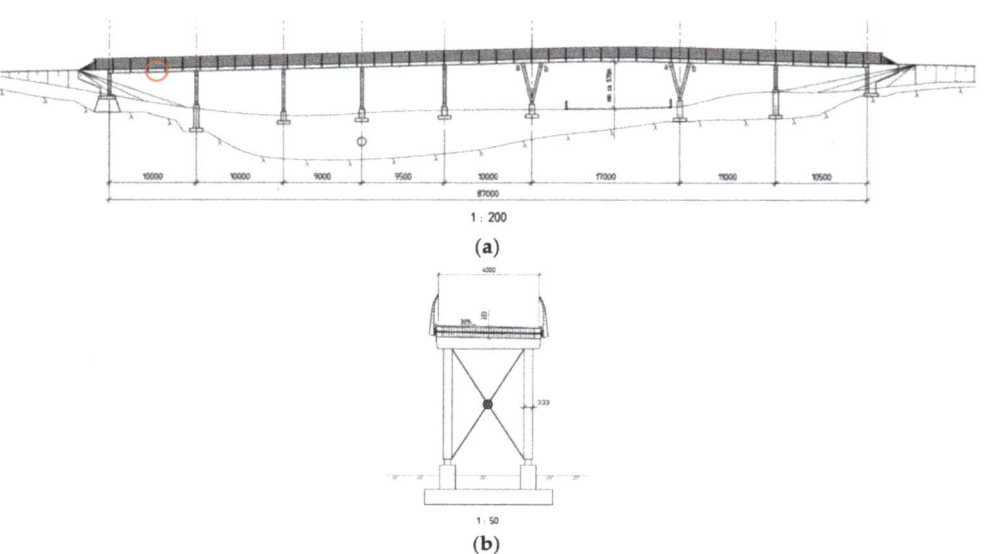

Figure 4. Sørliveien Bridge in Norway. (a) Side view with location of sensors (red circle). (b) Cross section showing the SLT deck with 333 mm thickness.

Figure 5. Sørliveien Bridge. Detail of the bottom deck with the monitoring equipment.

The monitoring equipment (Figure 5) collects data about the loading of the high-strength steel bars, temperature and humidity of the wood at different depths from the surface, and temperature and humidity of the air from the weather station positioned on the bridge. The collected data is processed directly on the embedded computing unit and regularly transmitted to the central monitoring server over the internet.

Three load cells were installed to the prestressing bars loaded to 227 kN, on the northeast side of the bridge. The measurements from load cells are not discussed in this paper, because they are not directly needed for the hygro-thermal simulations. Temperature and relative humidity were measured by ten integrated humidity-temperature sensors Vaisala Humitter 50Y [27]. This type of sensor has an operating range from −40 °C to +60 °C and from 0 to 100% of the RH. Its length is 70 mm and the diameter 12 mm. Nine sensors were installed in three different depths from the bottom surface (20 mm, 166 mm and 308 mm) and three different planks, and one additional sensor was measuring the temperature and relative humidity of the ambient air.

The FEM model for the Sørliveien deck is a 3D slice of the lamella, with the width of 48 mm, height 333 mm and thickness 5 mm. The weather data in Figure 6 was the primary information needed for the hygro-thermal simulation, because the boundary conditions of the model are based on the external RH and T. The numerical analysis is carried out from August 2005 until the end of January 2010, since in this period the measurements are continuous. The initial relative humidity and temperatures are equal to those of the air (RH_0 = 65%, T_0 = 26 °C) and the initial moisture content in equilibrium with RH_0 is MC_0 = 13.3%.

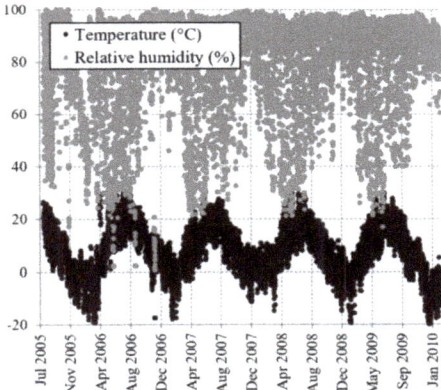

Figure 6. Sørliveien Bridge. Weather data measured between 2005 and 2010.

2.4. Case-Study: Tapiola Bridge

The highway overpass in the Tapiola district of Espoo, Finland, was built in the Spring of 2019 and was recently opened to traffic (Figure 8). Two of its three spans are stress-laminated timber decks compressed by steel bars in the transverse direction. This short-/mid-span highway crossing serves local car and bus transportation. In the width direction, the bridge deck is composed of 46 timber beams. The widths of the beams are 0.215 m, and the heights equal to the deck thicknesses are 0.765 m for the 13.45 m span and 1.035 m for the 22.13 m span. The width of the decks is 9.89 m, which is near to its useful width 9.79 m. The basic dimensions of the bridge are listed in Table 1.

Table 1. Tapiola Bridge decks.

Span	13.45 m	22.13 m	11.8 m
Material	stress-laminated timber	stress-laminated timber	concrete
Timber deck thickness	0.765 m	1.035 m	n/a
Useful deck width	9.79 m	9.79 m	13.13 m

Five integrated humidity-temperature sensors, two displacement and two force sensors were installed on the bridge. The displacement sensors monitor vertical and horizontal motion, while the force sensors are measuring the tension force variation in the steel bars. In addition, the monitoring unit cabinet has two thermocouples for tracking its inside and outside temperature. The sensors are described in Table 2. Figure 7 shows the locations of the sensors. More details on the sensor locations are provided in Figures A1 and A2 of Appendix B.

Table 2. The installed sensors in Tapiola Bridge deck.

No.	ID	Sensor Type	Model
1	KC1	Humidity and temperature	HMP110 [27]
2	KC2	Humidity and temperature	HMP110
3	KC3	Humidity and temperature	HMP110
4	KC4	Humidity and temperature	HMP110
5	KC5	Humidity and temperature	HMP110
6	V1	Force	C6A [28]
7	V2	Force	C6A
8	D1x	Slab longitudinal displacement	ELPC100 linear potentiometer of OPKON [29]
9	D2y	Slab vertical displacement	ELPC 100 linear potentiometer of OPKON

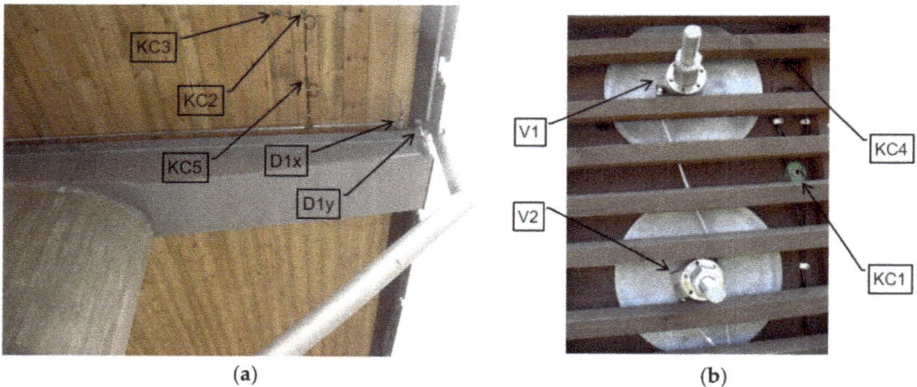

Figure 7. Tapiola Bridge. Photos of the sensor locations: (**a**) from the bottom; (**b**): from the lateral side. See the details about the locations of all sensors in Appendix B.

The sensors have wired connections to the monitoring unit. The unit itself is placed in the metal cabinet on the abutment of the bridge and it is connected to the electric grid and the internet. The devices and programmes of the unit are shown in Table 3.

Before the installation, the sensors were tested in the humidity-temperature controlled rooms at VTT Technical Research Centre of Finland Ltd (VTT). Only the temperature sensors (thermocouples), located inside and outside of the measurement enclosure were not calibrated, because they have lower precision requirements.

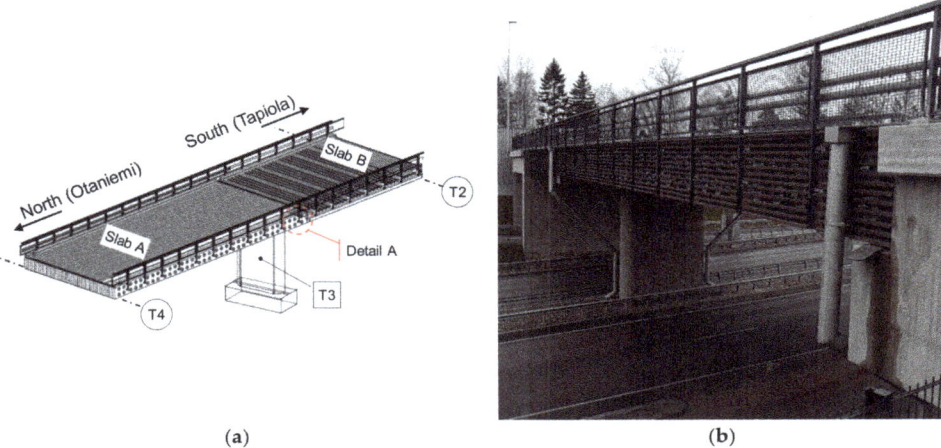

(a) (b)

Figure 8. Tapiola Bridge. (**a**) The scheme of the transverse prestressed glulam wooden slabs of the bridge where T2, T3 and T4 indicate the support, reproduced from [30] with permission from VTT publications. Detail A is shown in Appendix B. (**b**) A picture with the view of the bridge.

Table 3. The devices and software of the monitoring unit.

Industrial PC:	Advantech [31]
Measurement software:	Labview [32]
Remote desktop software:	DWAgent [33]
Data acquisition chassis:	NI cDAQ-9174 [32]
Data acquisition devices:	NI 9211 (thermocouple) [32]
	NI 9205 (temperature, moisture) [32]
	NI 9237 (Force) [33]
Power supply:	Quint Power [34]
Enclosure internal thermostat	
Enclosure heater	
Thermocouples	

The force sensor calibration was performed with the test rig of VTT, and a calibration factor of 0.95 was found. The current measurement system is able to record the relative force shift/change of the pre-tension bars. Displacement sensors were not explicitly calibrated, instead the manufacturer's instructions and precision requirements are followed [29].

The deck is protected with Valtti colour, an oil-based wood strain produced by Tikkurila [35]. According to the producer, this paint exhibits a low vapour resistance.

The FEM model for the deck is a 3D slice having a width of 107.5 mm (half of the lamination), height 1035 mm and thickness 5 mm. The numerical analysis of the Tapiola Bridge deck starts at the end of the construction time (April 2019) until October 2020, see the weather data in Figure 9, while the sensor-based monitoring started later (October 2019) and is on-going. The earlier starting of the numerical analysis demonstrates that the numerical models can assist the monitoring by predicting the hygro-thermal response of the SLTD also in the absence of measurements. The initial relative humidity and temperatures are equal to those of the air (RH_0 = 65%, T_0 = 0.85 °C) and the initial moisture content in equilibrium with RH_0 is MC_0 = 15.3%.

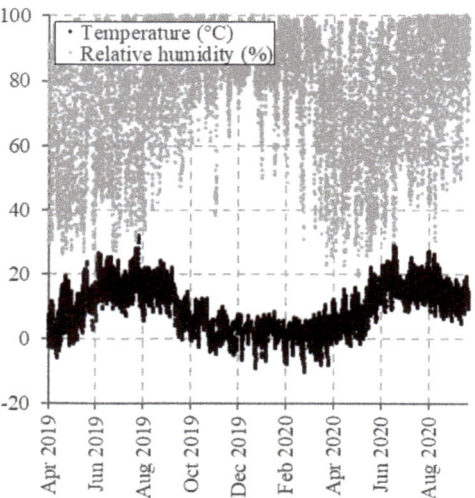

Figure 9. Tapiola Bridge. Weather data measured between 2019 and 2020.

3. Results

3.1. Hygro-Thermal Response of the Deck of Sørliveien Bridge

The outputs of the finite element model are the temperature, the moisture content and the vapour pressure in the wood material.

Since the RH and T in wood were measured directly by the monitoring system, reference results for the validation of the numerical model were available from all of the nine integrated humidity-temperature sensors installed in the wood lamellas. For the purpose of investigation of MCs and moisture gradients near the surface of untreated wood exposed to the external climate, data measured at 20 mm from the bottom surface were selected for comparison with the numerical results.

Figure 10 shows the comparisons in terms of vapour pressures between the measured and numerical data. The directly measured RH in wood was multiplied by the saturated vapour pressure by using the same Equation (9) adopted for the numerical model. Figure 11 presents the comparison between measured and numerical values of temperatures. The results of the FEM calculation show a good correlation to the yearly variation of the monitored temperatures and measurement-based vapour pressures.

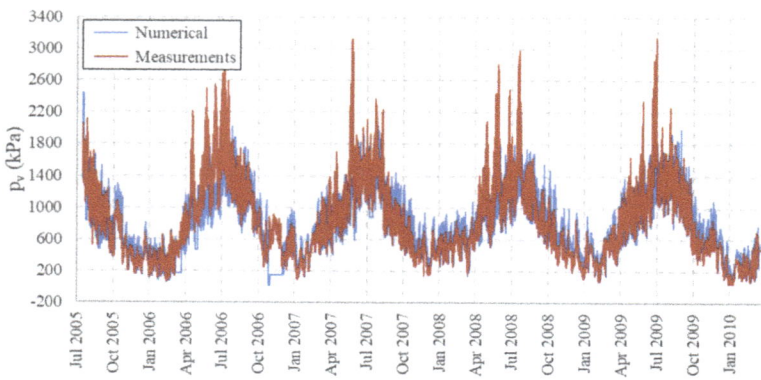

Figure 10. Sørliveien Bridge. Comparison between measurement-based and numerical vapour pressures in wood at 20 mm from the bottom surface.

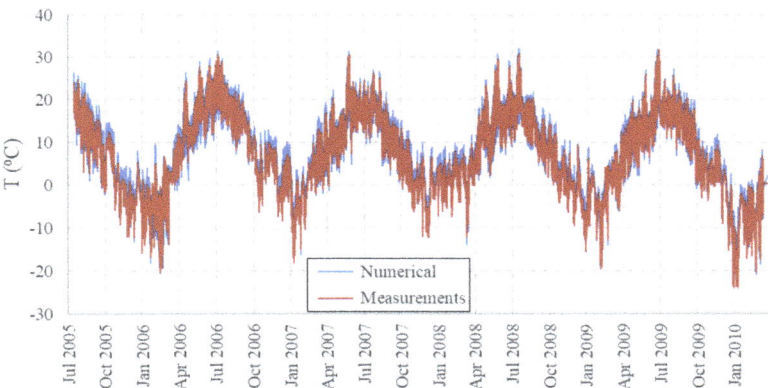

Figure 11. Sørliveien Bridge. Comparison between measured and numerical temperatures in wood at 20 mm from the bottom surface.

The numerical model assists the monitoring by allowing the evaluation of the *MC* from the bottom surface of the 3D slice until 20 mm, as shown in Figure 12. The numerical *MCs* close to the bottom surface are much higher than the ones at 20 mm from the surface that show small fluctuations (Figure 12a). The related moisture envelopes (minimum, maximum, average, 5th and 95th percentile), which show the trend of the moisture gradients, are presented in Figure 12b. A summary of the maximum and minimum *MC* values between the surface and 20 mm depth is shown in Table 4.

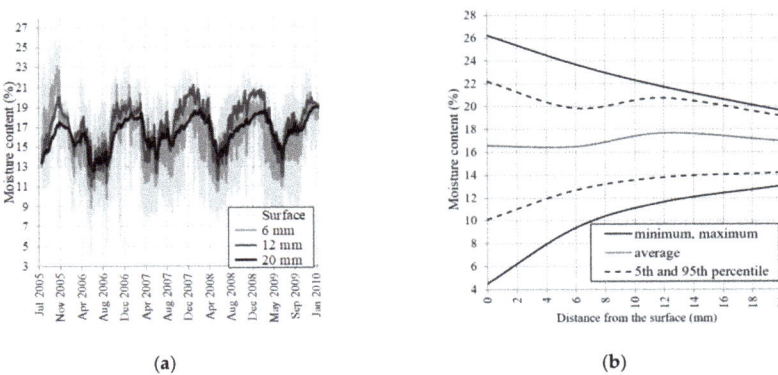

Figure 12. Sørliveien Bridge. Moisture content predicted by the finite element method (FEM) between the bottom surface and 20 mm depth. (**a**) *MC* vs. time. (**b**) Moisture envelopes (minimum, maximum, average, 5th and 95th percentile) from the external surface to 20 mm.

Table 4. Sørliveien Bridge. Numerical moisture content (MC) peaks at the bottom surface and 6, 12, 20 mm from the surface.

Distance from Bottom (mm)	Max MC (%)	Date	Min MC (%)	Date
0	25.8	31 October 2005	4	26 May 2006
6	23.2	12 November 2005	8.9	27 May 2006
12	21.2	19 January 2008	11.1	28 May 2006
20	19	26 February 2010	12.5	7 June 2006

3.2. Hygro-Thermal Response of the Deck of Tapiola Bridge

In this case-study, in this case-study, the results of the FEM analysis are in good agreement with the yearly variation of the temperatures and vapour pressures monitored at 60 mm from the bottom surface in sensors KC1 and KC2 (Figures 13 and 14). The larger temperatures measured in KC1 are because this sensor is located on the bridge side exposed to sun while KC2 is in the shadow. Since the model does not include the effect of solar radiation, the better comparison is with the data provided by sensor KC2 that is installed from the bottom of the deck.

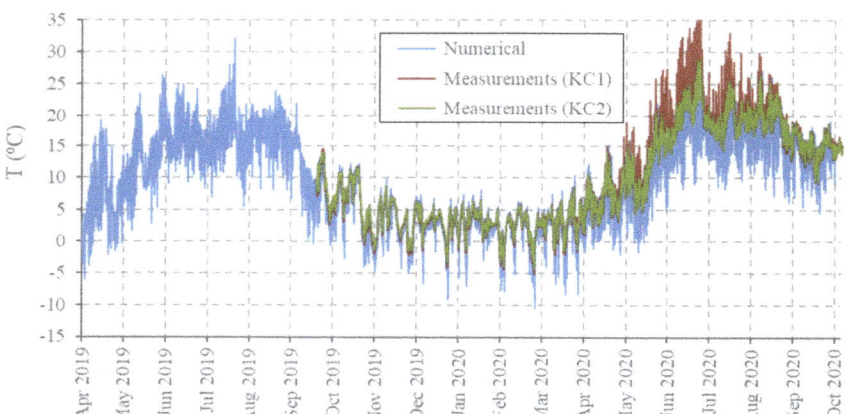

Figure 13. Tapiola Bridge. Comparison between numerical temperatures in wood and measurements in sensors KC1 and KC2 at 60 mm from the surface.

The *MC* history at 60 mm from the bottom surface shows very small daily fluctuations while the numerical results closer to the surface are larger (Figure 15a). Figure 15b shows the minimum and maximum moisture envelopes during the monitoring time, as well as the 5th and 95th percentile from the external bottom surface to 60 mm depth. For this thick deck, a summary of the maximum and minimum *MC* values between the surface and 400 mm depth is shown in Table 5.

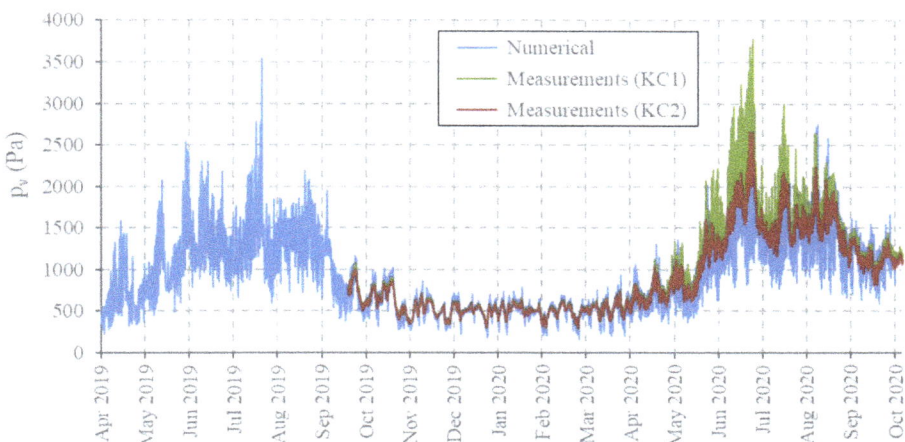

Figure 14. Tapiola Bridge. Comparison between numerical vapour pressures in wood and measurements in sensor KC2 at 60 mm from the bottom surface. In red the vapour pressure measurements in sensor KC1 at 60 mm from the lateral side exposed to the afternoon sun.

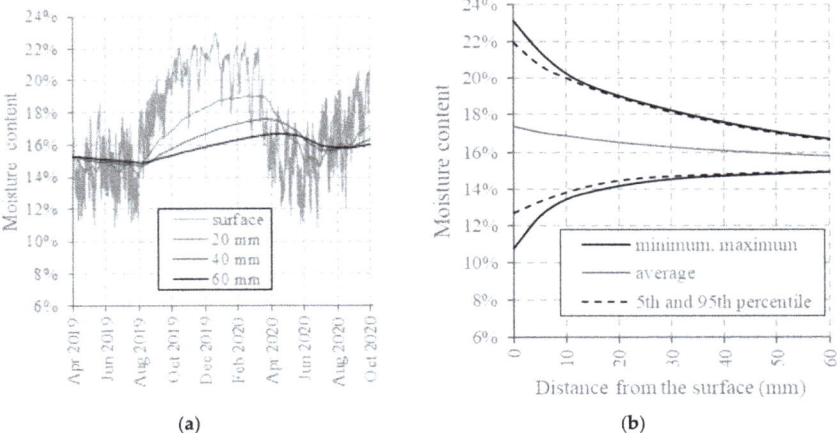

Figure 15. Tapiola Bridge. (**a**) Moisture contents from the external surface of the deck until 60 mm depth. (**b**) Moisture envelopes (minimum, maximum, average, 5th and 95th percentile) from the external surface to 60 mm.

Table 5. Tapiola Bridge. Numerical MC peaks at the bottom surface and 20, 60, 200 and 400 mm from the surface.

Distance from Bottom (mm)	Max MC (%)	Date	Min MC (%)	Date
0	23.1	25 December 2019	10.8	25 June 2020
20	19	15 March 2020	14.1	8 August 2019
60	16.7	19 April 2020	14.9	10 August 2019
200	15.4	30 September 2020	15.3	18 December 2019
400	15.3	12 April 2019	15.3	25 June 2020

4. Discussion

The combination of the sensor-based monitoring and numerical model presented in the previous sections, allowed analysing the hygro-thermal response of the uncoated SLTD of Sørliveien Bridge in Norway and the thick painted SLTD of Tapiola Bridge in Finland. In addition to the temperature T and vapour pressure p_v, the numerical model was able to provide quantitative values of the moisture content MC and the moisture gradients trends close to the surface that could be responsible of surface cracks, as shown in earlier works for large glulam beams of timber bridges [10–12] and in a research on the monitoring of large span timber structures [36]. In particular, we analysed the bottom part of decks which are sheltered from rain but subjected to both the continuously variable air humidity and temperature.

The uncoated deck of Sørliveien Bridge in Norway was analysed as a first case-study, and showed a relatively stable moisture behaviour after one year from the bridge erection. The main findings are listed below:

- Referring to Table 4, the MC values on the bottom surface span in a range between 4% (reached at the beginning of summer 2006) and 25.8% (autumn 2005), are highest during the first year of the analysis and more stable during the successive years (Figure 12). The minimum and maximum MC values at 20 mm from the bottom surface are 12.5% at the beginning of the summer 2006, and 19% at the end of winter 2010. The results show that no significant changes of the average moisture content would be expected after the first year of the bridge service life, i.e., from June 2006 to January 2010.
- High levels of MC over 20% were found only on the exposed surface at the bottom deck and in locations very close to this surface (Figure 12). These MC levels could be also critical for the wood durability [3], however the decay is a major problem mainly in the presence of liquid water due e.g., to rain and eventual water traps, and these cases were not investigated in the present work.

The painted and thick deck of Tapiola Bridge was analysed as second case-study, starting from an earlier stage after construction. The main findings are the following:

- Referring to Table 5, the moisture contents on the bottom surface varies between 10.8%, reached at the beginning of the summer, and 23% at the beginning of the winter. In the internal locations, the maximum and minimum moisture contents are reached earlier depending on the maximum moisture penetration depth, which is about 200 mm (see Table 5). The maximum and minimum MC values in the location of the humidity-temperature sensor at 60 mm from the bottom surface are 16.7%, at the beginning of spring, and 14.9% at the end of summer.
- High levels of MC (i.e., >20%) were found only on the exposed bottom surface and in locations very close to this surface. Compared to the MCs of Sørliveien Bridge, the peaks remained below 23% (Figure 15). This is because Tapiola Bridge is protected, even if the used paint has low vapour resistance (see Table A3 of Appendix A).

The following observations are based on the comparisons between the two case-studies:

- Considering the MC results of Sørliveien Bridge, it could be estimated that also for Tapiola Bridge the average values of MC will not change significantly during the successive years.
- The envelope curves shown in Figure 15b indicate similar levels of moisture gradients close to the surface as those of the Sørliveien Bridge's deck (Figure 12b). The average MCs are around 16% up to 20 mm depth, and remain at an almost constant level up until 60 mm depth in Tapiola Bridge's deck. Previous hygro-thermal models of SLTDs, based on single-phase moisture transport, found cupping deformations of around 16 mm and steel bar force losses of around 33% at these MC levels after 15 months [9].

The displacements and forces measured in the other sensors of the two monitoring systems can be simulated in future work by integrating the hygro-thermal analysis with a mechanical model for wood as in [9,12].

The embedded sensors and computational unit allow further expansion of the Internet of Things (IoT) network in order to efficiently exchange the monitoring data with passing vehicles and stationary objects of the road infrastructure. Moreover, combined with the results of the FEM simulation, the system can provide a comprehensive understanding of the bridge deck conditions in real time.

5. Conclusions

This paper proposed the use of an advanced multi-phase numerical model for wood below the fibre saturation point, previously introduced by some of the authors, to assist the monitoring of stress laminated timber decks by integrated humidity-temperature sensors.

The hygro-thermal simulation of a representative deck volume under Northern European climates supplements the sensor-based data. The simulation provides the distribution of the moisture content below the FSP, the temperature and the vapour pressure in the studied volume and allows to draw conclusions about the hygro-thermal response of the deck. However, the model does not include the effect of solar radiation and this is a task for future research. The modelling of the protective asphalt layer is also a topic for future work. The two analysed case-studies are sheltered from rain. To consider the effects of rain and possible water traps, the current model needs to be extended by introducing the variable concentration of free water in the lumens.

In future coupled hygro-thermo-mechanical models for SLTDs, the accurate evaluation of moisture contents is important for the prediction of moisture induced stresses which are responsible for surface cracking, cupping deformations and losses of the pre-stress force in steel bars.

The proposed method can be used to assist the monitoring techniques under Nordic climates contributing to maintenance cost reduction of timber bridge decks. FEM-assisted monitoring of bridges has a great potential to decrease the cost of instrumentation and increase safety. It can predict possible damages and communicate the results to other infrastructure components.

Author Contributions: Conceptualization, S.F., P.H., L.F.; methodology, S.F., P.H., H.B., K.K.; software, S.F., P.H., K.K.; validation, S.F., P.H., A.K.; formal analysis, S.F., P.H.; investigation, H.B., K.K.; resources, T.T.; data curation, P.H., K.K., H.B.; writing—original draft preparation, S.F., P.H.; writing—review and editing, P.H., L.F., T.T.; visualization, P.H., A.K.; supervision, S.F.; project administration, S.F.; funding acquisition, S.F., T.T. All authors have read and agreed to the published version of the manuscript.

Funding: This research was funded by WoodWisdom-Net project "Durable Timber Bridges" and by project "Delivering Fingertip Knowledge to Enable Service Life Performance Specification of Wood—Click Design", which is supported under the umbrella of ERA-NET Cofund ForestValue by the Ministry of the Environment of Finland. ForestValue has received funding from the European Union's Horizon 2020 research and innovation program. The Finnish Transport Infrastructure Agency (Väylävirasto) is a co-funder of the Click Design project.

Informed Consent Statement: Informed consent was obtained from all subjects involved in the study.

Data Availability Statement: The data presented in this study are available on request from the corresponding author. The data are not publicly available due to the agreements with the funding projects.

Acknowledgments: The authors wish to thank the Norwegian Public Road Administration for providing the monitoring data of Sørliveien Bridge. The Finnish Transport Infrastructure Agency (Väylävirasto) and the City of Espoo is acknowledged for supporting the monitoring of the Tapiola Bridge. The authors would like to warmly thank VTT colleagues Jukka Mäkinen, Mikko Kallio, Kalle Raunio, Pekka Halonen, Kari Korhonen for taking care of the on-going monitoring of Tapiola Bridge.

Conflicts of Interest: The authors declare no conflict of interest. Co-funder Väylävirasto (T.T.) participated in the interpretation of data, writing of the manuscript and in the decision to publish the results.

Appendix A

Table A1. Material parameters for the multi-Fickian model (all references can be found in [11]).

Water vapour diffusion tensor $\mathbf{D}_v = \boldsymbol{\xi}_v \left(2.31 \cdot 10^{-5} \left(\frac{p_{atm}}{p_{atm}+ p_v}\right) \left(\frac{T}{273}\right)^{1.81}\right)$ (m^{2}s^{-1}) • Atmospheric pressure $p_{atm} = 101325$ Pa • Vapour pressure $p_v = c_v RT/(\varphi M_{H2O})$ • Gas constant $R = 8.314$ $\left(\text{Jmol}^{-1}\text{K}^{-1}\right)$ • Molecular mass of water $M_{H2O} = 18.02 \times 10^{-3}$ $\left(\text{kgmol}^{-1}\right)$	Components of reduction factor $\boldsymbol{\xi}_v$ $\xi_{vL} = 0.9$ longitudinal $\xi_{vT} = 0.12$ transverse (*)
Bound water diffusion tensor $\mathbf{D}_b = \mathbf{D}_0 exp\left(-\frac{E_b}{RT}\right)$ (m^{2}s^{-1}) • Activation energy of bound water diffusion • $E_b = (38.5 - 29MC) \cdot 10^3$ $\left(\text{Jmol}^{-1}\right)$	Components of diagonal tensor $\mathbf{D}_0$ $D_{0L} = 17.5 \cdot 10^{-6}$ (m^{2}s^{-1}) longitudinal $D_{0T} = 7 \cdot 10^{-6}$ (m^{2}s^{-1}) transverse
Thermal conductivity tensor $\mathbf{K} = \boldsymbol{\xi}_H (G(0.2 + 0.38MC) + 0.024)$ $\left(\text{Wm}^{-1}\text{K}^{-1}\right)$ • Specific gravity of wood $G = 0.693 G_0/(0.653 + MC)$ • $G_0 = \varrho_0/\varrho_{0w} =$ dry wood density/water density	Components of reduction factor $\boldsymbol{\zeta}_H$ $\zeta_{HL} = 2.5$ longitudinal $\zeta_{HT} = 1$ transverse
Specific heat $c_w = \frac{0.0011T + MC - 0.0323}{1+MC}$ $\left(\text{Jkg}^{-1}\text{K}^{-1}\right)$ Wood density $\varrho = G(1 + MC)\varrho_w$ $\left(\text{kgm}^{-3}\right)$	
Enthalpy of bound water $h_b = -7.8955 \cdot 10^5 - 4.476206 \cdot 10^2 T + 2.274399 \cdot 10 T^2 - 4.9553577 \cdot 10^{-2} T^3 + 4.041035 \cdot 10^{-5} T^4$ $\left(\text{Jkg}^{-1}\right)$	
Enthalpy of water vapour $h_v = 1.891879 \times 10^6 + 2.56352 \times 10^3 T - 1.2360577 T^2$ $\left(\text{Jkg}^{-1}\right)$	
Sorption reaction rate function $H_c = \begin{cases} C_1 exp\left(-C_2 \left(\frac{c_b}{c_{bl}}\right)^{C_3}\right) + C_4 & c_b < c_{bl} \\ C_1 exp\left(-C_2 \left(2 - \frac{c_b}{c_{bl}}\right)^{C_3}\right) + C_4 & c_b > c_{bl} \end{cases}$	Constants $C_1 = 3.8 \cdot 10^{-4} (\text{s}^{-1})$ $C_2 = c_{21} exp(c_{22} RH) + c_{23} exp(c_{24} RH)$ $C_3 = 80.0$ $C_4 = 5.94 \cdot 10^{-7} (\text{s}^{-1})$ $c_{21} = 3.579$ $c_{22} = 2.21$ $c_{23} = 1.591 \cdot 10^{-3}$ $c_{24} = 14.98$
Anderson–McCarthy model for sorption isotherms $MC_{bl,\alpha} = -\frac{1}{f_{2\alpha}} \ln\left(\frac{\ln\left(\frac{1}{RH}\right)}{f_{1\alpha}}\right)$, $\alpha \in \{a, d\}$ • a and d refer to adsorption and desorption • $f_{i\alpha} = \sum_{j=0}^{n} b_{ij\alpha} T^j$, $i \in \{1, 2\}$	See Table A2
Permeances of the painted wood referred to wood temperature and air temperature: $k_v^w = \frac{1}{\frac{1}{k_w} + \frac{1}{k_p}} \frac{RT}{M_{H2O}}$, $k_v^a = \frac{1}{\frac{1}{k_w} + \frac{1}{k_p}} \frac{RT^a}{M_{H2O}}$ Thermal emission: $k_T = 20$ [W m^{-2} K^{-1}]	See Table A3

(*) selected in this paper.

Table A2. Shape parameters for the temperature dependent adsorption and desorption functions.

α	n	$b_{10\alpha}$ [-]	$b_{11\alpha}$ [K^{-1}]	$b_{20\alpha}$ [-]	$b_{21\alpha}$ [K^{-1}]
a	1	7.719	-0.011	5.079	0.046
d	1	9.739	-0.017	-13.419	0.100

Table A3. Permeances of weak paint and uncoated wood.

Paint	Permeance [kg/m^2 s Pa]
weak paint	4.0×10^{-9}
uncoated wood	5.0×10^{-9}

Appendix B

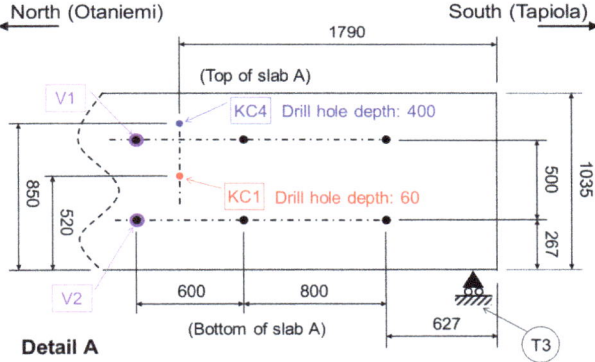

Figure A1. The locations of sensors in the side of the south-west corner of Slab A in the vicinity of the support T3 (see Figure 8a).

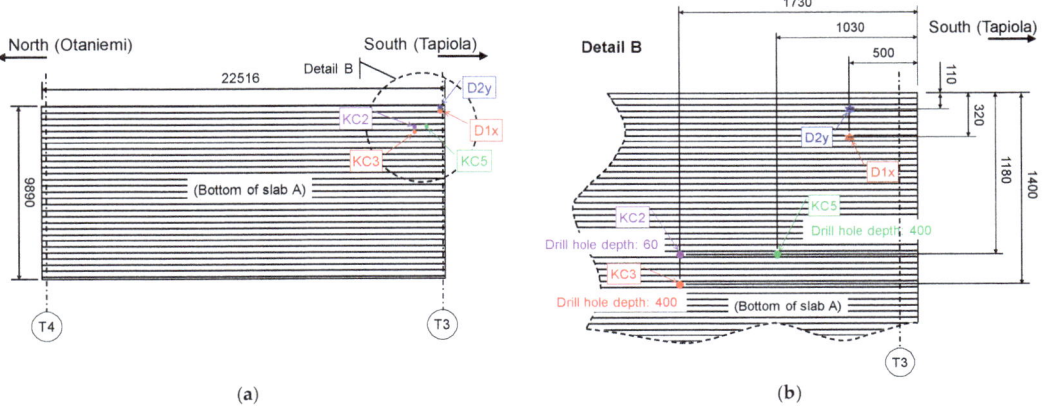

Figure A2. Tapiola Bridge: (**a**) the prestressed glulam wooden Slab A of the bridge. (**b**) The locations of sensors in the bottom of Slab A in the vicinity of the slab's south-west corner near the support T3 (see Figure 8a).

References

1. European Committee for Standardization. *EN 1990: Eurocode: Basis of Structural Design*; European Committee for Standardization: Brussels, Belgium, 2002.
2. Obataya, E. Characteristics of aged wood and Japanese traditional coating technology for wood protection. In Proceedings of the Actes de la Journée D'étude Conserver Aujourd'hui: Les "Vieillissements" du Bois, Cité de la Musique, Paris, France, 2 February 2007.
3. Brischke, C.; Meyer-Veltrup, L. Modelling timber decay caused by brown rot fungi. *Mater. Struct.* **2016**, *49*, 3281–3291. [CrossRef]
4. Wacker, J. *Cold Temperature Effect on Stress-Laminated Timber Bridges: A Laboratory Study*; Research Paper FPL–RP–605; U.S. Department of Agriculture, Forest Service, Forest Products Laboratory: Madison, WI, USA, 2003.
5. Scharmacher, F.; Müller, A.; Brunner, M. Asphalt surfacing on timber bridges. In Proceedings of the COST-Timber Bridges, Biel, Switzerland, 25–26 September 2014; Franke, S., Franke, B., Widmann, R., Eds.; Bern University of Applied Sciences: Bern, Switzerland, 2014.
6. Aasheim, E. Nordic timber bridge program—An overview. In Proceedings of the International Wood Engineering Conference, New Orleans, LA, USA, 21–28 October 1996; Gopu, V.K.A., Ed.; International Wood Engineering Conference: New Orleans, LA, USA, 1996.
7. Horn, H. *Rapport Oppdrag Nr. 310332: Monitoring Five Timber Bridges in Norway—Results 2012*; Norsk Treteknisk Institut: Oslo, Norway, 2013.
8. Ministry of the Environment, Department of the Built Environment. Wood Building Programme. Available online: https://ym.fi/en/wood-building (accessed on 14 November 2020).
9. Pousette, A.; Malo, K.; Thelandersson, S.; Fortino, S.; Salokangas, L.; Wacker, J. *Durable Timber Bridges—Final Report and Guidelines*; SP Report 25; Research Institutes of Sweden RISE: Skellefteå, Sweden, 2017.
10. Fortino, S.; Genoese, A.; Genoese, A.; Nunes, L.; Palma, P. Numerical modelling of the hygro-thermal response of timber bridges during their service life: A monitoring case-study. *Constr. Build. Mater.* **2013**, *47*, 1225–1234. [CrossRef]
11. Fortino, S.; Hradil, P.; Genoese, A.; Genoese, A.; Pousette, A. Numerical hygro-thermal analysis of coated wooden bridge members exposed to Northern European climates. *Constr. Build. Mater.* **2019**, *208*, 492–505. [CrossRef]
12. Fortino, S.; Hradil, P.; Metelli, G. Moisture-induced stresses in large glulam beams. Case study: Vihantasalmi Bridge. *Wood Mater. Sci. Eng.* **2019**, *14*, 366–380. [CrossRef]
13. Fragiacomo, M.; Fortino, S.; Tononi, D.; Usardi, I.; Toratti, T. Moisture-induced stresses perpendicular to grain in cross-sections of timber members exposed to different climates. *Eng. Struct.* **2011**, *33*, 3071–3078. [CrossRef]
14. Niklewski, J.; Fredriksson, M. The effects of joints on the moisture behaviour of rain exposed wood: A numerical study with experimental validation. *Wood Mater. Sci. Eng.* **2019**, 1–11. [CrossRef]
15. Florisson, S.; Vessby, J.; Mmari, W.; Ormarsson, S. Three-dimensional orthotropic nonlinear transient moisture simulation for wood: Analysis on the effect of scanning curves and nonlinearity. *Wood Sci. Technol.* **2020**, *54*, 1197–1222. [CrossRef]
16. Krabbenhøft, K. Moisture Transport in Wood: A Study of Physical-Mathematical Models and their Numerical Implementation. Ph.D. Thesis, Technical University of Denmark, Lyngby, Denmark, 2004.
17. Frandsen, H.L. Selected Constitutive models for simulating the hygromechanical response of wood. Ph.D. Thesis, Department of Civil Engineering Aalborg University, Aalborg, Denmark, 2007.
18. Eitelberger, J.; Hofstetter, K.; Dvinskikh, S.V. A multi-scale approach for simulation of transient moisture transport processes in wood below the fiber saturation point. *Compos. Sci. Technol.* **2011**, *71*, 1727–1738. [CrossRef]
19. Konopka, D.; Kaliske, M. Transient multi-Fickian hygro-mechanical analysis of wood. *Comput. Struct.* **2018**, *197*, 12–27. [CrossRef]
20. Huc, S.; Svensson, S.; Hozjan, T. Hygro-mechanical analysis of wood subjected to constant mechanical load and varying relative humidity. *Holzforschung* **2018**, *72*, 863–870. [CrossRef]
21. Autengruber, M.; Lukacevic, M.; Füssl, J. Finite-element-based moisture transport model for wood including free water above the fiber saturation point. *Int. J. Heat Mass Transf.* **2020**, *161*, 120228:1–120228:21. [CrossRef]
22. Svensson, S.; Turk, G.; Hozjan, T. Predicting moisture state of timber members in a continuously varying climate. *Eng. Struct.* **2011**, *33*, 3064–3070. [CrossRef]
23. Thalla, O.; Stiros, S.C. Wind-Induced Fatigue and Asymmetric Damage in a Timber Bridge. *Sensors* **2018**, *18*, 3867. [CrossRef] [PubMed]
24. Kepp, H.; Dyken, T. Thermal cction on timber bridges temperature variation measured in the deck of 3 timber bridges in Norway. In Proceedings of the International Conference Timber Bridges (ICTB2010), Lillehammer, Norway, 12–15 September 2010; Malo, K.A., Kleppe, O., Dyken, T., Eds.; Tapir Academic Press: Trondheim, Norway, 2010.
25. Dyken, T.; Kepp, H. Monitoring the moisture content of timber bridges. In Proceedings of the International Conference Timber Bridges (ICTB2010), Lillehammer, Norway, 12–15 September 2010; Malo, K.A., Kleppe, O., Dyken, T., Eds.; Tapir Academic Press: Trondheim, Norway, 2010.
26. Frandsen, H.L. *Modelling of Moisture Transport in Wood: State of the Art and Analytic Discussion*, 2nd ed.; Aalborg University: Aalborg, Denmark, 2005.
27. Vaisala Oyj Home Page. Available online: https://www.vaisala.com/ (accessed on 24 November 2020).
28. Hottinger Brüel & Kjaer GmbH Home Page. Available online: https://www.hbm.com/ (accessed on 24 November 2020).

29. Opkon Optik Elektronik Kontrol San. Tic. Ltd. Şti Home Page. Available online: https://www.opkon.com.tr/ (accessed on 24 November 2020).
30. Koski, K. *Instrumentation of Tapiolantien Risteyssilta (Bridge)*; Research Report VTT-R-00837-19; Technical Research Centre of Finland: Espoo, Finland, 2019.
31. Advantech, Co., Ltd. Home Page. Available online: https://www.advantech.com/ (accessed on 24 November 2020).
32. National Instruments. Home Page. Available online: http://www.ni.com/ (accessed on 24 November 2020).
33. DWS remote Control. Home Page. Available online: https://www.dwservice.net/ (accessed on 24 November 2020).
34. Phoenix Contact. Home Page. Available online: https://www.phoenixcontact.com/ (accessed on 24 November 2020).
35. Tikkurila Valtti Color Safety Data Sheet. Available online: https://tikkurila.com/sites/default/files/valtti-color-sds-en.pdf (accessed on 14 November 2020).
36. Dietsch, P.; Gamper, A.; Merk, M.; Winter, S. Monitoring building climate and timber moisture gradient in large-span timber structures. *J. Civil. Struct. Health Monit.* **2015**, *5*, 153–165. [CrossRef]

Article

Archetypal Use of Artificial Intelligence for Bridge Structural Monitoring

Bernardino Chiaia [1,2] and Valerio De Biagi [1,2,*]

[1] DISEG, Department of Structural, Geotechnical and Building Engineering, Politecnico di Torino, 10129 Torino, Italy; bernardino.chiaia@polito.it
[2] SISCON, Center for Safety of Infrastructures and Constructions, Politecnico di Torino, 10129 Torino, Italy
* Correspondence: valerio.debiagi@polito.it; Tel.: +39-011-0904842

Received: 31 August 2020; Accepted: 12 October 2020; Published: 14 October 2020

Abstract: Structural monitoring is a research topic that is receiving more and more attention, especially in light of the fact that a large part our infrastructural heritage was built in the Sixties and is aging and approaching the end of its design working life. The detection of damage is usually performed through artificial intelligence techniques. In contrast, tools for the localization and the estimation of the extent of the damage are limited, mainly due to the complete datasets of damages needed for training the system. The proposed approach consists in numerically generating datasets of damaged structures on the basis of random variables representing the actions and the possible damages. Neural networks were trained to perform the main structural monitoring tasks: damage detection, localization, and estimation. The artificial intelligence tool interpreted the measurements on a real structure. To simulate real measurements more accurately, noise was added to the synthetic dataset. The results indicate that the accuracy of the measurement devices plays a relevant role in the quality of the monitoring.

Keywords: structural health monitoring; damage detection; damage localization; hybrid approach; neural network

---

1. Introduction

Modern society makes large use of civil infrastructures. Hydroelectric power generation presupposes that dams are built along mountain valleys to store water. River and valleys are crossed by bridges and viaducts, tunnels are built under mountains. Any engineered product, in particular a structure such as a dam or a bridge, has a finite design working life. This means that after a certain amount of time (e.g., half a century), large maintenance works are required, otherwise, the structural safety reduces to unacceptable levels and the infrastructure must be destroyed or abandoned [1]. Large efforts have been made by researchers to understand the phenomena that occur on the infrastructures and the ageing processes acting on the structures, which reduce their bearing capacity [2]. After decades of tests and a larger and solid knowledge base, modern structural design philosophies account for the variability of the loads, the environmental effects on the structure, and the possibility of accidental phenomena (such as earthquakes or fire) and impose strict prescription in construction works to avoid the need of large maintenance works during the expected working life.

Nevertheless, it must be remembered that the larger part of the actual infrastructural heritage was built in the Sixties, and it has had more than fifty years of service. Such structures suffer from two major problems. First, it must be considered that loading and environmental scenarios not accounted for during the design phase could have occurred, thus, the infrastructures can have experienced a quicker ageing process. In fact, this results in costly maintenance or in abandonment before the end of the design working life. Second, the infrastructures cannot be replaced in an economical manner since service inconveniences and high construction costs must be considered before planning the demolition and the subsequent reconstruction [3]. Therefore, to lengthen the working period, a structural health monitoring (SHM) framework must be implemented to detect the location and the extent of damage on structures [3,4].

Briefly, SHM can be implemented through non-destructive or destructive techniques. The former presupposes that continuous or discrete measurements are carried out during a time period and modification of the trends are interpreted as the evolution of the damage. The latter implies that samples of material (concrete, steel, reinforcement bars) are taken from the structure and tested in in laboratories to determine the residual mechanical properties (uniaxial compression strength in concrete, tensile strength in steel) [5]. The resulting information is implemented in numerical finite element (FE) models of the structure, and the structural safety is assessed. Destructive investigations are also required when there is the possibility that components buried in the structure, e.g., post-tensioned tendons, are damaged. Referring to non-destructive methods, many approaches have been formulated, mainly based on the study of the evolution of the dynamical properties (vibration frequencies and damping) of the structure [6,7]. Details on the possible analysis techniques and their pros and cons can be found in the specific literature (see, e.g., [7–9]).

As reported by Farrar and Worden [3], machine learning is a useful tool in SHM as it can provide interesting insights into the following five hierarchical points: (i) detection of the damage, (ii) localization of the damage, (iii) classification of the type of damage, (iv) evaluation of the extent of the damage, and (v) prediction about the residual safety of the structure. Supervised and unsupervised learning approaches can be implemented into a machine learning framework. Referring to the former, it is mandatory that data from every conceivable damage situation should be available. In this sense, all the five points previously enumerated can be implemented into an artificial intelligence framework. In contrast, unsupervised learning can only be adopted for the detection of damage and, sometimes, for the localization of the damage.

The comparison between statistical patterns of recorded data provides information about the structural condition [10,11], provided that a calibration phase on an undamaged structure is performed [12]. Pattern recognition of extracted features has been found effective for detecting damages: various artificial intelligence techniques have been proposed, e.g., autoregressive models [13], artificial neural networks [14], or support vector machines [15]. Novelty detection or anomaly detection methods have been also largely adopted; some examples are provided in [16–19].

A recent review paper on the emerging artificial intelligence methods applied in structural engineering was written by Salehi and Burgueno [20]; it concluded that the use of artificial intelligence in feature extraction is a powerful tool in SHM [21].

The current approaches reveal that large datasets are needed to monitor a structure. In particular, to detect a damage, structural information from when the damage was not present is required. This reflects that the knowledge of at least one previous state of the system is needed to check for a change in structural behavior, which can be interpreted through the approaches previously illustrated. Obviously, this is not always possible since, usually, monitoring starts after the appearance of a defect on a structure. This evidence suggests two major problems. On one side, to encompass all the possible types of damage (location, type, and extent), a consistent amount of information related to a large number of monitored structures is required. On the other, adopting unsupervised learning approaches requires long time records on a single structure. The present paper proposes a hybrid approach that helps to address these problems. The method allows for numerical modeling to simulate all the possible damage configurations on a structure and for supervised learning to interpret the records of sensors on a structure. Exploiting just a small part of the power of machine learning algorithms, in the present paper, shallow neural networks (NN) [22] are adopted as learning machines for the analysis. To be implemented in a real case, the fact that the measurements can be affected by errors must be included in the analysis. Thus, random errors are included in the analysis.

2. Methods

The current investigation involved, first, the generation of a synthetic dataset in which all the parameters are known and, then, the training of learning tools to determine whether a structure is damaged and where the damage is located.

2.1. Generation of the Synthetic Dataset

The numerical evaluation of displacements and vibration frequencies of undamaged and damaged structural members onto which a moving load is acting was first performed. Such data served as information for the training of neural networks devoted to the identification and the localization of the damage. To this aim, a simple supported beam served as the reference structure for the test. The choice of this scheme was based on the fact that this type of support was largely adopted in historical bridge building and because of its inherent low damage tolerance as a statically determinate structure. The beam was 30 m long, had a rectangular cross section (0.4 m × 1.5 m), and was constituted by a material having elastic modulus E_0 and density equal to 30 GPa and 2500 kg/m^3, respectively. A moving load, namely P, whose position was identified through the variable x_P, acted on the beam. The magnitude of the load was not constant across the performed simulations: A uniform distribution between 50 kN and 100 kN was attributed to P to simulate a real scenario of heterogeneous traffic.

A set of 30k simulations with moving load were performed using Matlab (MathWorks, Natick, MA, USA) coupled to OpenSees solver [23]. A first subset of 10k simulations, namely Subset A, related to an undamaged beam: In each simulation, the position and the magnitude of the load were randomly modified. In the second subset of 10k simulations, Subset B, the damage was inserted. To this purpose, the beam was split into 20 parts (elements) of equal length (1.5 m each). For beams with stronger flexural resistance mechanisms, the parameter that rules the behavior of the system is the flexural inertia, which is the product between the inertia moment around the flexural axis and the elastic modulus of the material composing the cross section. Obviously, for elements made of materials with different mechanical properties, such as reinforced concrete, it is possible to determine equivalent inertia and elastic modulus values. Thus, the reduction of the cross section of the material, for example, due to corrosion, can be simulated with a reduction of either the moment of inertia or the elastic modulus [24]. A Lemaitre-Chaboche model [25] was adopted to simulate the damage. The damage model is herein intended as a reduction of the elastic modulus of the cross section. The elastic modulus of the damaged element is

$$E_d = (1 - d) E_0, \qquad (1)$$

where d is the damage parameter ($d = 0$ for undamaged, $d = 1$ for complete damage). For $d = 1$, a mechanism forms and, thus, the equilibrium cannot be guaranteed. The possible range of the damage parameter was set at [0;0.2]. Larger values are out of the scope of the present analysis since our interest is in incipient damage rather than on already evolved damage. For each simulation, the damaged element and the damage magnitude (parameter d) were randomly identified. The adopted approach encompasses all the possible damages that can occur on reinforced concrete elements and concrete elements with pretensioned tendons, which represent the major structural types of road bridge infrastructures.

The third subset of data (Subset C, 10k entries) was represented by five simulations in which the load P was located at different positions along the length of the damaged element once a damage was assigned. In detail, for each entry of the dataset, the damaged element and the damage magnitude were randomly identified, and the structure was solved for the load P at 1/5, 2/5, 3/5, and 4/5 of the beam length. An additional extra simulation accounting for the beam with just a gravity load was performed, and this condition is subsequently referred to as unloaded.

The structure was modeled as planar and discretized with 21 nodes and 20 beam-elements with 6 degrees-of-freedom each (3 for each end). The cross section and material properties were associated to the elements. The mass of the beam was attributed to each node considering its tributary length.

The mass associated to the moving load was not considered at this stage. Figure 1 illustrates the beam with its discretization.

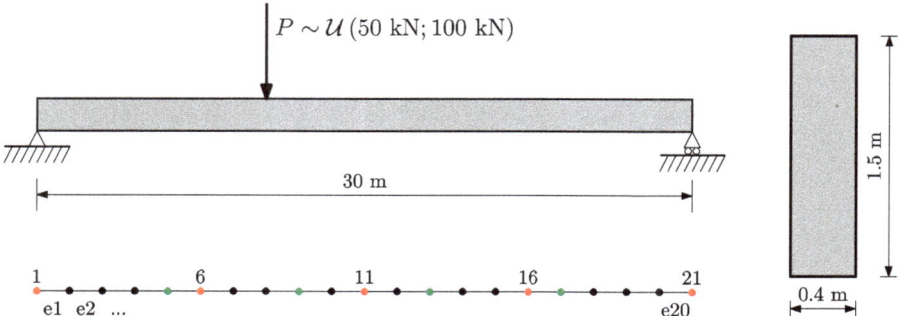

Figure 1. Sketch of the reference beam structure. The bottom scheme illustrates the discretization. Red nodes refer to the location of the inclinometers; green nodes refer to the location of the load P in the third simulations subset (Subset C).

An eigenvalue analysis was performed to determine the vibration modes, while a static analysis served for the vertical displacements. For each simulation, the frequencies associated to the first three vibration modes, namely f_1, f_2, and f_3, and the rotations at the beam ends, at midspan, and at quarter lengths were recorded (red points in Figure 1). The choice of measuring the rotations instead of the displacements follows the fact that on real structures inclinometers can be installed more easily than can displacement sensors. The presence (or not) of the damage, the position and the magnitude of the moving load, the location and the intensity (d) of the damage, and the recorded frequencies and rotation constitute the synthetic dataset.

2.2. Supervised Learning

To simulate the instrumental and post-process accuracy in real measurements, a white noise was applied to the data recorded in the simulations. It is supposed that the accuracy related to the frequency estimation differs from the accuracy in the measured rotations. In detail, a_{freq} is the half amplitude of the accuracy applied to each frequency and a_{rot} is the half amplitude of the accuracy applied to each measured rotation. The modified frequencies and rotations are, respectively,

$$\begin{aligned} f_i{}^n &= f_i + \text{rnd}(-a_{freq}; +a_{freq}) & i = 1,2,3 \\ \varphi{}^n &= \varphi + \text{rnd}(-a_{rot}; +a_{rot}) & i = 1,2,3 \end{aligned} \quad (2)$$

where f_i is a vibration frequency and φ a rotation, rnd($l; u$) is a random number generator in the range [$l; u$].

The obtained datasets served for the supervised learning of the neural networks (NN). The current investigation involved the training of three different neural networks to solve different problems in damage identification. The number of hidden neurons reflects the size of the input and output dataset [26], avoiding an excessive number of hidden units. The three considered networks were:

- Step 1: understanding whether the system is damaged or not. To this aim, a two-layers feed forward neural network-based classifier with 10 sigmoid hidden neurons and a softmax output neuron was built. Subset A and Subset B were used to train the neural network. In detail, the datasets were merged and shuffled. A two-variables state vector with values {−1; +1} identifies if the system is undamaged (−1) or damaged (+1). The supervised learning consisted in splitting the entire dataset (20k entries) into three groups: 14k served for training, 3k for validation, and 3k for testing. The Levenberg–Marquardt with Bayesian regularization algorithm was adopted to train the neural network [27].

- Step 2: identifying the location of the damage. To this aim, a two-layers feed forward neural network with 16 sigmoid hidden neurons and a linear output network was built. Subset C, only, was used to train the network. In detail, the subset was offset in such a way that the contribution of the imposed strains (due, for example, to thermal effects) was compensated for: The measured rotations when load P acts are subtracted from the rotations in the unloaded case. The input training dataset consisted of 7k entries, while the output consisted of a vector containing the location of the midpoint of the damaged element. The supervised training was validated and tested with the remaining 3k entries (1.5k for validation and 1.5k for testing). The Levenberg–Marquardt with Bayesian regularization algorithm was adopted to train the neural network.
- Step 3: identifying the magnitude of the damage. To this aim, a neural network similar to the one adopted for Step 2 was built and trained. The offset Subset C was split into three groups and the training consisted of fitting the output, i.e., the magnitude of the damage, with the measured inputs (frequencies and rotations).

Figure 2 summarizes the subsets and the performed analyses. The shallow neural networks were built and trained in a MATLAB environment adopting the built-in functions. To highlight the effects of the accuracy of the inputs, various half amplitudes, i.e., a_{freq} and a_{rot}, were tested. In detail, the term related to the vibration frequencies ranged between 0.001 Hz and 1 Hz, while the term related to the measured inclination ranged between 0.001° and 1°.

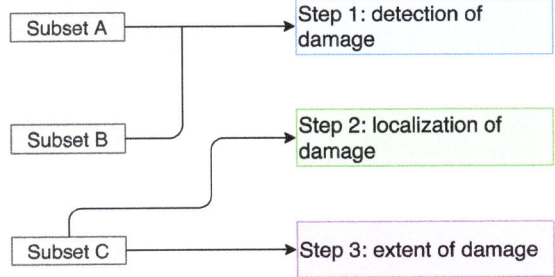

Figure 2. Scheme of the datasets adopted for the training of the neural networks for each step of the structural monitoring.

3. Results

The analyses showed a large variety of trends and dependencies on the accuracy of the estimated frequencies and measured rotations. The proposed parametric analysis only relates to the testing dataset of each neural network training process.

Referring to the identification of the presence of damage, the quality of the classification between damaged and undamaged was measured through the misclassification error. Figure 3 plots the confusion matrix for a_{freq} = 0.01 Hz and a_{rot} = 0.1°. The testing dataset was constituted by 1455 undamaged and 1545 damaged beams, for a total of 3000 structures. In the green cells, the number of true positives (the NN predicts damage on a structure that is really damaged) and true negatives (the NN predicts no damage on a structure that is really undamaged) are reported. The red cells highlight the false positives (the NN predicts damage on a structure that, on the contrary, is undamaged) and false negatives (the NN predicts no damage on a structure that, on the contrary, is damaged). The misclassification error, i.e., the percentage of false values (199) with the respect to the number of tested cases (3000), was 6.63% and is reported in red in the bottom-right cell. The misclassification error was adopted as an indicator of the performance of the capacity of the system to identify whether the beam is damaged.

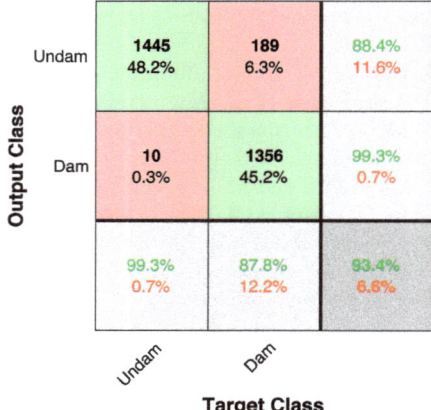

Figure 3. Example of a confusion matrix for $a_{freq} = 0.01$ Hz and $a_{rot} = 0.1°$. Dam and Undam refer to the damaged and undamaged structure, respectively. The percentages in the grey cells can be interpreted as "precision", "specificity", and so forth (check [28] for each single term).

The misclassification error was evaluated for various accuracies of the vibration frequencies and rotations. Figure 4 depicts a contour plot of the error for the considered ranges of variables. Contour lines, which identify an equal amount of error, parallel to one axis denote that the variable represented in the axis does not influence the performance of the classification. That is, for precise rotation measurements or for definite vibration frequencies estimation, the influence of the accuracy of the other data is negligible, as the contour lines are parallel to the X and Y axes, respectively.

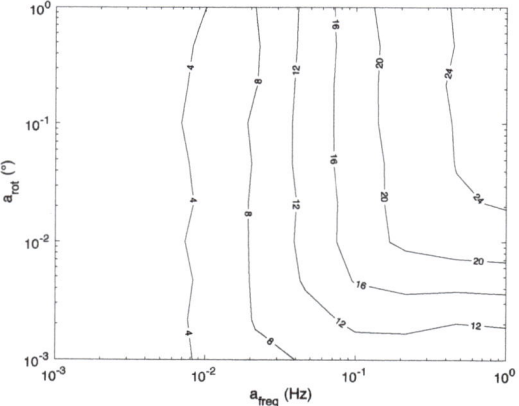

Figure 4. Misclassification error (in percentage) as a function of the accuracy in the estimation of the vibration frequency and in the measure of the rotation on log-log axes.

Referring to Step 2 and Step 3 simulations, the root-mean-squared error (RMSE) is the parameter that describes the precision in the location of the damage and its magnitude. The RMSE is defined as

$$\text{RMSE} = \frac{1}{N}\Sigma_{i=1}^{N}(O_i - T_i)^2, \quad (3)$$

where N is the number of testing entries (3k for the present case), T is the target value, and O is the output value from the trained neural network. Figure 5 illustrates the regression plot related to $a_{freq} = 0.001$ Hz and $a_{rot} = 0.001°$. Each circle corresponds to a couple $(T;O)_i$. The dashed red line relates to the perfect fitting, i.e., the output equals the target. The squared error tends to reduce for damage located at the midspan (target).

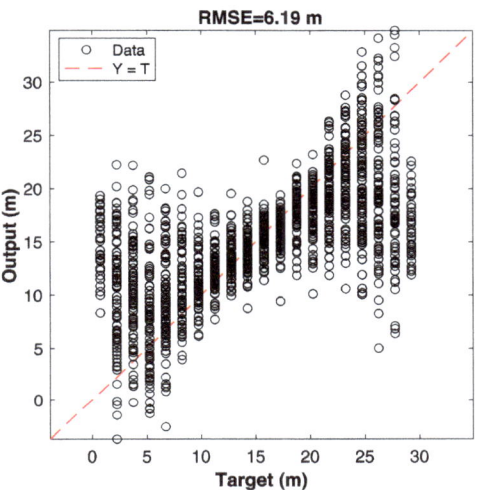

Figure 5. Regression plot related to $a_{freq} = 0.001$ Hz and $a_{rot} = 0.001°$. RMSE: root-mean-squared error.

The same analysis was repeated, varying the accuracy variables in the aforementioned range. As can be seen from Equation (2), the root-mean-squared error for damage location is a quantity with a physical dimension. The parameter related to damage location is in meters, while the parameter related to damage magnitude has the same physical dimension as d, i.e., it is dimensionless. Figure 6 reports the values of the errors for damage location and damage magnitude. It can be noted that damage location largely depends on the accuracy in the measure of the rotation of the beam, while it is roughly unaffected by the accuracy in the frequency estimation. The opposite consideration can be drawn for damage magnitude.

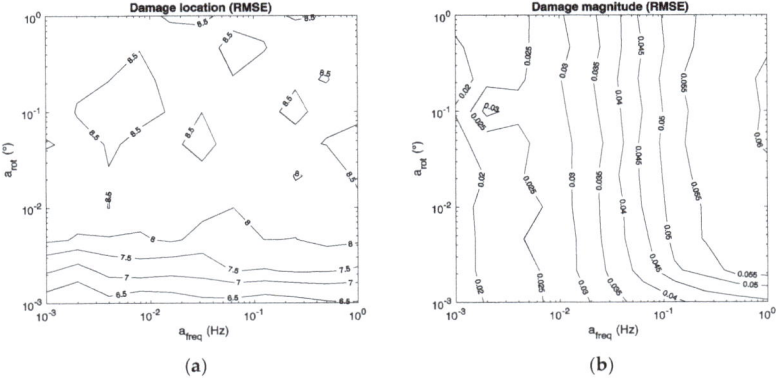

Figure 6. Root-mean-squared errors for the accuracy of the frequency in the range 0.001 Hz to 1 Hz and the accuracy of the rotation in the range $0.001°$ to $1°$. In (**a**) the root-mean-squared error, in meters, related to damage location is proposed. In (**b**) the root-mean-squared error related to damage magnitude is reported.

It should be mentioned that considering datasets without any errors, i.e., $a_{freq} = 0$ and $a_{rot} = 0$, the misclassification error drops to 2%, the root-mean-squared errors related to damage location and magnitude are 0.47 m and 0.001, respectively.

4. Discussions and Conclusions

Tests on a simply supported beam subjected to damage provided interesting insights into the possibility of implementing a hybrid method consisting of numerical simulations and real measurements for monitoring the state of conservation of a structure. Real measures can be, for example, recorded during the motion of a vehicle over the beam.

Referring to the detection of damage, Figure 4 details the relative importance between precision in the estimation of the vibration frequencies and the accuracy in the measured inclinations. In this sense, the performed analyses highlight that the most relevant information is provided by the dynamic properties of the system, i.e., the vibration frequencies if the accuracy in their evaluation is smaller than 5×10^{-2} Hz, otherwise both static (inclination) and dynamic information are needed.

With reference to the localization of the damage, the precision in the measurements of the inclination of the beam represents the key aspect for the determination of the position of the damage. For accurate measurements (around 1×10^{-3} degrees), it is expected that the accuracy of the localization is around 6 m, independent from the precision in the evaluation of the vibration frequencies. The accuracy is more or less 30% of the beam length. For more rough measurements, i.e., an accuracy of about 1×10^{-2} degrees or larger, the expected precision is around 40% of the beam length. This appears to be sufficient for a rough localization of the damage and a good input for other traditional techniques, say material sampling.

Referring to the extent of the damage, the finer the evaluation of the frequencies, the more precise the amount of damage. The accuracy of inclination measurements does not undermine the estimation of the extent of the damage.

Interesting comparisons can be drawn between the results of the present investigation with those of other studies found in the literature. The obtained results are in agreement with the studies performed by Neves and colleagues [24] on a numerical model of a railway bridge subjected to damage. In their case, they found that the accuracy of the trained classification neural network is very high if the damage extends along the beam for 3.5% of the beam length, roughly similar to the case herein analyzed (5%). The results, although applied to different structures, are similar. The dependency of the results on the dynamic properties is in accordance with Rageh et al. [29], who trained a neural network using a dataset of numerical time series solutions of a damaged bridge structure.

The importance of training the artificial intelligence system with processed data is a key aspect in damaged/undamaged classification. Cury et al. [30] determined an increase of the quality of the classification process by adopting modal data, i.e., processed information rather than raw accelerations. The same results were found in the present analysis: The classification error depicted in Figure 4 shows that the accuracy of the system was sensitive to the precision of the estimated frequencies, which are processed data, rather than to the measured inclinations, which are raw measures. The importance of a full analysis of the influence of sensors precision on the accuracy of the system improved the results of Yan et al. [31], who limited their analysis to 1% and 3% noise relative amplitude.

The possibility of using computational mechanics methods for building datasets for training the artificial intelligence system aims at overcoming the problem of having sets of measures on damaged structures. This point was precisely pointed out by Cheung et al. [21], who showed that the autoregressive algorithm provides precise indication, provided that the system previously experienced damage. In civil engineering structures, this is not possible, since there is no possibility of damaging a structure without causing its destruction. An example of modelling the damage on a recoverable structural element is proposed by Shahsavari et al. [32]. Although interesting and a harbinger of suggestions, their experimental campaign is tailored for the single tested steel element, rather than to a complete frame structure. Hence, the present approach is a tentative implementation of existing techniques for wider use of artificial intelligence for structural health monitoring.

As highlighted by Salehi and Burgueno [20], the majority of the studies focusing on the use of pattern recognition for SHM serve to detect damage. The localization of the damage and its extent are not possible since information on the evolution of the damage is not known a priori on the structure. This is the result of the fact that the structural behavior is directly dependent on damage location. In this sense, artificial intelligence can play an integrated role with structural numerical modeling.

The novelty of the proposed approach consists in the fact that the dataset that servers for training the learning algorithm comes from numerical analyses, rather than from observations on the structure. A set of damage configurations (location and extent) are modelled and the resulting structural displacements and dynamical properties are used for supervised training, in the present case, of a shallow neural network. In this sense, the use of machine learning can be enlarged to damage location and damage extent. The proposed approach was aimed at overcoming the limitations that emerged in the previous studies. Although theoretical, the approach can be applied for studying structures that are very similar (in construction engineering two structures cannot be identical since there is always human intervention in the construction process). This is the case, for example, of the overpasses of Highway A21 "Torino-Piacenza-Brescia" in the Northwest of Italy. Here, there are several overpasses that are coeval and have the same structural scheme, details, and techniques (Figure 7). The design of such infrastructures was performed by Dott. Ing. Gervaso [33]. The hybrid approach herein proposed fits well with monitoring such types of structures, for which an in-depth preliminary study and a detailed numerical modeling can be performed. The simulated structural dataset can be considered for a large number of similar real structures. Future studies on a real monitored structure are planned, as well as tests on more complex structural types, for example, portal frames.

Figure 7. Street view of one of the overpasses of the Italian Highway A21 (source Google Maps).

Author Contributions: Conceptualization, B.C. and V.D.B.; methodology, software, validation, formal analysis, writing—original draft preparation, V.D.B.; writing—review and editing, B.C. and V.D.B. All authors have read and agreed to the published version of the manuscript.

Funding: This research received no external funding.

Conflicts of Interest: The authors declare no conflict of interest.

References

1. *EN 1990:2002 Eurocode 0—Basis of Structural Design*; Comité Européen de Normalisation: Brussels, Belgium, 2002.
2. Wang, C.; Li, Q.; Ellingwood, B.R. Time-dependent reliability of ageing structures: An approximate approach. *Struct. Infrastruct. Eng.* **2016**, *12*, 1566–1572. [CrossRef]

3. Farrar, C.R.; Worden, K. *Structural Health Monitoring: A Machine Learning Perspective*; John Wiley & Sons: London, UK, 2013.
4. Farrar, C.R.; Worden, K. An introduction to structural health monitoring. *Philos. Trans. R. Soc. Lond. Ser. A* **2007**, *365*, 303–315. [CrossRef] [PubMed]
5. Seo, J.; Hu, J.W.; Lee, J. Summary review of structural health monitoring applications for highway bridges. *J. Perform. Constr. Facil.* **2016**, *30*, 04015072. [CrossRef]
6. Cawley, P.; Adams, R.D. The location of defects in structures from measurements of natural frequencies. *J. Strain Anal. Eng. Des.* **1979**, *14*, 49–57. [CrossRef]
7. Salawu, O.S. Detection of structural damage through changes in frequency: A review. *Eng. Struct.* **1997**, *19*, 718–723. [CrossRef]
8. Chen, H.-P. *Structural Health Monitoring of Large Civil Engineering Structures*; Wiley Blackwell: New York, NY, USA, 2018.
9. Deraemaeker, A.; Worden, K. (Eds.) *New Trends in Vibration Based Structural Health Monitoring*; Springer: Wien, Austria, 2010.
10. Sohn, H.; Farrar, C.R.; Hunter, N.F.; Worden, K. Structural health monitoring using statistical pattern recognition techniques. *J. Dyn. Syst. Meas. Control* **2001**, *123*, 706–711. [CrossRef]
11. Noman, A.S.; Deeba, F.; Bagchi, A. Health monitoring of structures using statistical pattern recognition techniques. *J. Perform. Constr. Facil.* **2012**, *27*, 575–584. [CrossRef]
12. Gul, M.; Catbas, F.N. Statistical pattern recognition for structural health monitoring using time series modeling: Theory and experimental verifications. *Mech. Syst. Signal Process.* **2009**, *23*, 2192–2204. [CrossRef]
13. Andre, J.; Kiremidjian, A.; Liao, Y.; Georgakis, C.T. Structural health monitoring approach for detecting ice accretion on bridge cables using the autoregressive model. In *Eighth International Conference on Bridge Maintenance, Safety and Management*; International Society for Optics and Photonics: Foz do Iguacu, Brazil, 2016; p. 431.
14. Lam, H.F.; Ng, C.T. The selection of pattern features for structural damage detection using an extended Bayesian ANN algorithm. *Eng. Struct.* **2008**, *30*, 2762–2770. [CrossRef]
15. Liu, C.-C.; Liu, J. Damage identification of a long-span arch bridge based on support vector machine. *Zhendong Yu Chongji J. Vib Shock* **2010**, *29*, 174–178.
16. Bishop, C.M. Novelty detection and neural network validation. *IEEE Proc. Vis. Image Signal Process.* **1994**, *141*, 217–222. [CrossRef]
17. Worden, K. Structural fault detection using a novelty measure. *J. Sound Vib.* **1997**, *201*, 85–101. [CrossRef]
18. Markou, M.; Singh, S. Novelty detection—A review. Part I: Statistical approaches. *Signal Process.* **2003**, *83*, 2481–2497. [CrossRef]
19. Markou, M.; Singh, S. Novelty detection—A review. Part II: Neural network based approaches. *Signal Process.* **2003**, *83*, 2499–2521. [CrossRef]
20. Salehi, H.; Burgueno, R. Emerging artificial intelligence methods in structural engineering. *Eng. Struct.* **2018**, *171*, 170–189. [CrossRef]
21. Cheung, A.; Cabrera, C.; Sarabandi, P.; Nair, K.K.; Kiremidjian, A.; Wenzel, H. The application of statistical pattern recognition methods for damage detection to field data. *Smart Mater. Struct.* **2008**, *17*, 065023. [CrossRef]
22. Flach, P. *Machine Learning: The Art and Science of Algorithms that Make Sense of Data*; Cambridge University Press: Cambridge, UK, 2012.
23. Mazzoni, S.; McKenna, F.; Scott, M.H.; Fenves, G.L. OpenSees command language manual. *Pac. Earthq. Eng. Res. (Peer) Cent.* **2006**, *264*, 1–465.
24. Neves, A.C.; Gonzalez, I.; Leander, J.; Karoumi, R. Structural health monitoring of bridges: A model-free ANN-based approach to damage detection. *J. Civ. Struct. Health Monit.* **2017**, *7*, 689–702. [CrossRef]
25. Lemaitre, J.; Chaboche, J.L. *Mechanics of Solid Materials*; Cambridge University Press: Cambridge, UK, 1994.
26. Sheela, K.G.; Deepa, S.N. Review on methods to fix number of hidden neurons in neural networks. *Math. Probl. Eng.* **2013**, *2013*, 425740. [CrossRef]
27. Foresee, F.D.; Hagan, M.T. Gauss-Newton Approximation to Bayesian Learning. In Proceedings of the International Conference on Neural Networks (ICNN'97), Houston, TX, USA, 9–12 June 1997; Volume 3, pp. 1930–1935.
28. Fawcett, T. An Introduction to ROC Analysis. *Pattern Recognit. Lett.* **2006**, *27*, 861–874. [CrossRef]

29. Rageh, A.; Linzell, D.G.; Azam, S.E. Automated, strain-based, output-only bridge damage detection. *J. Civ. Struct. Health Monit.* **2018**, *8*, 833–846. [CrossRef]
30. Cury, A.; Crémona, C.; Diday, E. Application of symbolic data analysis for structural modification assessment. *Eng. Struct.* **2010**, *32*, 762–775. [CrossRef]
31. Yan, B.; Cui, Y.; Zhang, L.; Zhang, C.; Yang, Y.; Bao, Z.; Ning, G. Beam Structure Damage Identification Based on BP Neural Network and Support Vector Machine. *Math. Probl. Eng.* **2014**, *2014*, 850141. [CrossRef]
32. Shahsavari, V.; Chouinard, L.; Bastien, J. Wavelet-based analysis of mode shapes for statistical detection and localization of damage in beams using likelihood ratio test. *Eng. Struct.* **2017**, *132*, 494–507. [CrossRef]
33. Gervaso, A. Calcavia prefabbricati in cemento armato precompresso per l'autostrada Torino-Piacenza. *L'industria Ital. del. Cem.* **1969**, *39*, 449–470.

Publisher's Note: MDPI stays neutral with regard to jurisdictional claims in published maps and institutional affiliations.

© 2020 by the authors. Licensee MDPI, Basel, Switzerland. This article is an open access article distributed under the terms and conditions of the Creative Commons Attribution (CC BY) license (http://creativecommons.org/licenses/by/4.0/).

Article

Structural Reliability Estimation with Participatory Sensing and Mobile Cyber-Physical Structural Health Monitoring Systems

Ekin Ozer * and Maria Q. Feng

Civil Engineering and Engineering Mechanics, Columbia University, 500 W 120th Street, 610 Mudd, New York, NY 10027, USA
* Correspondence: eo2327@columbia.edu

Received: 26 May 2019; Accepted: 9 July 2019; Published: 16 July 2019

Abstract: With the help of community participants, smartphones can become useful wireless sensor network (WSN) components, form a self-governing structural health monitoring (SHM) system, and merge structural mechanics with participatory sensing and server computing. This paper presents a methodology and framework of such a cyber-physical system (CPS) that generates a bridge finite element model (FEM) integrated with vibration measurements from smartphone WSNs and centralized/distributed computational facilities, then assesses structural reliability based on updated FEMs. Structural vibration data obtained from smartphones are processed on a server to identify modal frequencies of an existing bridge. Without design drawings and supportive documentation but field measurements and observations, FEM of the bridge is drafted with uncertainties in the structural mass, stiffness, and boundary conditions (BCs). Then, 2700 FEMs are autonomously generated, and the baseline FEM is updated by minimizing the error between the crowdsourcing-based modal identification results and the FEM analysis. Furthermore, using 151 strong ground motion records from databases, the bridge response time history simulations are conducted to obtain displacement demand distribution. Finally, based on reference performance criteria, structural reliability of the bridge is estimated. Integrating the cyber (FEM analysis) and the physical (the bridge structure and measured vibration characteristics) worlds, this crowdsourcing-based CPS can provide a powerful tool for supporting rapid, remote, autonomous, and objective infrastructure-related decision-making. This study presents a new example of the emerging fourth industrial revolution from structural engineering and SHM perspective.

Keywords: structural reliability estimation; structural health monitoring; modal identification; finite element model updating; cyber-physical systems; crowdsourcing

1. Introduction

Deriving economical, sustainable, and practical solutions without a compromise in infrastructure safety and integrity is a broad challenge in civil and structural engineering disciplines. The unpredictable nature of hazardous events combined with limited resources lead the current practice to inherit performance-based criteria in structural design and evaluation. Therefore, controlling the extent of structural damage rather than exclusively avoiding it, is the trending principle in up-to-date engineering codes and regulations [1,2].

Observing the changes in vibration characteristics of structures with the state-of-the-art sensing and processing tools, structural health monitoring (SHM) technologies attract significant attention in research and industry in the last three decades [3–6]. On the other hand; instrumentation, cabling, operation, and maintenance of SHM systems require labor work, knowhow, and financing; declining the growth rate of SHM use in practice. Especially in the past decade, these drawbacks lead researchers to focus on innovative methods such as noncontact vibration measurement techniques [7–9], wireless sensor network (WSN) and distributed sensor network (DSN) systems [10–16], as well as smart [17–19], mobile [20–23], and multisensory [24–27] sensing platforms. Eventually, smartphones are adopted into SHM such that their built-in sensors, operating systems, computation, and wireless communication capabilities can perform as structural vibration measurement devices [28–31].

The authors' previous works present the first vibration-based SHM system (CS4SHM) using crowdsourcing power [32] and offer multisensory solutions to citizen-induced errors by considering spatiotemporal [33] and directional [34] uncertainties. Without any prior engineering education and background, citizens as uncontrolled SHM device operators can provide a central server system with ubiquitous vibration data. The acquired data is autonomously processed for modal identification which is an important indicator of structural vibration characteristics. Unlike conventional SHM systems, CS4SHM points out unorthodox monitoring issues which are concurrently discussed in the upcoming technological boom "Industry 4.0", the latter phase of digital revolution [35,36]. Collecting the distributed crowd sensed information through a central server and conducting modal identification autonomously, civil infrastructures as physical objects are connected with server-side computing in a massive scale forming a CPS [37–41], or in some cases, an Internet of Things system [42–45]. This highlights a significant potential to evolve from pure theoretical structural response simulation (FEM) to experiment-aided and calibrated models (model updating) in massive scales. In other words, with the help of autonomous, connected, scalable cyber networks; citizen-engaged sensing; digital (FEM predictions) and physical (field measurements) civil infrastructure representations; monitoring systems can be adopted to the upcoming technological innovations.

Aforementioned hybrid models can be used for large-volume analysis to retrieve quick evaluation of structural status. This can be performed by utilizing the modal identification results, calibrating mathematical models, and obtaining the probabilistic failure distribution under a wide range of strong ground motions. Eventually, using identification results as model calibration tools, civil infrastructures' seismic response and structural reliability can be estimated [46,47] to provide the decision makers with the necessary information.

This study presents the last section of a PhD dissertation dedicated to cyber-physical system applications from structural engineering perspective [48]. The paper extends the outcomes of a crowdsourcing-based modal identification platform by modifying FEMs constructed with limited information. FEMs as cyber representatives of building behavior are updated to minimize the error between the simulated models and the identification results obtained from the "physical objects". Then, the updated models, which represent the actual vibration characteristics to a better extent, are used to simulate structural response under different earthquake motion scenarios. Finally, collecting the simulation results obtained from numerous time history analyses, the demand distribution is evaluated according to the exemplary code and regulation criteria. To summarize, the developed platform presents an innovative mobile CPS by converting the very initial physical vibrations into highly abstracted decision-making information (i.e., service close, retrofitting, and reconstruction) through a digital and multiphase mathematical information processing framework.

Section 2 discusses the experimental and theoretical phases of the methodology followed throughout the study. These phases include information about the testbed bridge structure, the CPS adaptation, model updating, and reliability estimation methodologies. Section 3 presents the application to the testbed and presents the monitoring results including objective function minimization and determination of structural reliability. Finally, Section 4 reviews the overall work, introduces the future goals, and highlights the concluding remarks.

2. Materials and Methods

The methodology presented in this study connects the experimental data obtained from civil infrastructures with the advanced mathematical modeling and analysis procedures. The following subsections introduce the testbed structure, modal identification, FEM updating, and reliability estimation processes as a CPS framework. From sensing to decision making, Figure 1 represents an idealized cyber-physical information processing scheme, with a comparison of the current CS4SHM system.

The up-to-date platform is capable of receiving vibration measurements from citizens and conduct modal identification on the server-side. Then, the identification results are collected to set the reference modal analysis values for FEM updating and reliability estimation procedures. These phases are currently conducted through a scripted Matlab and OpenSees [49] loop, and the ultimate goal is to handle these cyber procedures through cloud computing. Nevertheless, in both cases, the decision makers can be provided with the quantitative information regarding structural status. Depending on the changes made to the structural system, the effects will be reflected on the future vibration characteristics which completes the cyclic information processing scheme.

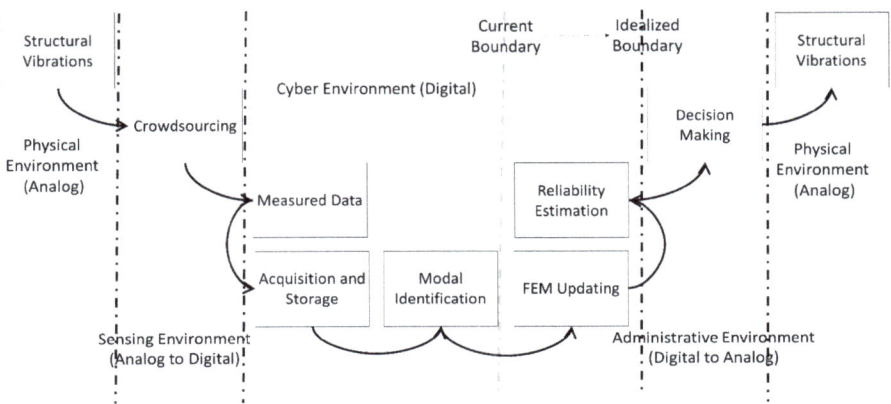

Figure 1. Cyber-physical processes of CS4SHM system.

It should be noted that the probabilistic nature given here is mainly concentrated on input ground motion and FE model updating. In fact, the effect of participatory sensing is an indispensable aspect of the system proposed in this paper since citizen engagement brings numerous uncertainties into the measurements. Previously, it has been shown that actual crowdsourcing results (results from uncontrolled citizens) matched well with the reference identification results [32]. These uncertainties are extensively studied in [48] including spatiotemporal variation of a citizen sensor [33], phone orientation which is subjected to change before, during, and after the measurements [34], and biomechanical distortions caused by human nature [50]. Therefore, in this paper, the main focus is on uncertainties induced by ground motions and FE model parameters.

2.1. Testbed Structure

In order to select a testbed structure with crowdsourcing potential, a link bridge with pedestrian access is implemented. The bridge is a steel frame structure connecting two adjacent buildings in Columbia University Morningside Campus, namely, Mudd and Schapiro Buildings. Mudd Schapiro Link Bridge, shown in Figure 2, is an arch structure with rigid connections spanning approximately 10.5 m. Using the known dimensions of a window and a vision-based scaling procedure, the structural dimensions are approximated without any supplementary documents and design drawings. These dimensions constitute the baseline for the mathematical model, which later on, will be updated with the information from crowdsourced vibration data.

Figure 2. Mudd Schapiro Link Bridge.

2.2. Cyber-Physical System

As a new emerging technology, CPSs attract significant attention from numerous research and industry fields in the last decade. The link and coordination between physical objects and computational resources set the fundamental system goal, which in return, brings different disciplines such as computer, control, electronic, and mechanical systems together [51]. Combining multilayered computer architectures [52] with embedded systems, sensors and control [53], or expanding WSNs to take action in the physical world [54], CPSs present a diverse interpretation of the up-to-date existing technologies.

The overall motivation of the CPS platform presented in this study is to connect the physical, cyber, and sensor system objects through a multilayered information processing SHM framework. The physical object formulated in this scheme is the bridge structure which represents the outer layer of the developed CPS system, as shown in Figure 3. The physical parameter of interest is the bridge vibrations, which can be gathered by smartphone accelerometers with the help of pedestrian volunteers. Moreover, the sensing process is enhanced by the hybrid foundation of pedestrians and sensors, composing the citizen sensor layer. Eventually, the bridge vibrations sensed by smartphone accelerometers are submitted to the server where the signal processing and data analytics take place. With the help of the cloud-based acceleration record manager system, which is the innermost layer, the vibration data can be stored, viewed, reprocessed, and their results can directly be extracted by the system administrators. Interconnecting these components successively forms the two core elements (sensor networks and application platforms) with transactions (sensing and knowledge) of a typical CPS and produce the actuation information with intelligent decision systems to complete the cyber-physical loop [39]. To summarize, citizen sensors provide the binding components of the smartphone-based SHM network by integrating the civil infrastructure with the cloud services, and the numerical representations of the bridge (FEMs) can be fed with the actual bridge response through the cyber-physical SHM system phases.

Formerly, the bridge is registered to the CS4SHM online server system and database to store, process, and monitor its structural vibrations. An iOS application is developed as a data acquisition interface to enable smartphone users gather vibration data from the bridge and submit it to the server. Pedestrians with bridge access are assigned as the test group and submitted 135 vibration measurements in total. The data is processed through the online server system, and modal identification results are recorded. These identification results can be used to calibrate the mathematical model of the bridge by following the FEM updating procedure. Figure 4 shows an exemplary citizen sample in the time and the frequency domains. Based on the whole set of submission records, first, second, and third modal frequencies are identified in [32] as 8.5, 19, and 30 Hz, respectively.

Figure 3. Conceptual CPS scheme for smartphone-based SHM.

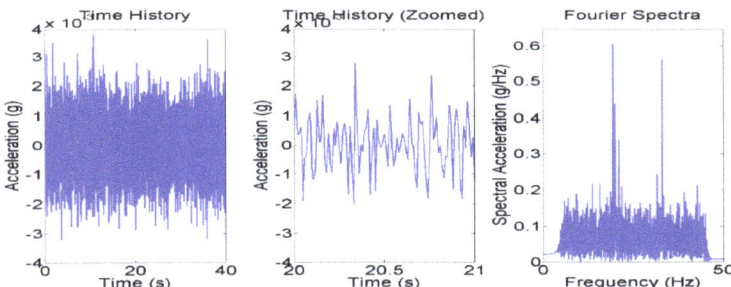

Figure 4. Exemplary crowdsourced submission time histories and Fourier spectra.

2.3. Finite Element Model Updating

In order to predict the structural response accurately, the available information should be effectively used such that FEM parameters can be determined to the best extent. In this modeling example, the design drawings and material properties are unavailable, therefore, the initial FEM is based on site observations and estimations. The observations include the length and the outer diameter of structural members by scaling the pixel values with respect to the known dimensions (i.e., window size). Although the outer diameter can be determined using bridge photographs, the cross-sectional thickness or the inner diameter is unknown. Likewise, support restraints are set as uncertain parameters with possible realizations such as fixed, pinned, or roller. Other than these, contribution of the nonstructural components is difficult to estimate, therefore, mass sources are assigned based on crude assumptions. The analysis primarily considers displacement demand in the vertical direction which is influential on glass façade safety, and the paper presents time history analysis results in z-direction. The model consists of 48 nodes and 98 beam-column elements. Figure 5 shows the node and element tags for the baseline model.

Figure 6 shows the modeling uncertainties taking place without the necessary documentation. To summarize, tubular structural member section dimensions, distributed mass due to non-structural components, and support restraints all contribute to the modeling uncertainties and will be determined throughout the FEM updating process.

The proposed FEM updating method consists of generating a large number of models changing in uncertain parameters, comparing the modal analysis results of each FEM with the experimental data, and selecting the model which minimizes the error between the simulation (model) and the reality (identification). In order to establish an autonomous parameter study and FEM updating procedure, an OpenSees-Matlab integration loop is pursued. Specifically, OpenSees scripts are simultaneously generated, run, and evaluated by a controller Matlab code. As mentioned previously, three different parameters are selected to create different FEM batches. These are the boundary conditions, element stiffness values, and nodal masses, respectively. For each boundary condition combination changing in fixity definitions, a set of models with varying stiffness and mass values are generated. Each of the model batches are evaluated according to the difference between the first, second, and the third FEM and identification results. This is conducted by developing an objective function quantifying the error between a model and the reference modal parameter values.

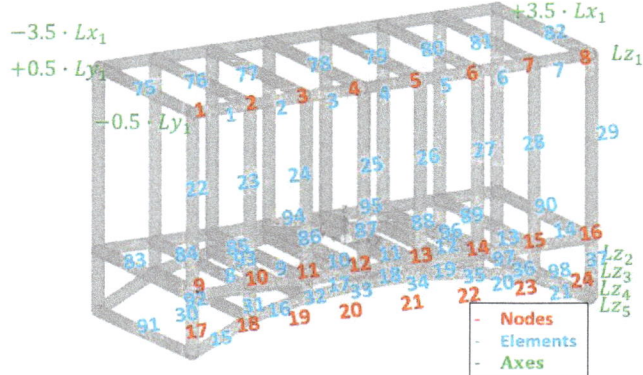

Figure 5. Finite element model nodes, elements and axes.

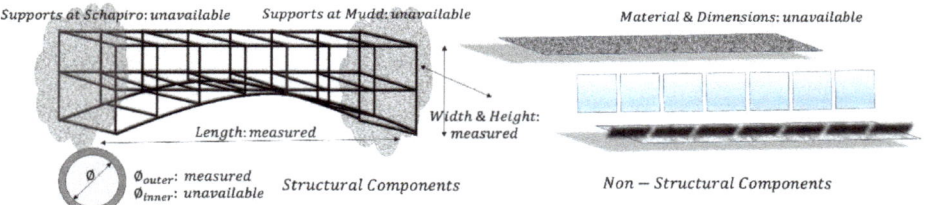

Figure 6. Finite element modeling uncertainties.

In the former studies, the authors adopted Least Square Method (LSM) to formulate the objective function [46,47] whereas different approaches are present in the literature [55,56], in this study, the objective function is structured in terms of error between the simulation and the experimental results and multiplication of multiple modal parameter errors. To specify, the objective function, which is a function of the support restraints, stiffness and mass distributions, is formulated as

$$\text{Obj}(BC, K, M) = \prod_{\text{mode}=1}^{3} (|f_{\text{mode FEM}} - f_{\text{mode EXP}}| / f_{\text{mode EXP}})$$

where BC, K, M represent changing FEM parameters such as boundary condition (BC), member stiffness, and mass values, respectively. Each boundary condition, stiffness, and mass combination corresponds to a different set of first, second, and third modal frequencies represented with $f_{\text{mode FEM}}$ term, and the model accuracy is determined based on the deviation from experimental values represented with $f_{\text{mode EXP}}$. At the end of the loop analyses, the optimal model which minimizes the error between the simulated and identified values becomes the updated model. Afterwards, this model can be used as a baseline for seismic response simulations and reliability estimation.

To summarize the updating process, Figure 7 shows the relationship between the OpenSees and Matlab platforms. The finite element model generation and updating process consists of two integrated platforms which are Matlab and OpenSees. Matlab basically works as a commander, it manipulates finite element parameters stored in OpenSees script files and utilizes the time history analysis outputs from OpenSees. OpenSees functions as a script-based modeling program (suitable for automated batch analysis), is used to conduct modal analysis or time history analysis and is controlled by Matlab. The two pieces of software work in harmony to conduct a large number of automated analysis with a common baseline model but differences in updated parameters.

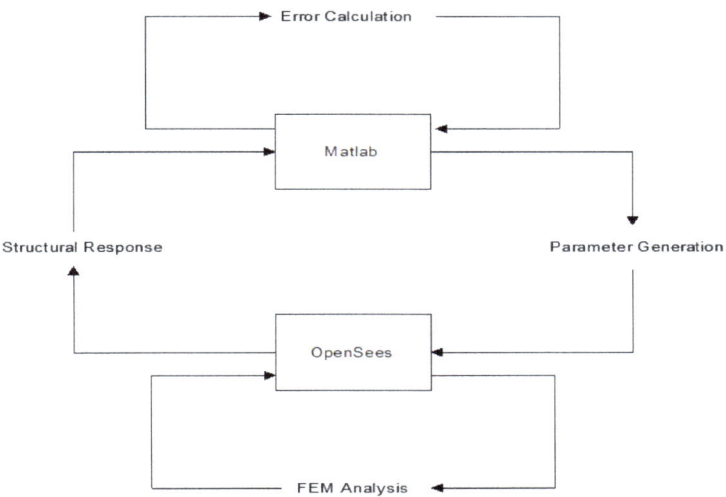

Figure 7. Integration scheme of Matlab and OpenSees software.

2.4. Structural Reliability Estimation

In the authors' previous studies, SHM-integrated reliability estimation is performed by generating fragility curves of different performance levels taking peak ground acceleration (PGA) as the random variable [46,47]. This method can result in high computational cost as the number of available seismic ground motions increases. Compatible with the smartphone-based identification procedure presented in this study, it is expected that ground motion demands under a seismic event can be determined by a dense seismic network composed of smartphone seismometers [57]. Besides, as the number of input ground motions increases like in a mobile CPS scenario, accuracy of the fragility curve parameters may run into obstacles due to truncation and round-off errors. Therefore, in this study, the probabilistic structural response is directly obtained from log-normal distribution of the time history analysis results. Damping term in equation of motion is modeled with Rayleigh Damping where the associated matrix is a combination of stiffness and mass proportional damping. Alpha and beta coefficients are determined based on 2% damping ratio at first and third modes.

For each ground motion taken from the 1994 Northridge Earthquake, a time history analysis is conducted, and the simulated response is obtained. Because the bridge considered in this study is a high frequency structure compared to the low frequency character of Northridge Earthquake records, it is assumed that the structure undergoes linear behavior and its response can be simulated with linear time history analysis. In this case, secondary performance indicators such as maximum drift or displacement become important as they are decisive in the basic engineering mechanics assumptions.

Therefore, the response from each seismic event is collected in terms of maximum deflection and finally, the distribution demand under the given set of earthquakes is obtained. Looking at the distribution demand as well as the reference code and regulation criteria, it can be predicted whether the structural response will exceed certain performance thresholds. In conclusion, with the proposed reliability estimation framework, the high computational cost of fragility curve development is swapped with a simpler approximation, provided that the ground motion response distribution matches well with log-normal type distribution features.

It should be noted that 1994 Northridge Earthquake records perform as an exemplary dataset thanks to the high number of stations and well-distributed strong motion parameters, however, they do not necessarily mirror testbed site conditions presented in this paper. In other words, they are referred for demonstration purposes. In contrast with California, seismicity in the state of New York possesses a lower risk and lacks a comprehensive dataset available to public. However, in the event of ground motion record lack, synthetic ground motions can be generated for a particular site which has designated earthquake spectra [58]. What is more, in a futuristic scenario where there is seismic activity in urban areas, smartphones have shown feasible performance as low-cost seismometers which can be used to detect input ground motion imposed on civil infrastructure [57].

Considering the similar geometry and accordingly similar dynamic behavior of adjacent buildings, out of phase motions are not taken as a primary source of seismic damage. Therefore, forcing function at the boundaries are assumed to be uniform rather than multi-support excitation. It should be noted that changes in boundary conditions can also be monitored with the help of the proposed model updating procedure. Besides, given that the bridge lays on top of campus area with complete occupant access, glass façade integrity can also add additional life safety concerns in case of a seismic event. To incorporate that, vertical displacement behavior is taken as an exemplary performance parameter for reliability analysis.

3. Results and Discussion

Following the outline presented in the methodology, the testbed bridge data is used for modal identification, FEM updating, and reliability estimation with the updated model. The results obtained throughout the analysis are presented with two subsections discussing objective function minimization (FEM Updating) and simulation of seismic response (Reliability Estimation), respectively.

3.1. Objective Function Minimization

In order to predict the structural performance under hazardous events accurately, a well-tuned baseline model is essential. With limited modeling information due to lack of design drawings and reports, an approximate model may deviate from the actual behavior of the structure. Based on the field observations, mass estimations, and fixed BCs, modal analysis results of the initial non-updated model are 8.98, 14.41, and 22.05 Hz for first, second, and third modes, respectively. Comparing these results with the actual dynamic response obtained from the identification results, one can see there is a significant mismatch in second and third modes. Therefore, such modeling discrepancy should be diminished to improve the accuracy of the baseline model.

For this purpose, the FEM updating procedure explained in the previous section is adopted. The updating procedure is composed of three loops each manipulating one modeling variable to generate multiple FEM instances. These three parameters are related to the support restraints, member thicknesses, and distributed mass over the entire span. Looking at the support restraints of the bridge, there are two different types of BCs. The first type is anchored to the adjacent buildings, and the second type is bolted connections. The support details observed through visual inspection show that the bolted connections are only used for the arch restraints, and the rest of the connections are most likely anchored to the structure. To decrease the number of parameter updates, considering that anchored connections form rigid supports, the bolted connection type is considered as an updating parameter which leads to three different combinations such as fixed-fixed, fixed-pinned, and fixed-roller. For each

BC case, 900 FEM instances are created ranging in stiffness (K), and mass (M) parameters. The objective function error between the FEMs and the modal identification results are computed to find the optimal parameter combination. Figure 8 shows the error surfaces of the fixed-fixed, fixed-pinned, and fixed-roller cases.

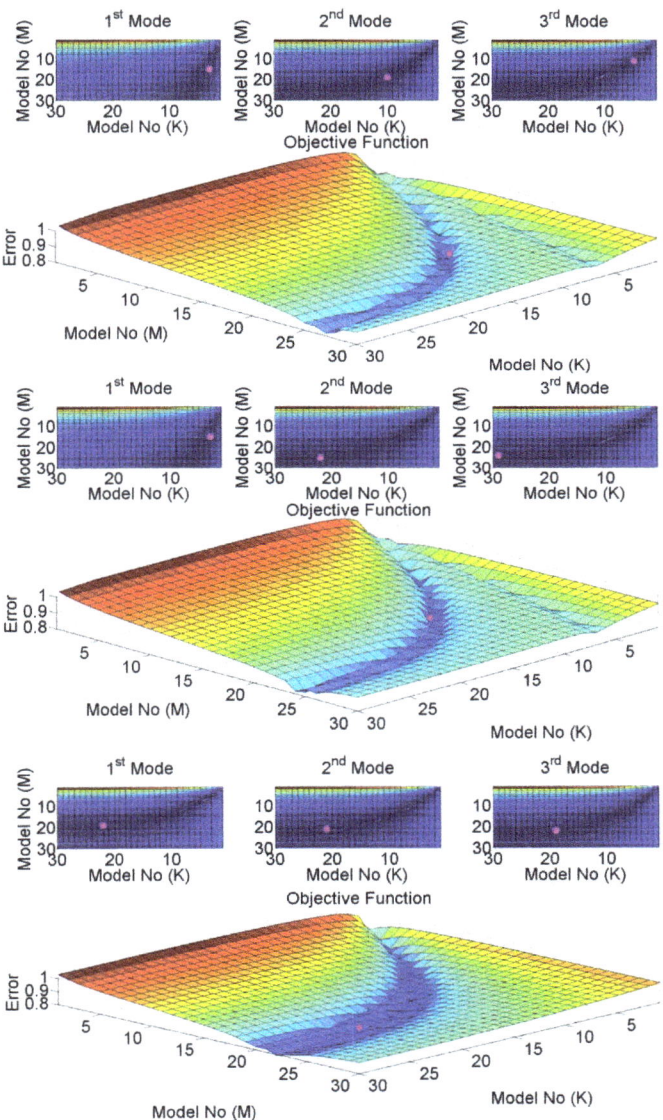

Figure 8. Modal frequency error surfaces for fixed-fixed, fixed-pinned, fixed-roller BCs.

According to Figure 8, the uppermost three figures of each BC case shows the error due to each individual modal frequency, whereas the three-dimensional figures show the combination of these individual components as the objective function product. For visualization purposes, the error between FEM and identification results is demonstrated with colored surfaces. The error surface ranges between red and blue where red corresponds to maximum dispersion from physical reality (modal frequencies from accelerometer data) and blue corresponds to minimal difference between mathematical model and identified modal parameters. Other colors (e.g., orange, yellow, green, turquoise) lay between maximum and minimum error based on the objective function calculations. The magenta spots on each subfigure points out the optimal combination of updating parameters. The overall behavior shows that the model accuracy is very sensitive to the BCs. In other words, combinatory results as well as individual modal frequency errors heavily rely on the modeling of the support restraints.

Stiffness and mass domains are meshed into 30 pieces when candidate models are developed. So, each dimension consisting of 30 individual values represents the uncertainty range within the minimum and maximum values. To explain, Table 1 presents the modal frequency errors obtained from different BC cases and Figure 9 presents the modal parameters of optimal combination cases for each BC case. Stiffness parameters vary from $1.20 \times 10^{-5} m^4$ to $11.5 \times 10^{-5} m^4$ for moment of inertia and $2.1 \times 10^{-3} m^2$ to $36.8 \times 10^{-3} m^2$ for cross-sectional area of a single element. Meanwhile, mass per unit area ranges between $5.7 \times 10^{-2} t/m^2$ and $1.7 t/m^2$. Table 1 implies that for fixed-fixed and fixed-pinned cases, the optimal solutions from each mode varies significantly, and the objective function is either dominated by one of the modes or an irregular combination of them. Fixed-roller case, on the other hand, is contradictory with the first two BC cases. Optimal combinations obtained from first, second, and third modes are evidently similar with each other (ranging around 21th model number), as well as the optimal objective function solution.

Table 1. Optimal models for different BCs.

Boundary Conditions	Optimal	Parameter	Combinations	<K, M>
	Mode 1	Mode 2	Mode 3	Objective
Fixed-Fixed	<3, 15>	<10, 19>	<5, 11>	<10, 19>
	$I_{member} = 3.33 \times 10^{-5} m^4$	$I_{member} = 8.3 \times 10^{-5} m^4$	$I_{member} = 5.1 \times 10^{-5} m^4$	$I_{member} = 8.3 \times 10^{-5} m^4$
	$A_{member} = 6.0 \times 10^{-3} m^2$	$A_{member} = 1.8 \times 10^{-2} m^2$	$A_{member} = 1.0 \times 10^{-2} m^2$	$A_{member} = 1.8 \times 10^{-2} m^2$
	$M_{per\ area} = 0.86\ t/m^2$	$M_{per\ area} = 1.09\ t/m^2$	$M_{per\ area} = 0.63\ t/m^2$	$M_{per\ area} = 1.09\ t/m^2$
Fixed-Pinned	<3, 15>	<22, 25>	<29, 24>	<11, 18>
	$I_{member} = 3.33 \times 10^{-5} m^4$	$I_{member} = 11.2 \times 10^{-5} m^4$	$I_{member} = 11.5 \times 10^{-5} m^4$	$I_{member} = 8.7 \times 10^{-5} m^4$
	$A_{member} = 0.6 \times 10^{-2} m^2$	$A_{member} = 3.2 \times 10^{-2} m^2$	$A_{member} = 3.6 \times 10^{-2} m^2$	$A_{member} = 1.9 \times 10^{-2} m^2$
	$M_{per\ area} = 0.86\ t/m^2$	$M_{per\ area} = 1.43\ t/m^2$	$M_{per\ area} = 1.38\ t/m^2$	$M_{per\ area} = 1.03\ t/m^2$
Fixed-Roller	<22, 19>	<21, 21>	<19, 22>	<21, 21>
	$I_{member} = 11.2 \times 10^{-5} m^4$	$I_{member} = 11.1 \times 10^{-5} m^4$	$I_{member} = 10.9 \times 10^{-5} m^4$	$I_{member} = 11.1 \times 10^{-5} m^4$
	$A_{member} = 3.2 \times 10^{-2} m^2$	$A_{member} = 3.1 \times 10^{-2} m^2$	$A_{member} = 2.9 \times 10^{-2} m^2$	$A_{member} = 3.1 \times 10^{-2} m^2$
	$M_{per\ area} = 1.09\ t/m^2$	$M_{per\ area} = 1.20\ t/m^2$	$M_{per\ area} = 1.26\ t/m^2$	$M_{per\ area} = 1.20\ t/m^2$

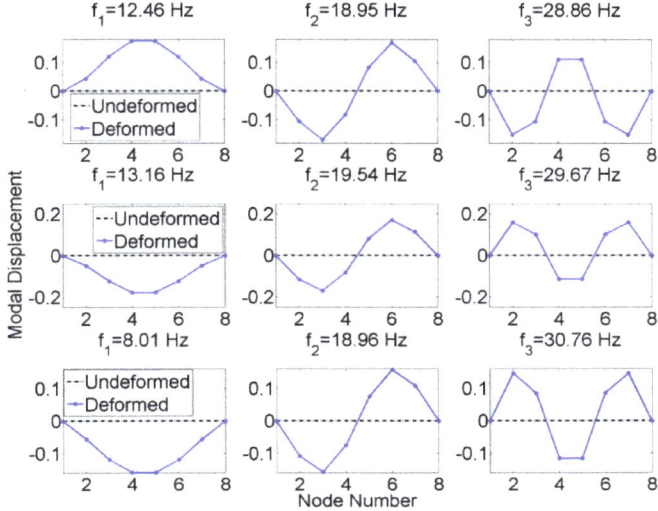

Figure 9. Updated FEM modal parameters for fixed-fixed, fixed-pinned, fixed-roller BCs.

To understand the difference between the fixed-roller case and the other cases, the modal frequencies obtained from each case are investigated. Looking at the first modal frequency of the updated models, it can be observed that the fixed-fixed and fixed-pinned cases have very high errors (47%, 55%), although the second (0.3%, 2.8%) and the third (0.5%, 1.1%) modal frequencies are represented well. In contrast, fixed-roller case represents all three modes with a fair and even accuracy such as 5.8%, 0.2%, and 2.5%. These results show that the arch support fixities are decisive to set the proportion between the first modal frequency and the others, and the fixed-roller case performs significantly better than the other BC cases.

According to Figure 9, comparing the ratio between the modal frequencies, it is seen that the BCs qualitatively do not have a significant effect on the updated mode shapes. On the other hand, without the correct proportion between modal frequencies, even if one or two modes are accurately identified, the remaining mode will have a very high error value. This phenomenon can be proven with a sensitivity study, yet it is the beyond of the scope, and therefore is not addressed further in this paper. Specifically, releasing the arch support fixities in the longitudinal direction can tremendously increase the accuracy of the FEM modal frequencies. Conclusively, an accurate FEM is developed with the presented model updating procedure, and such model can be used to simulate the seismic performance of the structure.

3.2. Simulation of Seismic Response and Reliability

After the optimal modeling parameters are determined and the FEM with limited information is updated, the resultant model can be used as a baseline to predict structural performance under hazardous events. Specifically, in this study, seismic response is scoped, yet similar analysis procedure can be extended to other damaging events. The PEER Strong Motion Database have an extensive set of real earthquake records, therefore, one of these largest sets, 1994 Northridge Earthquake is taken as an exemplary structural demand due to a seismic event [59]. Table 2 shows the overall information about the ground motion dataset features and strong motion parameters.

One hundred and fifty-one earthquake ground motion records are taken from the Northridge Earthquake dataset and used as structural input for time history analyses. With the time history analysis of the baseline model under different earthquake ground motions, the structural response can be probabilistically simulated. Figure 10 shows an example of these analyses illustrating the time and the frequency content of the structural input and outputs.

Table 2. Ground motion dataset summary.

Ground Motion Parameter	Range 1	Range 2	Range 3	Range 4
Frequency (Hz)	0–3	3–6	6–10	10–50
Number	112	37	2	0
PGA (g)	0–0.1	0.1–0.4	0.4–1	1–2
Number	50	70	29	2
Duration (s)	0–25	25–50	50–75	75–100
Number	17	120	14	0

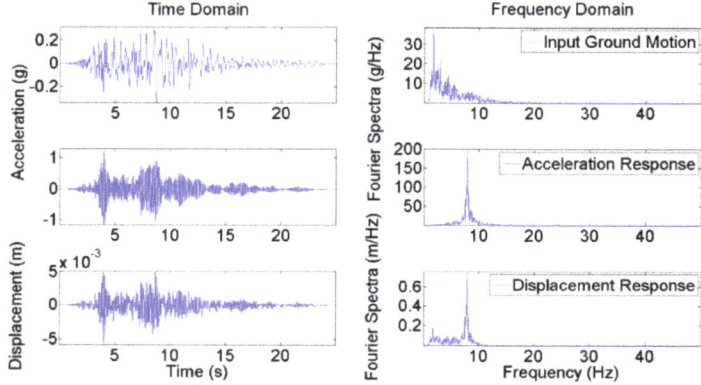

Figure 10. Exemplary input ground motion and simulated structural response.

According to Figure 10, it can be observed that the frequency content of the input ground motion is dominated in low frequencies (below 5 Hz), whereas the structural response peaks around 8–9 Hz. The mode with the lowest frequency, the first mode, is excited more than the second and third modes, and therefore, the response peaks are observed around the first frequency range. This is due to the fact that the higher structural frequencies (e.g., 8.5, 19, 29 Hz) are very far away from relatively low frequency seismic activity. For these reasons, the seismic response is expected to have less structural damage compared with the low frequency civil infrastructure. As a result, the structure behaves in the linear range, yet, it should still be checked whether the bridge maximum deformations exceed certain regulations. One reason is, the nonstructural earthquake damage losses still compose a significant percentage of overall losses [60]. Likewise, even slight damages following a seismic event might result in functionality losses [61]. Besides, it is seen that the low-frequency sensitive displacement response still includes the effects of seismic input, whereas these effects vanish in case of the acceleration response. Finally, and the most important of all, excessive displacements occurring at façade components can lead to glass failure, which possesses safety threat for campus occupants nearby the bridge during the catastrophic incident [62,63].

To summarize the overall dataset results, Figure 11 shows the maximum acceleration and displacement response values indexed according to the strong motion parameters amplitude, frequency, and duration [64], respectively. The analysis results are obtained considering the excitation in the vertical direction. Location of the output corresponds to the absolute maximum displacement value observed on multiple bridge deck nodes throughout the time history analyses.

Time history analysis results are recorded and the maximum response values from each analysis are collected to form a distribution demand. Figure 12 shows the distributed and the cumulative maximum displacement distribution obtained from 151 analysis results. Assuming that the distribution type is log-normal, if the probability density function (PDF) and cumulative distribution function (CDF) are plotted, one can see that the current dataset is a good representative of such type.

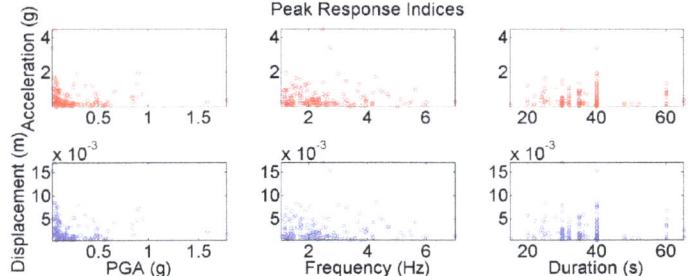

Figure 11. Peak responses indexed according to the strong motion parameters.

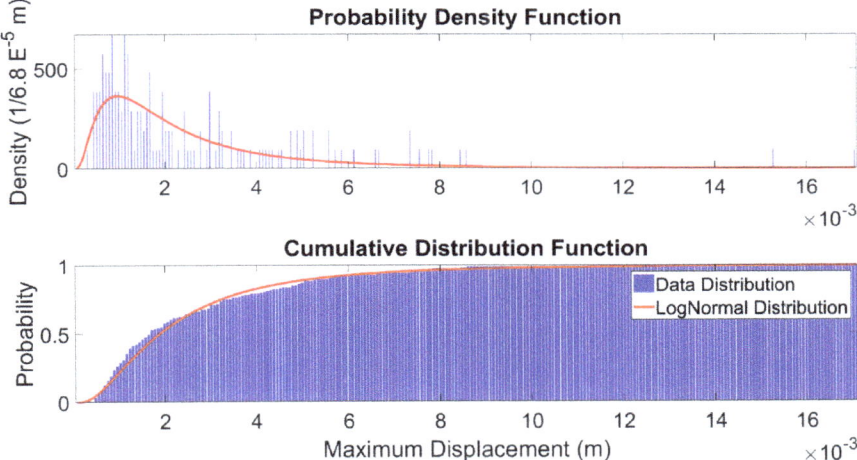

Figure 12. Maximum displacement demands based on Northridge Earthquake records.

The relationship between the arithmetic and logarithmic means can be established with the following relationships,

$$\sigma_y^2 = \ln\left(\frac{\sigma_x^2}{\mu_x^2} + 1\right)$$

$$\mu_y = \ln(\mu_x) - \frac{1}{2}\sigma_y^2$$

where μ and σ corresponds to mean and standard deviation, whereas y and x subscripts correspond to the normal and lognormal distributions. Distribution obtained from the 151 analysis results is treated as log-normal distribution with the specified mean and standard deviation values, rather than following a fragility curve fitting procedure described in [46,47]. Nevertheless, the red plots show that the log-normal distribution assumption is a good representative of the discrete data distribution obtained from time history analyses. Looking at these CDF values of a particular displacement demand, one can determine the structural reliability under that particular threshold.

After the CDF is determined, the bridge performance can be evaluated according to the reference criteria. An example corresponding to an alter load case is that the US pedestrian steel bridges under live loads are limited by a maximum deflection value of L/1000 [65]. Likewise, allowable live load deflection limit for the bridges in Japan ranges between L/2000 (L shorter than 10 m) and L/500 (L longer than 40 m) depending on the main span length [66]. Considering Mudd-Schapiro Bridge dimensions, L/1000 and L/2000 values correspond to approximately 0.01 and 0.005 m. Static deflection limits for the Ontario highway bridges with pedestrian sidewalks are formulated as a function of the first flexural frequency, and the allowable threshold for 10 Hz is equal to 0.002 m [67]. Although, these parameters are indirectly related to structural damage, extreme relative displacements can possess non-structural threats to the community as well. As mentioned above, Mudd-Schapiro Bridge and its glass facades lay above campus area which has pedestrian access 24 h a day. Quantifying exceedance of certain deflection values is therefore beneficial practice for occupant safety.

Finally, the exceedance probabilities of exemplary reference criteria are investigated according to the CDF values. Considering 0.010, 0.005, and 0.002 m as the performance thresholds, structural reliability values of the data distribution are 0.987, 0.868, and 0.576, respectively. Likewise, log-normal distribution reliability values of the same performance thresholds are within a close range such as 0.981, 0.887, and 0.533, respectively. In general, based on similar reliability values under Northridge Earthquake example, the authorities can take action for pre-event preparation. These can include exemplary decisions such as claiming the structure's safety, service shutdown, initiating a retrofitting process, destruction if the performance thresholds are unachievable and reconstruction needed. Yet, it should be noted that for a different set of earthquake records with different frequency character, the structural performance is likely to be different. In the future, this issue can be further investigated with ground motion simulation using site-specific spectra (theory-driven), utilize location-aware smartphone seismic networks (measurement-driven), or both. Automated, remote, and computer-aided survey approaches will be more and more important for civil infrastructure systems which is in line with advances in measurement techniques and building information modeling. Basically, imagery data such as point clouds obtained from aerial or terrestrial tools can be converted into FEMs [68,69]. In fact, terrestrial laser scanning is recently linked to FEM updating process, and therefore, SHM [70]. In addition, the advent of drone technologies combined with photogrammetry made it possible to collect aerial information for building inspection [71]. Such complementary tools can also take part in the development of future cyber-physical infrastructure and collocated usage of similar systems is likely to happen in the near future. Nevertheless, in summary, with the multilayered and detailed analysis procedure presented in this paper, response distributions to different datasets can autonomously be performed by a well-structured cyber-physical SHM system.

4. Conclusions

In this study, present and possible future implementations of a crowdsourcing-based mobile cyber-physical SHM system are presented. Civil infrastructure as physical objects are connected to a cyber-structural model and a response simulation scheme, and the real vibration data obtained from smartphone users are used to calibrate these model parameters. This procedure includes a number of information processing phases such as mobile, server, and administrative components. The mobile platform digitizes structural vibrations via accelerometers and submits it to the server. The server conducts modal identification, returns, and stores the analysis results. The identification results obtained from smartphone sensors are used to update the FEM and increase its accuracy by minimizing the error between the model and the identified modal parameters, which is formerly created with limited information and modeling uncertainties.

Using the updated model as a baseline, structural responses subjected to 151 earthquake records are simulated by time history analyses. The displacement demand distribution obtained from the time history analysis results is evaluated according to the exemplary maximum allowed deflection criteria. Finally, for an earthquake scenario with a wide set of records, one can determine the structural reliability according to the desired performance levels. This information can provide the decision makers with a good foundation for risk assessment, preparedness, and mitigation. Based on the evaluation results of this cyber-physical information flow, the bridge service can be interrupted, structural members can be retrofitted, or the existing structure can be demolished if there is no feasible maintenance scenario. As the volume of invisible operations in computational zone increases, the cyber loops will become more remote and automated.

The framework is demonstrated on an actual pedestrian bridge structure, and the results are presented. The results show that even with limited information, accurate FEMs can be developed with the help of a model updating procedure. Besides, the necessary information is provided by smartphone sensor data and crowdsourcing which solely relies on participatory sensing and pure citizen contribution. Once the physical information is extracted from the sensors, the corresponding data can be combined with a deep mathematical process without any human intervention. Automation, connectivity, scalability, and mobility of the presented platform has a great potential for future mobile cyber-physical SHM systems. Especially, as the seismic monitoring arrays become dense and abundant (e.g., smartphone seismometers), seismic performance of a structure can be simultaneously evaluated with ubiquitous data according to the reference code regulations and standards.

Author Contributions: Conceptualization, E.O. Data curation, E.O.; Formal analysis, E.O.; Investigation, M.Q.F.; Methodology, E.O.; Supervision, M.Q.F.; Visualization, E.O.; Writing—original draft, E.O.; Writing—review & editing, M.Q.F.

Funding: This research received no external funding.

Acknowledgments: The authors would like to acknowledge Demosthenes Long from Public Safety and Daniel Held from Facilities, Columbia University, for their support throughout on-campus pedestrian bridge tests.

Conflicts of Interest: The authors declare no conflict of interest.

References

1. ASCE. *Seismic Evaluation and Retrofit of Existing Buildings (ASCE/SEI 41-13)*; American Society of Civil Engineers: Reston, VA, USA, 2014.
2. AASHTO. *LRFD Bridge Design Specifications*; American Association of State Highway and Transportation Officials: Washington, DC, USA, 2015.
3. Doebling, S.W.; Farrar, C.R.; Prime, M.B. A summary review of vibration-based damage identification methods. *Shock Vib. Dig.* **1998**, *30*, 91–105. [CrossRef]
4. Carden, E.P.; Fanning, P. Vibration based condition monitoring: A review. *Struct. Health Monit.* **2004**, *3*, 355–377. [CrossRef]
5. Brownjohn, J.M. Structural health monitoring of civil infrastructure. *Philos. Trans. R. Soc. Lond. A Math. Phys. Eng. Sci.* **2007**, *365*, 589–622. [CrossRef] [PubMed]
6. Farrar, C.R.; Worden, K. An introduction to structural health monitoring. *Philos. Trans. R. Soc. Lond. A Math. Phys. Eng. Sci.* **2007**, *365*, 303–315. [CrossRef] [PubMed]
7. Wahbeh, A.M.; Caffrey, J.P.; Masri, S.F. A vision-based approach for the direct measurement of displacements in vibrating systems. *Smart Mater. Struct.* **2003**, *12*, 785. [CrossRef]
8. Lee, J.J.; Shinozuka, M. A vision-based system for remote sensing of bridge displacement. *Ndt E Int.* **2006**, *39*, 425–431. [CrossRef]
9. Feng, D.; Feng, M.Q.; Ozer, E.; Fukuda, Y. A vision-based sensor for noncontact structural displacement measurement. *Sensors* **2015**, *15*, 16557–16575. [CrossRef]
10. Farrar, C.R.; Park, G.; Allen, D.W.; Todd, M.D. Sensor network paradigms for structural health monitoring. *Struct. Control Health Monit.* **2006**, *13*, 210–225. [CrossRef]
11. Lynch, J.P.; Loh, K.J. A summary review of wireless sensors and sensor networks for structural health monitoring. *Shock Vib. Dig.* **2006**, *38*, 91–130. [CrossRef]
12. Gao, Y.; Spencer, B.F.; Ruiz-Sandoval, M. Distributed Computing Strategy for Structural Health Monitoring. *Struct. Control Health Monit.* **2006**, *13*, 488–507. [CrossRef]
13. Lynch, J.P. An overview of wireless structural health monitoring for civil structures. *Philos. Trans. R. Soc. Lond. A Math. Phys. Eng. Sci.* **2007**, *365*, 345–372. [CrossRef] [PubMed]
14. Aygün, B.; Cagri Gungor, V. Wireless sensor networks for structure health monitoring: Recent advances and future research directions. *Sens. Rev.* **2011**, *31*, 261–276. [CrossRef]

15. Prasad, P. Recent trend in wireless sensor network and its applications: A survey. *Sens. Rev.* **2015**, *35*, 229–236. [CrossRef]
16. Farhey, D.N. Integrated virtual instrumentation and wireless monitoring for infrastructure diagnostics. *Struct. Health Monit.* **2006**, *5*, 29–43. [CrossRef]
17. Liu, S.C.; Tomizuka, M.; Ulsoy, G. Strategic issues in sensors and smart structures. *Struct. Control Health Monit.* **2006**, *13*, 946–957. [CrossRef]
18. Spencer, B.F.; Ruiz-Sandoval, M.E.; Kurata, N. Smart sensing technology: Opportunities and challenges. *Struct. Control Health Monit.* **2004**, *11*, 349–368. [CrossRef]
19. Jeong, M.J.; Koh, B.H. A decentralized approach to damage localization through smart wireless sensors. *Smart Struct. Syst.* **2009**, *5*, 43–54. [CrossRef]
20. Taylor, S.G.; Farinholt, K.M.; Flynn, E.B.; Figueiredo, E.; Mascarenas, D.L.; Moro, E.A.; Park, G.; Todd, M.D.; Farrar, C.R. A mobile-agent-based wireless sensing network for structural monitoring applications. *Meas. Sci. Technol.* **2009**, *20*, 045201. [CrossRef]
21. Chen, B.; Liu, W. Mobile agent computing paradigm for building a flexible structural health monitoring sensor network. *Comput.-Aided Civ. Infrastruct. Eng.* **2010**, *25*, 504–516. [CrossRef]
22. Zhu, D.; Yi, X.; Wang, Y.; Lee, K.M.; Guo, J. A mobile sensing system for structural health monitoring: Design and validation. *Smart Mater. Struct.* **2010**, *19*, 055011. [CrossRef]
23. OBrien, E.J.; Keenahan, J. Drive-by damage detection in bridges using the apparent profile. *Struct. Control Health Monit.* **2015**, *22*, 813–825. [CrossRef]
24. Sun, H.; Büyüköztürk, O. Identification of traffic-induced nodal excitations of truss bridges through heterogeneous data fusion. *Smart Mater. Struct.* **2015**, *24*, 075032. [CrossRef]
25. Lienhart, W. Challenges in the analysis of inhomogeneous structural monitoring data. *J. Civ. Struct. Health Monit.* **2013**, *3*, 247–255. [CrossRef]
26. Chatzi, E.N.; Smyth, A.W. The unscented Kalman filter and particle filter methods for nonlinear structural system identification with non-collocated heterogeneous sensing. *Struct. Control Health Monit.* **2009**, *16*, 99–123. [CrossRef]
27. Cho, S.; Giles, R.K.; Spencer, B.F. System identification of a historic swing truss bridge using a wireless sensor network employing orientation correction. *Struct. Control Health Monit.* **2015**, *22*, 255–272. [CrossRef]
28. Morgenthal, G.; Höpfner, H. The application of smartphones to measuring transient structural displacements. *J. Civ. Struct. Health Monit.* **2012**, *2*, 149–161. [CrossRef]
29. Oraczewski, T.; Staszewski, W.J.; Uhl, T. Nonlinear acoustics for structural health monitoring using mobile, wireless and smartphone-based transducer platform. *J. Intell. Mater. Syst. Struct.* **2015**. [CrossRef]
30. Zhao, X.; Han, R.; Ding, Y.; Yu, Y.; Guan, Q.; Hu, W.; Ou, J. Portable and convenient cable force measurement using smartphone. *J. Civ. Struct. Health Monit.* **2015**, *5*, 481–491. [CrossRef]
31. Feng, M.; Fukuda, Y.; Mizuta, M.; Ozer, E. Citizen Sensors for SHM: Use of Accelerometer Data from Smartphones. *Sensors* **2015**, *15*, 2980–2998. [CrossRef]
32. Ozer, E.; Feng, M.Q.; Feng, D. Citizen Sensors for SHM: Towards a Crowdsourcing Platform. *Sensors* **2015**, *15*, 14591–14614. [CrossRef]
33. Ozer, E.; Feng, M.Q. Synthesizing spatiotemporally sparse smartphone sensor data for bridge modal identification. *Smart Mater. Struct.* **2016**, *25*, 085007. [CrossRef]
34. Ozer, E.; Feng, M.Q. Direction-sensitive smart monitoring of structures using heterogeneous smartphone sensor data and coordinate system transformation. *Smart Mater. Struct.* **2017**, *26*, 045026. [CrossRef]
35. Lasi, H.; Fettke, P.; Kemper, H.G.; Feld, T.; Hoffmann, M. Industry 4.0. *Bus. Inf. Syst. Eng.* **2014**, *6*, 239. [CrossRef]
36. Lee, J.; Bagheri, B.; Kao, H.A. A cyber-physical systems architecture for industry 4.0-based manufacturing systems. *Manuf. Lett.* **2015**, *3*, 18–23. [CrossRef]
37. Lee, E.A. The past, present and future of cyber-physical systems: A focus on models. *Sensors* **2015**, *15*, 4837–4869. [CrossRef] [PubMed]
38. Schirner, G.; Erdogmus, D.; Chowdhury, K.; Padir, T. The future of human-in-the-loop cyber-physical systems. *Computer* **2013**, *46*, 36–45. [CrossRef]
39. Wu, F.J.; Kao, Y.F.; Tseng, Y.C. From wireless sensor networks towards cyber physical systems. *Pervasive Mob. Comput.* **2011**, *7*, 397–413. [CrossRef]

40. Kim, K.D.; Kumar, P.R. Cyber–physical systems: A perspective at the centennial. *Proc. IEEE* **2012**, *100*, 1287–1308.
41. Hu, X.; Chu, T.; Chan, H.; Leung, V. Vita: A crowdsensing-oriented mobile cyber-physical system. *Emerg. Top. Comput. IEEE Trans.* **2013**, *1*, 148–165. [CrossRef]
42. Atzori, L.; Iera, A.; Morabito, G. The internet of things: A survey. *Comput. Netw.* **2010**, *54*, 2787–2805. [CrossRef]
43. Gubbi, J.; Buyya, R.; Marusic, S.; Palaniswami, M. Internet of Things (IoT): A vision, architectural elements, and future directions. *Future Gener. Comput. Syst.* **2013**, *29*, 1645–1660. [CrossRef]
44. Gershenfeld, N.; Krikorian, R.; Cohen, D. The Internet of things. *Sci. Am.* **2004**, *291*, 76. [CrossRef]
45. Miorandi, D.; Sicari, S.; De Pellegrini, F.; Chlamtac, I. Internet of things: Vision, applications and research challenges. *Ad Hoc Netw.* **2012**, *10*, 1497–1516. [CrossRef]
46. Özer, E.; Soyöz, S. Vibration-based damage detection and seismic performance assessment of bridges. *Earthq. Spectra* **2015**, *31*, 137–157. [CrossRef]
47. Ozer, E.; Feng, M.Q.; Soyoz, S. SHM-integrated bridge reliability estimation using multivariate stochastic processes. *Earthq. Eng. Struct. Dyn.* **2015**, *44*, 601–618. [CrossRef]
48. Ozer, E. Multisensory Smartphone Applications in Vibration-Based Structural Health Monitoring. Ph.D. Thesis, Columbia University, New York, NY, USA, 2016.
49. McKenna, F. OpenSees: A framework for earthquake engineering simulation. *Comput. Sci. Eng.* **2011**, *13*, 58–66. [CrossRef]
50. Ozer, E.; Feng, M.Q. Biomechanically influenced mobile and participatory pedestrian data for bridge monitoring. *Int. J. Distrib. Sens. Netw.* **2017**, *13*, 1550147717705240. [CrossRef]
51. Suh, S.C.; Tanik, U.J.; Carbone, J.N.; Eroglu, A. *Applied Cyber-Physical Systems*; Springer: New York, NY, USA, 2014.
52. Liu, C.H.; Zhang, Y. *Cyber Physical Systems: Architectures, Protocols and Applications*; CRC Press: Boca Raton, FL, USA, 2015.
53. Hu, F. *Cyber-Physical Systems: Integrated Computing and Engineering Design*; CRC Press: Boca Raton, FL, USA, 2013.
54. Siddesh, G.M.; Deka, G.; Srinivasa, K.G.; Patnaik, L.M. *Cyber-Physical Systems: A Computational Perspective*; CRC Press: Boca Raton, FL, USA, 2016.
55. Ghanem, R.; Shinozuka, M. Structural-system identification. I: Theory. *J. Eng. Mech.* **1995**, *121*, 255–264. [CrossRef]
56. Shinozuka, M.; Ghanem, R. Structural system identification. II: Experimental verification. *J. Eng. Mech.* **1995**, *121*, 265–273. [CrossRef]
57. Dashti, S.; Bray, J.D.; Reilly, J.; Glaser, S.; Bayen, A.; Mari, E. Evaluating the reliability of phones as seismic monitoring instruments. *Earthq. Spectra* **2014**, *30*, 721–742. [CrossRef]
58. Shinozuka, M.; Deodatis, G. Simulation of stochastic processes by spectral representation. *Appl. Mech. Rev.* **1991**, *44*, 191–204. [CrossRef]
59. Chiou, B.; Darragh, R.; Gregor, N.; Silva, W. NGA project strong-motion database. *Earthq. Spectra* **2008**, *24*, 23–44. [CrossRef]
60. Kircher, C.A.; Reitherman, R.K.; Whitman, R.V.; Arnold, C. Estimation of earthquake losses to buildings. *Earthq. Spectra* **1997**, *13*, 703–720. [CrossRef]
61. Nielson, B.G.; DesRoches, R. Analytical seismic fragility curves for typical bridges in the central and southeastern United States. *Earthq. Spectra* **2007**, *23*, 615–633. [CrossRef]
62. GANA, B. *GANA Glazing Manual*; Glass Association of North America: Topeka, KS, USA, 2004.
63. O'Brien, W.C., Jr.; Memari, A.M.; Kremer, P.A.; Behr, R.A. Fragility curves for architectural glass in stick-built glazing systems. *Earthq. Spectra* **2012**, *28*, 639–665. [CrossRef]
64. Kramer, S.L. *Geotechnical Earthquake Engineering*; Prentice Hall: Upper Saddle River, NJ, USA, 1996.
65. Roeder, C.W.; Barth, K.E.; Bergman, A. Effect of live-load deflections on steel bridge performance. *J. Bridge Eng.* **2004**, *9*, 259–267. [CrossRef]
66. Nishikawa, K.; Murakoshi, J.; Matsuki, T. Study on the fatigue of steel highway bridges in Japan. *Constr. Build. Mater.* **1998**, *12*, 133–141. [CrossRef]
67. Billing, J.R. Dynamic loading and testing of bridges in Ontario. *Can. J. Civ. Eng.* **1984**, *11*, 833–843. [CrossRef]

68. Hinks, T.; Carr, H.; Truong-Hong, L.; Laefer, D.F. Point cloud data conversion into solid models via point-based voxelization. *J. Surv. Eng.* **2012**, *139*, 72–83. [CrossRef]
69. Castellazzi, G.; D'Altri, A.; Bitelli, G.; Selvaggi, I.; Lambertini, A. From laser scanning to finite element analysis of complex buildings by using a semi-automatic procedure. *Sensors* **2015**, *15*, 18360–18380. [CrossRef]
70. Yang, H.; Xu, X.; Neumann, I. Laser scanning-based updating of a finite-element model for structural health monitoring. *IEEE Sens. J.* **2015**, *16*, 2100–2104. [CrossRef]
71. Rakha, T.; Gorodetsky, A. Review of Unmanned Aerial System (UAS) applications in the built environment: Towards automated building inspection procedures using drones. *Autom. Constr.* **2018**, *93*, 252–264. [CrossRef]

© 2019 by the authors. Licensee MDPI, Basel, Switzerland. This article is an open access article distributed under the terms and conditions of the Creative Commons Attribution (CC BY) license (http://creativecommons.org/licenses/by/4.0/).

Article

An Efficient Pipeline to Obtain 3D Model for HBIM and Structural Analysis Purposes from 3D Point Clouds

Massimiliano Pepe [1,*]**, Domenica Costantino** [1] **and Alfredo Restuccia Garofalo** [2]

1. Polytechnic of Bari, via E. Orabona 4, 70125 Bari, Italy; domenica.costantino@poliba.it
2. AESEI Spin-off - Polytechnic of Bari, via S. Eligio 1/L, 74015 Martina Franca (Taranto), Italy; ingrestucciachristian@gmail.com
* Correspondence: massimiliano.pepe@poliba.it

Received: 2 January 2020; Accepted: 10 February 2020; Published: 12 February 2020

Abstract: The aim of this work is to identify an efficient pipeline in order to build HBIM (heritage building information modelling) and create digital models to be used in structural analysis. To build accurate 3D models it is first necessary to perform a geomatics survey. This means performing a survey with active or passive sensors and, subsequently, accomplishing adequate post-processing of the data. In this way, it is possible to obtain a 3D point cloud of the structure under investigation. The next step, known as "scan-to-BIM (building information modelling)", has led to the creation of an appropriate methodology that involved the use of Rhinoceros software and a few tools developed within this environment. Once the 3D model is obtained, the last step is the implementation of the structure in FEM (finite element method) and/or in HBIM software. In this paper, two case studies involving structures belonging to the cultural heritage (CH) environment are analysed: a historical church and a masonry bridge. In particular, for both case studies, the different phases were described involving the construction of the point cloud and, subsequently, the construction of a 3D model. This model is suitable both for structural analysis and for the parameterization of rheological and geometric information of each single element of the structure.

Keywords: scan-to-BIM; point cloud; HBIM; FEM; Rhinoceros

1. Introduction

Modern surveying technologies in cultural heritage (CH) offer new perspectives of application both as regards the acquisition of metric data and the representation or analysis of objects of historical and artistic interest [1]. In this way, it is possible to obtain a digital representation of objects or structures belonging to the CH environment in terms of position, shape, geometry and description of each element. Geomatics surveys are the primary step in the process of conservation, enhancement and management of CH. A geomatics survey can be performed using image-based 3D modelling (IBM) or range-based modelling (RBM).

IBM methods use 2D images (generated by passive sensor) measurements in order to obtain 3D models. In the last few years, a very successful approach in the construction of 3D models has been that based on the structure from motion (SfM) and multi-view stereo (MVS) algorithms. Using these approaches, a 3D model or 2D orthophotos can be obtained in a rapid and automatic way using photogrammetric software. In general, the several processing steps that lead to the construction of the model are: (*i*) alignment of the images; (*ii*) building a dense point cloud (PC); (*iii*) building mesh and; (*iv*) building an orthomosaic. Furthermore, the passive sensors used in the IBM method may be used even on mobile platforms (such as cranes, unmanned aerial vehicles (UAVs), hot-air balloons, etc.). In this way, it is possible to acquire data even in big, complex and inaccessible structures, such as upper parts of buildings, aqueducts, bridges etc.

Range-based modelling is based on active sensors, which provide a highly detailed and accurate representation of a 3D object or structure. An example of active sensor is the terrestrial laser scanner (TLS). TLS is a ground-based method that rapidly acquires accurate 3D dense point clouds of a scene through laser range-finding [2].

While in the past these two techniques have been often treated as two separate methodologies, comparing them in terms of accuracy, cost and flexibility [3], only in recent times, they have started to be considered as complementary [4]. The benefit of integrating these two technologies is to take advantage of the TLS capability to directly acquire a dense coloured cloud, with the flexibility of photogrammetry to operate even in exceptional conditions.

By an adequate post-processing of geomatics survey data, it is possible to obtain a georeferenced point cloud of the structure (or object) under investigation.

Next, it is necessary to transform the point cloud into objects for BIM (building information modelling) and FEM (finite element method) analysis. Recently, many studies have been focusing on the possibility of managing point clouds within BIM or structural analysis software and/or identifying a suitable pipeline in order to obtain 3D model for these purposes [5]. While allowing the import of data, the current BIM and structural analysis software does not provide flexible and manageable procedures such as transforming them into models suitable for subsequent processing. Indeed, this is the main challenge pertaining to modelling, as it is necessary to develop simple methods to obtain BIM or HBIM (historic building information modelling) models that still guarantee accuracy, precision and quality of representation consistent with the acquired data. In addition, the model must be enriched with data and information that are not strictly geometric, such as historical information, analysis of degradation or deformation, and levels of detail not granted by the complete model.

1.1. Related Works

HBIM for the integration of contemporary technology and the BIM approach in the field of CH documentation was introduced by Murphy et al., 2009 [6]. The purpose of this research was to identify a new methodology for creating full engineering models from laser scan and image survey data for historic structures. Therefore, the identification of a suitable procedure able to obtain a BIM model from the survey is key, especially in the management of structures of particular historical-architectural interest. A comprehensive review of the several BIM software types for CH is reported in López et al., 2018 [7] where some information, such as functionality, tools, object structure, interoperability and links are addressed.

Fregonese et al., 2015 [8] developed a procedure to obtain a 3D model for BIM purposes. Once the model from the 3D survey is obtained, solid model software was recreated directly in Autodesk Revit, where each single element was modelled using a system family or "Model in Place". This BIM software has allowed to model historical and complex elements in a parametric way which allowed it to be connected with a database. However, due to the limitations of BIM commercial software, the authors have developed software for the management and planning of restoration operations.

Barazzetti et al., 2015 [9] have showed a procedure for BIM generation from point clouds via BIM parameterization of NURBS (non-uniform rational B-spline) curves and surfaces using Revit software. In the case study, the authors suggest a procedure that provides BIM objects of complex elements by using the NURBS surface turned into specific BIM families. Using this approach, some problems were found in the modelling of complex objects and in the building of the layer-based reconstruction from the intrados to the extrados.

Eigenraam et al., 2016 [10] presents a method in order to obtain free-form shell structures from point cloud to finite element model. In the paper, special attention is given to the geometric accuracy, considering that shape and force interact. The method was applied to Heinz Isler's models for reverse engineering purposes.

Furno et al., 2017 [11] compared two different modelling methods: one based on the use of NURBS and the parametric one on BIM objects, using Rhinoceros and Revit software. The "direct" modelling of Rhinoceros made it possible to process the survey data and obtain a model divided into blocks, with the possibility of modifying the intrinsic parameters of the individual elements using the Grasshopper plug-in (included in Rhinoceros). However, the model obtained in this way does not add information of any kind to the elements. For this reason, the modelling of the same structure was also performed with the Revit software and applied to Milan Cathedral in Italy.

León-Robles et. al, 2019 [12] discussed HBIM applied to a masonry bridge using Revit commercial BIM software, but they encountered great difficulties in doing so because only a few families of libraries are dedicated to the modelling of complex civil constructions such as bridges. Moreover, in this case study, an analysis of the deformations between the designed model of the bridge and that surveyed was carried out.

Bassier et al., 2019 [13] suggest a fast and accurate procedure to capture the spatial information required using FEM. The workflow involves two parallel methods: the former converts the point cloud to a complex FEM mesh (through a series of semi-automated procedures) while the second extracts crack information and enhances the FEM mesh to incorporate the crack geometry.

1.2. Organization of the Article

This paper is organized as follows. The first part describes the several approaches used in order to reconstruct the surface of the object from a point cloud generated through geomatics surveys. Subsequently, after describing the method that allows obtaining a 3D model for HBIM and FEM from 3D a point cloud, two case studies are discussed. In particular, the method developed is applied to a historical church featuring a rather simple shape, and an old masonry bridge with a complex structure. Conclusions are summarized at the end of the paper.

2. Surface Reconstruction from Point Cloud

2.1. Three-Dimensional (3D) Surface

To generate a surface model from a point cloud, the reconstruction technique implemented in dedicated software conventionally uses: tessellation, 3D reconstruction using Delaunay triangulation or NURBS surfaces. This procedure may show poor accuracy near the edges or with sudden surface changes from normal. Furthermore, the representation of such surfaces could require numerous pieces and, consequently, greater computational capabilities. By decimating the triangulation, information could be lost on the geometry of the structure under examination. This is the reason why many commercial software products are not able to use mesh models, but use precise analytical models in which surfaces are represented mathematically [14]. In the following sections, the triangular irregular network (TIN) and NURB are described in detail.

2.2. Triangular Irregular Network (TIN)

TIN generation is a way to obtain surface reconstruction. Triangulation may be performed in two or in three dimensions, in accordance with the geometry of the input data. TIN utilizes the original sample points to create many non-overlapping triangles that cover the entire region according to a set of rules. The surface is described (approximately) with these triangles [15]. The computer graphics community tends to call this polygonal model "*mesh*". A mesh contains vertices, edges and faces and its easiest representation is a single face. For triangular meshes, an indexed face list consists of an array of vertices each having three coordinates, and an array of faces each having three indices in the vertex array [16]. The criterion for triangulation division is often used to construct the non-overlapping triangles based on the discrete sampling points. Delaunay is the most common triangulation algorithm.

2.3. Non-Uniform Rational B-Spline (NURBS)

The NURBS are mathematical representations of 3D geometry that accurately define a generic geometric entity, such as simple or more complex shapes. The NURBS curve is mathematically defined by the following equation:

$$C(u) = \frac{\sum_{i=0}^{n} N_{i,p}(u)\, w_i\, P_i}{\sum_{i=0}^{n} N_{i,p}(u)\, w_i} \qquad (1)$$

where the w_i are the weights, the P_i are the control points, and the $N_{i,p}(u)$ are the normalized B-Spline basis functions of degree p recursively as [17,18]:

$$N_{i,0}(u) = \begin{cases} 1 & \text{if } u_i \leq u \leq u_{i+1} \\ 0 & \text{otherwise} \end{cases}$$
$$N_{i,p}(u) = \frac{u - u_i}{u_{i+p} - u_i} N_{i,p-1}(u) + \frac{u_{i+p+1} - u}{u_{i+p+1} - u_{i+1}} N_{i+1,p-1}(u) \qquad (2)$$

where u_i are the knots forming the knot vector $U = \{u_0;\, u_0;\, \ldots;\, u_{n+p+1}\}$.

Therefore, a NURBS curve is defined by four characteristics: the degree, the control vertices, the knot vectors and the weights. The degree of the NURBS (a positive integer) defines mathematically the piecewise polynomial blending function. The higher the degree of the polynomial, the more flexible the curve and surface. The control vertices are a row of points at least equal to (degree + 1). The knot vectors define how the polynomial pieces are blended together with the proper smoothness. Generally, there are two kinds of knot vector definition: uniform (i.e., with constant spacing between the knots) and non-uniform (i.e., with varying spacing between knot vectors).

A weight is associated with each control point (i.e., its ability to attract the curve). Excluding some exceptions, weights are positive numbers. When the control points of a curve all have the same weight (usually equal to 1), the curve is called "non-rational" and the NURBS curve is reduced to a B-spline curve; otherwise, it is called "rational." For this reason, the letter R of the acronym NURBS stands for "rational" and indicates that a NURBS curve can be rational.

3. Method

The creation of surfaces suitable for the modelling of objects or structures starting from a 3D "dense point cloud" model (obtained through geomatics surveys) can take place in different ways. Several pipelines have been examined [1,19]: the most efficient of these (in terms of linearity of the method, accuracy, processing times) can thus be schematized (Figure 1). In fact, should the model generation take place in Revit, it would require the generation of families responding to the geometric characteristics of the object. The generation of the same model in Rhinoceros is "semi-automatic" because it requires the adaptation of any complex surface from the point cloud. This task can be carried out using the different plug-ins within the Rhinoceros software. The processing times for model generation in Revit are considerably longer than those required in Rhinoceros, primarily because complex surfaces do not always find adaptive models in BIM, while in Rhinoceros surfaces can be generated to adapt to the point cloud.

As showed in the pipeline (Figure 1), the first step, after performing geomatics surveys, is to import the point cloud into the Rhinoceros software. Through the Arena4D plug-in, implemented in Rhinoceros software, it is possible to obtain optimal management of the point cloud. In other words, this plug-in creates a series of filters on the point cloud such as the elimination of outliers, etc.

In Rhinoceros software, it was also possible to create detailed profiles in the specific part of the structure and, consequently, to build complex and irregular shapes according to NURBS-type geometries. In this way, it is possible to differentiate the several elements of a structure, such as that of a bridge (geometry of the pylons, vaults, retaining walls, etc.). The characterization of each structural element allows each of them to be assigned a specific material.

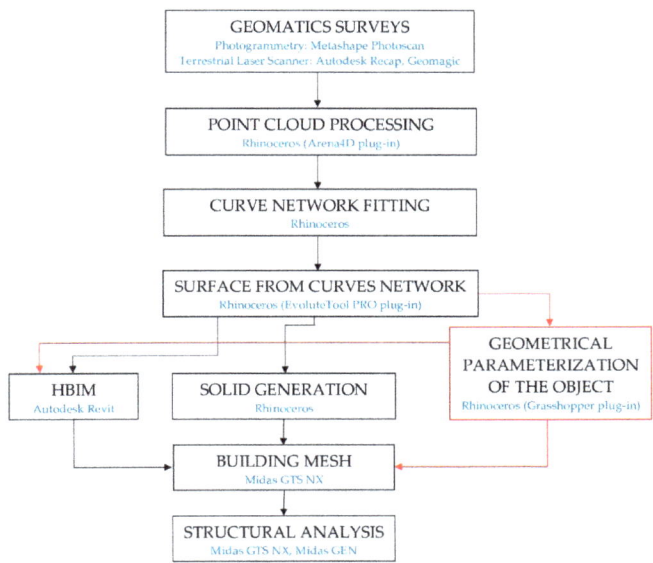

Figure 1. Pipeline of the developed method.

If the structure shows irregular geometries, it is possible to use an additional plug-in developed in Rhinoceros, called "EvoluteTools PRO", which is able to generate highly complex and sophisticated NURBS surfaces.

Subsequently, the surfaces can be imported into the software of HBIM or structural analysis. In the latter case, NURBS surfaces cannot be imported directly into the software, but it is necessary to build solids. As a result, each NURBS surface can be transformed into a solid through modelling in Rhinoceros. Once solid geometric objects are exported into Midas GTS NX software, the structural mesh can be built.

The transformation from NURBS into solid is performed through solid generation commands such as "offset surface", "loft evolut", "revolution", "extrusion" (i.e., Boolean commands)". Obviously, this phase can be achieved knowing the thicknesses of the structural elements that have been detected and identified through the use of multi-sections on the structure. Consequently, structural objects can be constrained and subjected to loads (permanent and accidental); in this way, it is possible to perform the analysis of stresses and deformations of the structure taken into consideration. However, depending on the structure under investigation, it is possible to use the Grasshopper plug-in, implemented within the Rhinoceros software. This plug-in allows the problem of repeatability of similar objects to be overcome or "parameterization" from time to time of structural elements that have similar geometric characteristics. The programming in Grassopher starts from the insertion of the (surveyed) surfaces generated by the point cloud and adapted in Rhinoceros, which then allows the geometric parameterization. The latter allows us to define any geometric parameter of the object (length, height, thickness, etc.). These geometric elements can be modified and managed according to the space-time use (duplication of the object, comparison with temporal deformations, cracking) using commands such as "number slider" or "Nurbs Curve" (insertable and manageable within the "canvas"). A further advantage of using the Grasshopper plug-in is the possibility to parameterize any type of surface. This is particularly useful in the 4D monitoring activity since it is possible to update the parameterized model according to the deformations detected in different eras. Therefore, the different structural elements generated in this way can be imported into HBIM software or used in structural analysis, as previously described.

Furthermore, all plug-ins and software used in this paper require (commercial) user licenses and support interchange formats. In the Revit environment, the processing requires more manual interventions on the part of the operator than the semi-automatism provided by Rhinoceros. Moreover, if the goal of the process is the complete geometrical parameterization of the object (up to foreseeing temporal modifications or similarities between the objects) it is necessary to have more programming knowledge (Grasshopper) and, consequently, greater manual intervention on behalf of the operator.

4. Case Studies

4.1. Brief History and Location of the Structures Taken under Investigation

4.1.1. The Church of San Nicola in Montedoro

The church of San Nicola in Montedoro is one of the oldest in the town of Martina Franca in the province of Taranto (Italy) (see Figure 2a,b). The construction of the church presumably dates back to the 14th century, the period of the Angevin foundation of the city. It is located in the Montedoro district, hence "San Nicola di Montedoro" [20]. The church preserves its original structure, despite the internal transformations of the seventeenth century. The structure is characterized by a modest rectangular hall and late medieval architectural elements visible especially on the outside. The simplicity of the external façade is embellished only by the roof with raised pitches that intersect and form two gables with cladding made with the typical "*chiancarelle*" (a type of limestone slab). The portal is surmounted by a lunette and a small radial rose window, while on the tympanum of the main façade stands a graceful bell tower (Figure 2c). The interior consists of a single room and has two baroque altars in stone. On the walls, frescoes are visible, painted on two layers (Figure 2d) [1].

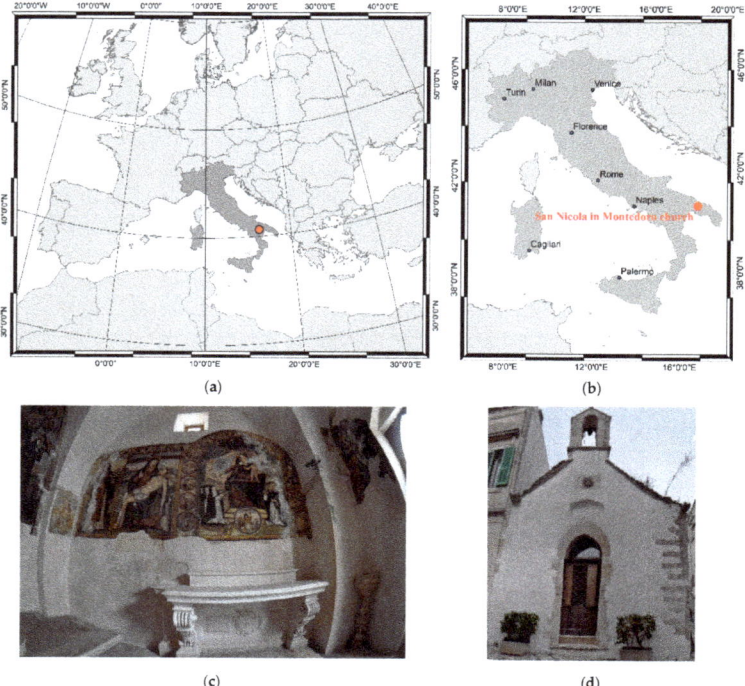

Figure 2. The church of San Nicola in Montedoro: location (**a**,**b**); image of the internal of the church acquired with fisheye lens (**c**); image of the external of the church (**d**).

4.1.2. San Cono Bridge

San Cono bridge spans the Bianco river located in the municipality of Buccino, in southern Italy (Figure 3a,b). As reported by the inscription on the bridge, the construction of San Cono bridge can be dated to the Augustan age (Figure 3c).

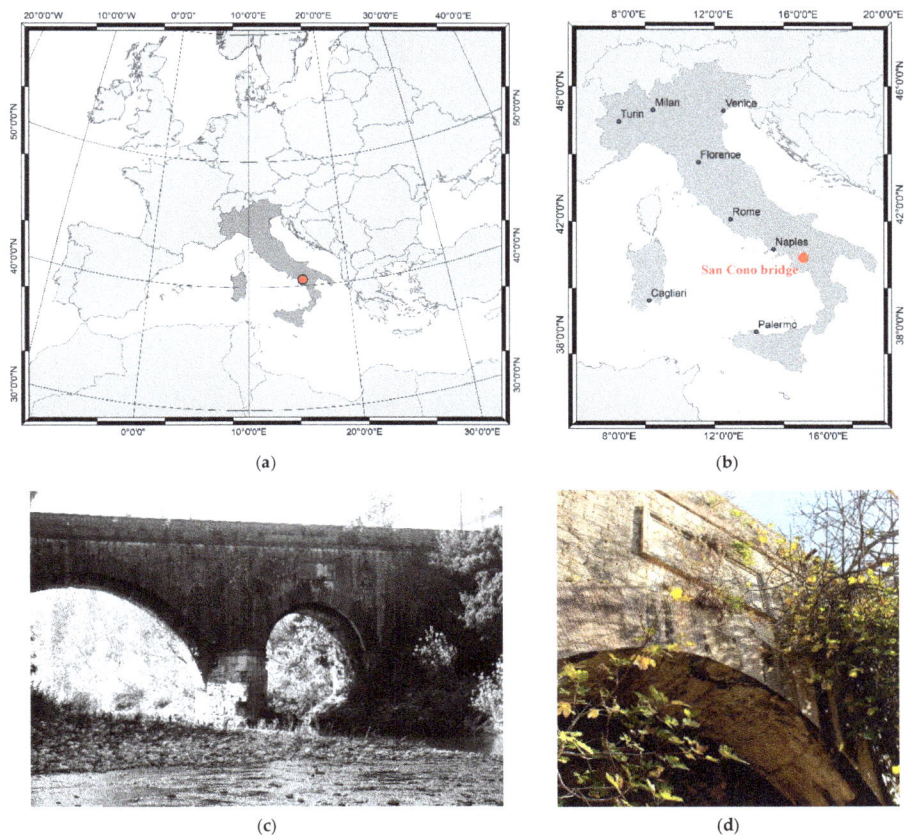

Figure 3. San Cono bridge: location (**a**,**b**) and panoramic images of masonry bridge (**c**,**d**).

Originally, the bridge had a pronounced donkey-back profile with two shoulders and a steep slope at the ends and a pylon with a triangular rostrum [20]. Now, the current shape of the bridge is incorporated into a new bridge, which in 1872 levelled the road and widened the site (taking it from 3.20 m to 6.45 m), covering it, so as to leave only the original arches visible, below the new ones. In this way, the intervention represented an exceptional example of respect for the ancient monument. As for the bridge architecture, it has two spans of unequal light, for a total length of 40 m. Part of the ancient arches can still be seen below the nineteenth-century one, which changes its profile. The central round arch has a light of 17.3 m and at the base there are five projecting brackets with three others at a higher altitude to complete the support of the rib; the minor arc has a light of 5.9 m with three shelves.

The original vestments of the tympani were in square work; today they are inserted in the new 19th-century vestments, with an upper parapet that modifies the original donkey back profile [20] (Figure 3d).

4.2. Three-Dimensional Point Cloud of San Nicola in Montedoro Church

4.2.1. Three-Dimensional Survey of the Church

The survey of the church was carried out through the use and integration of active and passive sensors, terrestrial and aerial. In particular, the external façade was surveyed using a TLS, the inner part using a digital single-lens reflex (DLSR) camera with fish-eye lens and the upper part of the building (i.e., the roof and other architectural elements not visible through a terrestrial survey) through the use of a camera mounted on a UAV platform.

Before performing the surveys with photogrammetric techniques and laser scanners, a survey with a total station was performed. The survey was carried out by TS30 Leica Geosystems. This total station allows discrete points to be acquired with an angular precision of 0.5" (0.15 mgon) and to acquire distance with prism (precision of 0.6 mm + 1 ppm) and without prism (2 mm + 2 ppm).

In this case study, the survey was carried out by two base stations. In this way, it was possible to obtain horizontal and vertical angular observations of the ground control points (GCPs). The GCPs, inside and outside the building, were chosen so as to be easily recognizable even on the image (Figure 4). The post processing of the data was carried out in LGO (Leica Geo Office) developed by the Leica Geosystem company.

Figure 4. Images of some ground control points used for the georeferencing of the point cloud.

4.2.2. Survey of the Terrestrial Laser Scanner of the External Part of the Structure

Regarding the generation of the model for the external part of the church, the survey was carried out by a terrestrial laser scanning survey. In this case study, FARO FocusS 350 instruments were used because specially designed for outdoor applications. HDR imaging and HD photo resolution (overlay up to 165-megapixel colour) ensure true-to-detail scan results with high data quality (distance accuracy up to ± 1 mm). The main features of this scanner are summarized in the following Table 1:

Table 1. Main Features of the FARO FocusS 350 terrestrial laser scanner (TLS).

Distance Accuracy	Range	Measurement Rate	Laser Class	Integrated Colour Camera	Operating Temperature
up to ± 1 mm	0.6 m to 350 m	up to 976,000 points/s	1	yes	+5 °C to + 40 °C

In order to cover the entire external surface of the church, three acquisition stations were built.

The post-processing of the TLS scans was performed in Autodesk Recap software. This software, where the word "Recap" stands for Reality Capture, allows a fully automatic recording of the scans. In the case the procedure is partially successful, the software allows manual identification of targets and natural homologous points, to reduce distance among contiguous scans, improving their alignment using the iterative closest point (ICP) algorithm [21].

4.2.3. Unmanned Aerial Vehicle (UAV) Photogrammetry to Obtain the 3D Point Cloud of the Upper Part of the Church

The aerial survey was carried out using a Parrot Anafi, a UAS (unmanned aerial system) quadcopter equipped with a Sony Sensor® 1/2.4" 21MP (5344 × 4016) CMOS (complementary metal-oxide semiconductor), which allows obtaining, thanks also to a 3-axis stabilizer, clear and detailed images (Figure 5a). The distance between the UAV and the building was really close due to the presence of many obstacles in the old town where the church is located. Consequently, the images were acquired with high geometric resolution (Figure 5b). In any case, the photogrammetric survey was carried out with a high degree of overlap between the images. In addition, by varying the tilt angle of the camera, it was possible to acquire images of every part of the building. In this way, it was possible to build a network of the 97 images with a high degree of overlap and convergent image configuration (Figure 5c).

Figure 5. Image acquisition by unmanned aerial vehicle (UAV): (**a**) Parrot Anafi and Parrot Sky-controller; (**b**) acquisition step; (**c**) an aerial image acquired by UAV platform.

Taking into account 5 GCPs, the root mean square error (RMSE) for spatial coordinates, evaluated on the cameras used in this dataset, was of 0.009 m; in particular, this RSME refers to the georeferencing process of the images and not to the resolution of the model. In addition, the modelling of the roof was obtained through the insertion of its own and accidental load.

4.2.4. Photogrammetry of the Internal Part of the Structure Using a Fisheye Lens

For the interior of the church, since there are also frescoes of great historical and cultural value and considering the rather restricted environment, a photogrammetric survey was carried out using a Nikon D5000 DSLR camera with a calibrated fisheye lens (focal length 10 mm). The fisheye is a wide-angle photographic lens that allows a wide scene to be observed. This type of lens has been used successfully in the photogrammetry field, as shown in Kannala and Brandt, 2006 [22], especially in narrow spaces.

In self-calibration mode, the dataset of the 22 images was processed in Agisoft Metashape software. The total error, i.e., standard deviation evaluated on 6 GCPs, was 0.003 m.

Considering the high value of the frescoes and the architecture of the small altars inside the structure, orthophotos of each single façade and floor were taken. In order to carry out this task, it was necessary to build a mesh of the interior of the structure. Subsequently, identifying the planes of the single façade, the orthophotos with a geometric resolution of 0.1 mm of the interior of the church were built (Figure 6).

Figure 6. Orthogonal projection of the inside of the structure: orthophotos (in very high resolution) of the single façade and floor of the church.

4.2.5. Merging of the Datasets (Point Clouds)

Through the survey activity and post processing of the data obtained either with IBM or RBM methods, it was possible to obtain three datasets, as shown in Table 2.

Table 2. Point Clouds Obtained in the Several Datasets.

	Dataset	Method		Point Cloud
1	Outside (exterior façade)	Terrestrial	RBM	4,871,426
2	Inside	Terrestrial	IMB	7,047,448
3	Outside (top of the structure)	Aerial	IMB	3,700,522

The several point clouds were merged in a single point cloud on the base of common point. This task was carried out in 3DF Zephyr environment, which is a commercial photogrammetry software, developed and marketed by the Italian software house 3DFLOW. A representation of the whole structure according to point cloud is shown in Figure 7.

Figure 7. Three-dimensional (3D) point cloud of San Nicola in Montedoro.

4.3. Three-Dimensional Point Cloud of San Cono Bridge

In order to build the 3D model of the bridge, the photogrammetric survey was divided into an aerial and a terrestrial one. Taking into account the scale of representation (SR) and the aim of the project, a Ground Sample Distance (GSD) equal to 1 cm was chosen as reference for the survey. The terrestrial survey was carried out in order to survey the lower part of the bridge using a Canon EOS 100D DSLR camera (Charged Coupled Device -CCD size = 4.29 µm) with a focal length of 18 mm. A total amount of 400 terrestrial images was acquired. As regards the aerial survey, this was carried out using a UAS Xiaomi Mi 4K, a multi-copter rotary wing weighing less than 1.5 kg and whose declared maximum speed is 18 m per second (about 65 km/h). This UAV was developed and produced by Flymi, a company of Mi Ecosystem. The photogrammetric features of the camera mounted on UAV platform were: CCD size = 4.29 µm and focal length of 3.5 mm. The aerial survey was designed using a software called Mission Planner, which is developed by Oborne for the open-source APM autopilot project. The flight plan was designed with the following characteristics [23]: 80% longitudinal (end-lap) and 60% transversal overlap (sidelap). In addition, flight lines (FLs) inclined at 30° and 45° were designed in a direction longitudinal to the bridge in order to increase the rigidity of the aerial photogrammetric block and, at the same time, to increase the redundancy of information with the data obtained from the terrestrial survey. In total, 285 images were taken during the aerial survey.

The post-processing of terrestrial and aerial images was carried out using Agisoft Metashape software. In this case study, two separate chunks were built: one involving aerial (UAV) surveying and another involving terrestrial surveying. To evaluate the quality of image matching (alignment step), the number of the projections and the error achieved on the single chunk were taken into account. Table 3 shows the high quality of the image matching and, consequently, the correctness in the phase of acquisition, for both the aerial and the terrestrial surveys.

Table 3. Report on image matching for the two datasets.

Dataset	Projections (#)		Error (Pixel)	
	min	max	min	max
Aerial	1831	2426	0.524	0.8427
Terrestrial	4031	5114	0.622	0.247

According to the photogrammetric pipeline, a dense point cloud was built for both datasets. Consequently, in order to obtain the model of the bridge under investigation, it was necessary to integrate the two datasets on the basis of common points. In total, the final 3D point cloud consisted of approximately 8 million points (Figure 8). Subsequently, the model was scaled using 12 Ground Control Points obtained through a traditional topographic survey.

Figure 8. Three-dimensional point cloud of San Cono bridge (visualization in Agisoft Metashape software).

4.4. Three-Dimensional Reconstruction of the Models

The point cloud obtained from the geomatics surveys must be classified in objects which the structure under examination consists of. The processes necessary to perform this task must take into consideration several parameters, such as noise, occlusions, the association between faces of neighbouring objects, etc. We carried out this task in Rhinoceros software because it has more tools and plug-ins for 3D modelling. The key point of this software application is the possibility of generating a profile of the structure and, especially, to build a surface that can be adapted to the point cloud obtained in geomatics surveys. Once the point cloud was imported into Rhinoceros, it was possible to reanalyse it using the Arena plug-in. In this way, the density of the points of the PC was decreased and, consequently, it was possible to assess if there were any holes in the 3D model. Within the Rhinoceros software, the tools available to users are quality, point size and the visual analysis tools (render, ratio, opacity). This allowed for editing the point cloud of the structure. Subsequently, the point cloud was dissected into several planes in space. This operation allowed sections in strategic points of the structure, such as the arches of the bridge (Figure 9), to be performed.

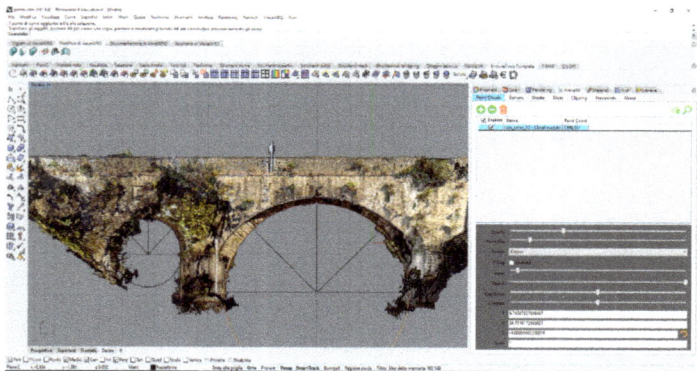

Figure 9. Scheme of the position of the sections.

The plug-in allowed saving the sections in a specific layer. As a result, the sections were displayed as "construction plans" (see Figure 10).

Sections that are transverse and longitudinal to the structure were used to create NURBS. Using the EvoluteTools PRO plug-in, it was possible to generate NURBS surfaces (Figure 11a). This plug-in allowed us to shape NURBS surfaces on objects of the structure, exploiting both the sections and the point cloud through an appropriate algorithm developed within this plug-in. For example, the bridge pillar was modelled using an adaptive NURBS (Figure 11b).

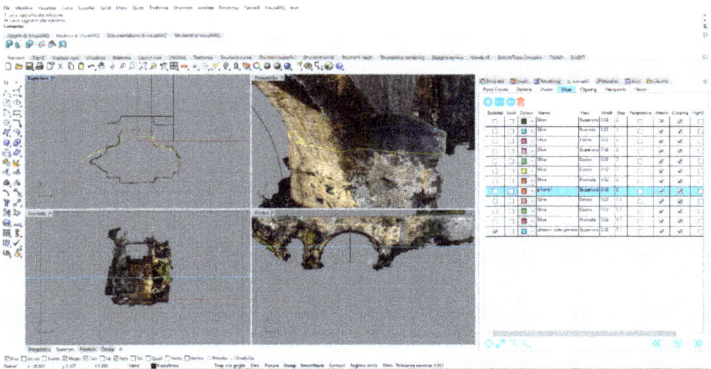

Figure 10. Construction planes within Rhinoceros software.

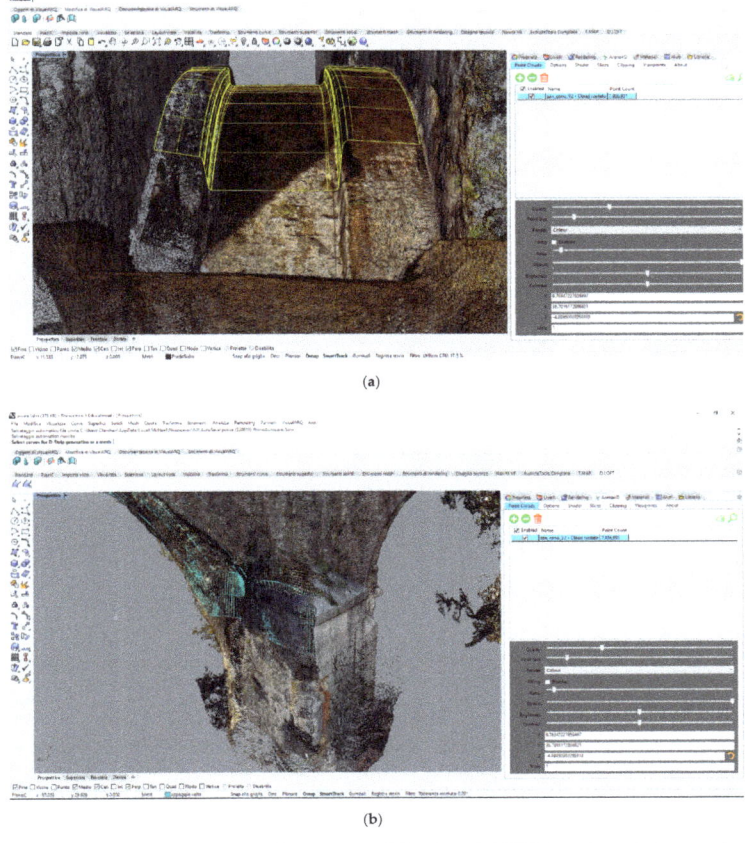

Figure 11. Adaptation of the non-uniform rational B-spline (NURBS) to the existing surface of the masonry bridge: detail of the arches (**a**) and of a pylon (**b**).

Of course, the time of the clustering task was related to the complexity of the structure. In this way, it was possible to create surfaces that represent the elements of the structure (vault, stack, retaining walls and superstructure of the bridge), as shown in Figure 12a. Using the same procedure just described for the masonry bridge, it was possible to build a 3D model of the San Nicola in Montedoro church too (Figure 12b).

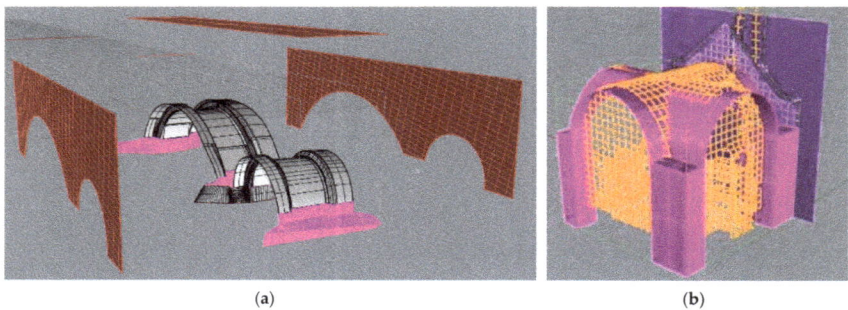

Figure 12. Three-dimensional model created in Rhinoceros environment: exploded structural elements of the bridge (**a**); 3D model of the San Nicola in Montedoro church (**b**).

Lastly, thanks to the development of the Grasshopper plug-in, it was possible to model similar structural elements (or parts of them) in 3D. Thus, it was possible to parameterize both from the geometric point of view and from the point of view of the type of material. For example, Figure 13 shows the parameterization of the arch of the bridge using the tools developed in Grasshopper.

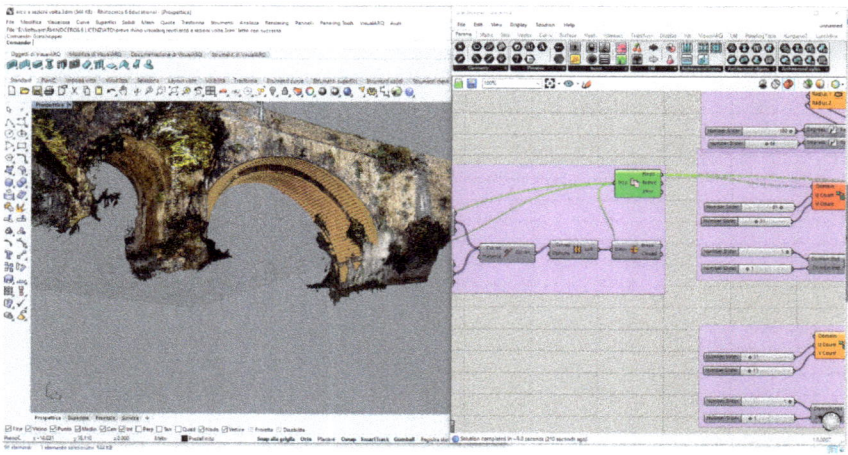

Figure 13. Three-dimensional model of the masonry bridge using Grasshopper software.

4.5. Building Information Modelling (BIM)

Many commercial BIM software products are available on the market. One of the most efficient is Autodesk Revit. The original software was developed by Charles River Software, founded in 1997, renamed Revit Technology Corporation in 2000, and acquired by Autodesk in 2002. Autodesk Revit allows users to design a building and structure and its components in 3D, annotate the model with 2D drafting elements and access building information from the building model's database. Modelling in the BIM environment of the two case studies was carried out using Autodesk Revit software.

In both case studies, the resulting mesh surface obtained in Rhinoceros software in 3D ACIS Modeler (ACIS) format (*.sat) was imported into the BIM Revit software. In this way, the surface created can be quickly opened by the BIM software and can be easily manipulated with rotations and translations. The high detail of the polysurface allowed the precise determination of the levels for the creation of BIM objects. Screenshots of the modelling and management of the information in Revit software, both of the masonry bridge and of the church, are shown below (Figure 14 a,b).

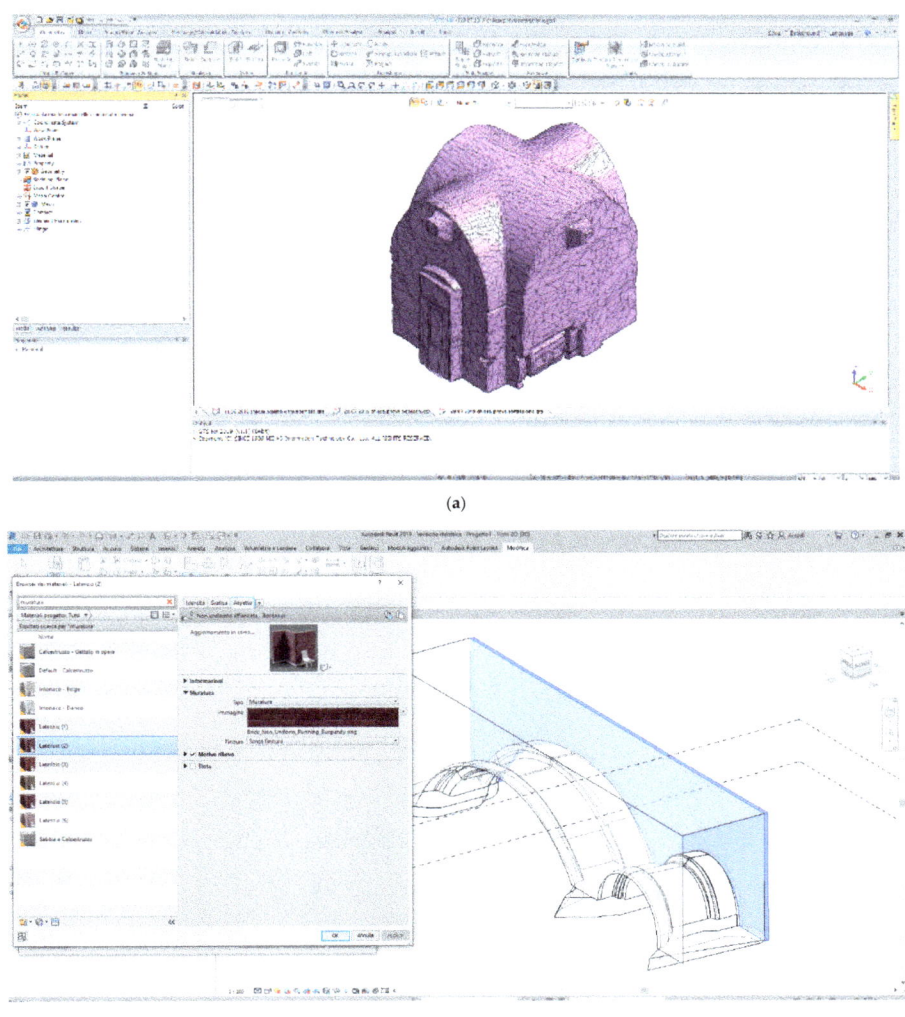

Figure 14. Visualization of the structures in Revit environment: San Nicola in Montedoro church (**a**) and San Cono masonry bridge (**b**).

4.6. Structural Analysis

The 3D model obtained in Rhinoceros was used in structure analysis software based on the FEM method. The finite element method is the most widely used method for solving problems of engineering and mathematical models, such as structural analysis, heat transfer, fluid flow, etc.

In this paper, an FEM model was used for structural analysis. In particular, Midas GTS NX software, developed by MIDAS Information Technology Co, was used for the several structural analyses. Midas GTS NX is a comprehensive finite element analysis software package that is equipped to handle the entire range of structural design applications.

The procedure that allowed structural information to be generated, starting from the 3D model, is quite simple. In fact, once the surface is imported into the Midas GTS NX structural software, it was possible to create structural meshes. Subsequently, the conditions of external, internal, deadweight (structural elements weight) and accidental loads constraint were assigned to the structure.

As for the materials, the customized information of each of them can be assigned within Rhinoceros through the "VisualARQ" plug-in. The styles of objects with such customize can be exported, in IFC (Industry Foundation Classes) format, in Revit software. These objects, recognized in Revit according to their style and custom information (material properties, costs per unit and custom metric information associated with any object in the model), are further enriched through the Revit libraries with the appropriate material characteristics useful for volumic, thermal, computational-maintenance elaborations. In order to use the advanced structural constitutive relation it is necessary to use FEM calculation software.

The object created in Rhinoceros was imported into Midas GTS NX through: "step" and "parasolid" format. The imported object is congruent and all its structural parts are correctly connected. The object, however, represents a single solid of a single material. Through specific Boolean operations, such as "divide solid", it is possible to divide and auto-connect the different surfaces. Therefore, each of the structural parts generated will be given the appropriate structural material. The materials are characterized by the appropriate constitutive relations (Mohr–Coulomb, Drucker–Prager, Von Mises). The elastic modulus, friction angle, Poisson coefficient etc. were indicated in the software. Once the correct materials were assigned, through the congruence of the structural elements, linear and non-linear seismic analisys can be performed.

For example, Figure 15 shows a view of the results in terms of deformation of the San Nicola in Montedoro church.

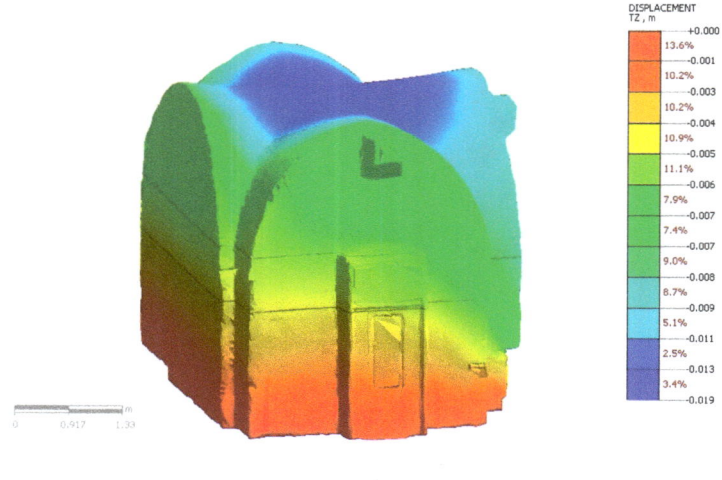

Figure 15. Static analysis: results of deformations of the structure. The maximum value achieved was 1.8 mm (blue) while the minimum was 1.2 mm (red).

Of course, the same approach, but using a different method related to the load of the structure and the constraints, was used for the masonry bridge. Specifically, the Mohr–Coulomb constitutive relation was used to assign the materials to the masonry bridge. This constitutive relation allows for linear and non-linear seismic analysis. This task was carried out within the structural software (see Figure 16).

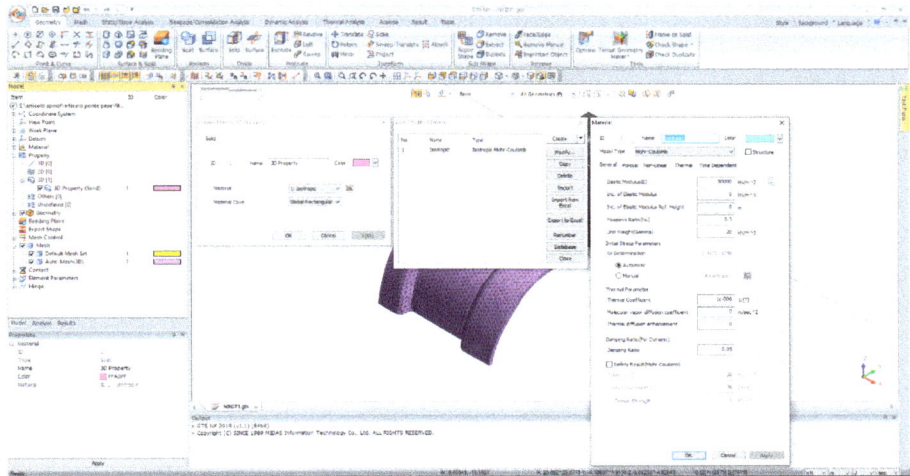

Figure 16. Materials and constitutive relation assigned to the masonry arch of the bridge in Midas GTS NX software.

As a result, it was possible to analyse the deformation state of the masonry bridge. However, it is necessary to clarify that the analysis performed on the structures represents a test based on the evidence of the correctness of the structural model within the software. Therefore, in order to define a model of deformation closer to reality, it would be necessary to take into consideration further investigations of the dynamic effects, the geotechnical-geological characteristics of the soil, the hydraulic effects (in the case of the bridge), etc. However, the consideration of the latter aspects goes beyond the scope of this paper, whose goal was to identify a specific procedure that we considered more suitable to switch from a 3D point cloud representation (obtained through a geomatics survey) to a 3D model manageable in BIM and FEM environments.

5. Conclusions

This paper reports an effective procedure to obtain 3D models for HBIM and FEM environments. In addition, using the procedure described herein, it was also possible to model structures (as shown in the case study of the masonry bridge) that had thick vegetation covering part of them. However, the procedure required several manual steps and the use of multiple softwares. At present, no single software has been developed that allows this process to be tackled directly from a geomatics survey to modelling and subsequent transformation into an object useable in BIM or FEM.

In the construction of 3D models, a key role is played by geomatics surveying. In fact, the higher the quality with which a model is built (in terms of precision and structure details) the more suitable the model will be to be implemented within BIM and FEM software.

Lastly, parametric modelling with the Grasshopper tool (implemented in the Rhinoceros software) allowed us to efficiently parameterize the elements of the analysed structures. A further potential of this tool is related to the possible updating of the static condition of the structure. In other words, Grasshopper allows building suitable models for structural verification over time, i.e., in 4D. In addition, this tool allows creating surfaces capable of representing existing structures; therefore, once a model is obtained, it is possible to build structural reinforcements that can be applied to the structure.

Author Contributions: Conceptualization, M.P., D.C. and A.R.G.; Methodology, M.P., D.C. and A.R.G.; Validation, M.P., D.C. and A.R.G.; Writing–review and editing, M.P., D.C. and A.R.G. (all authors contributed equally to the research and writing of the manuscript). All authors have read and agreed to the published version of the manuscript.

Funding: This research was conducted with funds from the DICATECh of the Polytechnic of Bari (Italy).

Acknowledgments: We want to thanks the reviewers for their careful reading of the manuscript and their constructive remarks. This research was carried out in the project: PON "Ricerca e Innovazione" 2014-2020 A. I.2 "Mobilità dei Ricercatori" D.M. n. 407-27/02/2018 AIM "Attraction and International Mobility" (AIM1895471 – Line 1).

Conflicts of Interest: The authors declare no conflict of interest.

References

1. Costantino, D.; Carrieri, M.; Restuccia Garofalo, A.; Angelini, M.G.; Baiocchi, V.; Bogdan, A.M. Integrated survey for tensional analysis of the vault of the church of San Nicola in Montedoro. *Intern. Arch. Photogramm. Remote Sens. Spat. Inf. Sci.* **2019**, *4211*, 455–460. [CrossRef]
2. Artese, S.; Achilli, V.; Zinno, R. Monitoring of bridges by a laser pointer: Dynamic measurement of support rotations and elastic line displacements: Methodology and first test. *Sensors* **2018**, *18*, 338. [CrossRef] [PubMed]
3. Kersten, T.; Lindstaedt, M.; Maziull, L.; Schreyer, K.; Tschirschwitz, F.; Holm, K. 3D recording, modelling and visualisation of the fortification kristiansten in Trondheim (Norway) by photogrammetric methods and terrestrial laser scanning in the framework of Erasmus programmes. *Int. Arch. Photogramm. Remote Sens. Spat. Inf. Sci.* **2015**, 225–230. [CrossRef]
4. Kadobayashi, R.; Kochi, N.; Otani, H.; Furukawa, R. Comparison and evaluation of laser scanning and photogrammetry and their combined use for digital recording of cultural heritage. *Int. Arch. Photogramm. Remote Sens. Spat. Inf. Sci.* **2004**, *35*, 401–406.
5. Bassier, M.; Hardy, G.; Bejarano-Urrego, L.; Drougkas, A.; Verstrynge, E.; Van Balen, K.; Vergauwen, M. Semi-automated Creation of Accurate FE Meshes of Heritage Masonry Walls from Point Cloud Data. In *Structural Analysis of Historical Constructions*; Springer: Cham, Germany, 2019; pp. 305–314.
6. Murphy, M.; McGovern, E.; Pavia, S. Historic building information modelling (HBIM). *Struct. Surv.* **2009**, *27*, 311–327. [CrossRef]
7. López, F.J.; Lerones, P.M.; Llamas, J.; Gómez-García-Bermejo, J.; Zalama, E. A review of heritage building information modeling (H-BIM). *Multimodal Technol. Interact.* **2018**, *2*, 21. [CrossRef]
8. Fregonese, L.; Achille, C.; Adami, A.; Fassi, F.; Spezzoni, A.; Taffurelli, L. BIM: An integrated model for planned and preventive maintenance of architectural heritage. *2015 Digit. Herit.* **2015**, *2*, 77–80.
9. Barazzetti, L.; Banfi, F.; Brumana, R.; Previtali, M. Creation of parametric BIM objects from point clouds using NURBS. *Photogramm. Rec.* **2015**, *30*, 339–362. [CrossRef]
10. Eigenraam, P.; Borgart, A. Reverse engineering of free form shell structures; From point cloud to finite element model. *Heron* **2016**, *61*, 193.
11. Furno, F.L.; Pietrucci, F.; Tommasi, C.; Mandelli, A. Un modello informativo parametrico per il Duomo di Milano-Test e sperimentazioni. *Archeomatica* **2017**, *7*, 22–25.
12. León-Robles, C.A.; Reinoso-Gordo, J.F.; González-Quiñones, J.J. Heritage building information modeling (H-BIM) applied to a stone bridge. *ISPRS Intern. J. Geo-Inf.* **2019**, *8*, 121.
13. Bassier, M.; Vergauwen, M. Clustering of Wall Geometry from Unstructured Point Clouds Using Conditional Random Fields. *Remote Sens.* **2019**, *11*, 1586. [CrossRef]
14. Mineo, C.; Pierce, S.G.; Nicholson, P.I.; Cooper, I. Introducing a novel mesh following technique for approximation-free robotic tool path trajectories. *J. Comput. Des. Eng.* **2017**, *4*, 192–202. [CrossRef]
15. Dardari, D.; Luise, M.; Falletti, E. (Eds.) *Satellite and Terrestrial Radio Positioning Techniques: A Signal Processing Perspective*; Academic Press: Cambridge, MA, USA, 2012.
16. Pajarola, R. Advanced 3D Computer Graphics. Available online: http://mat-web.upc.edu (accessed on 10 October 2019).
17. De Boor, C. On calculating with B-splines. *J. Approx. Theory* **1972**, *6*, 50–62. [CrossRef]
18. Piegl, L. On NURBS: A survey. *IEEE Comput. Graph. Appl.* **1991**, *11*, 55–71. [CrossRef]

19. Pepe, M.; Costantino, D.; Crocetto, N.; Garofalo, A.R. 3D modeling of roman bridge by the integration of terrestrial and UAV photogrammetric survey for structural analysis purpose. *Int. Arch. Photogramm. Remote Sens. Spat. Inf. Sci.* **2019**, *42*, 249–255. [CrossRef]
20. Liuzzi, G. *Il Santacroce E Il Distretto Di Martina Compassato Nell'antica Selva Di Monopoli*; Umanesimo della pietra: Martina Franca TA, Italy, 2011; Volume 34, pp. 3–16. Available online: https://www.umanesimodellapietra.it/UmanesimoManager//File/pubblicazioni/000019/allegati//riflessioni_2011.pdf (accessed on 1 September 2019).
21. Ronchi, D.; Limongiello, M.; Ribera, F. Field work monitoring and heritage documentation for the conservation project. The" Foro Emiliano" in Terracina (Italy). *Int. Arch. Photogramm. Remote Sens. Spat. Inf. Sci.* **2019**, *4215*, 1031–1037. [CrossRef]
22. Kannala, J.; Brandt, S.S. A generic camera model and calibration method for conventional, wide-angle, and fish-eye lenses. *IEEE Trans. Pattern Anal. Mach. Intell.* **2006**, *28*, 1335–1340. [CrossRef] [PubMed]
23. Pepe, M.; Fregonese, L.; Scaioni, M. Planning airborne photogrammetry and remote-sensing missions with modern platforms and sensors. *Eur. J. Remote Sens.* **2018**, *51*, 412–436. [CrossRef]

© 2020 by the authors. Licensee MDPI, Basel, Switzerland. This article is an open access article distributed under the terms and conditions of the Creative Commons Attribution (CC BY) license (http://creativecommons.org/licenses/by/4.0/).

Article

Categorization of the Condition of Railway Embankments Using a Multi-Attribute Utility Theory

Meho Saša Kovačević [1], Mario Bačić [1,*], Irina Stipanović [2] and Kenneth Gavin [3]

1. Faculty of Civil Engineering, University of Zagreb, 10000 Zagreb, Croatia; msk@grad.hr
2. Faculty of Engineering Technology, University of Twente, 7500 AE Enschede, The Netherlands; i.stipanovic@utwente.nl
3. Faculty of Civil Engineering and Geosciences, TU Delft, 2628 CN Delft, The Netherlands; k.g.gavin@tudelft.nl
* Correspondence: mbacic@grad.hr; Tel.: +385-1-4639-636

Received: 9 October 2019; Accepted: 22 November 2019; Published: 25 November 2019

Abstract: In the current economic climate, it is crucial to optimize the use of all resources regarding railway infrastructure maintenance. In this paper, a multi-attribute decision support framework is applied to categorize railway embankments in order to prioritize maintenance activities. The paper describes a methodology to first determine the current condition of embankments using a combination of ground penetrating radar (GPR) surveys, visual inspection, and historical data about maintenance activities. These attributes are then used for the development of a multi-attribute utility theory model, which can be used as a support for decision making process for maintenance planning. The methodology is demonstrated for the categorization of 181 km of railway embankments in Croatia.

Keywords: railway embankment; condition assessment; ground penetrating radar; multi-attribute utility theory

1. Introduction

A significant part of the major infrastructure on European railway networks was built in the 19th century, prior to the advent of modern design standards and specifications. Increased axle loading, aging, and climate impacts are known stressors to this infrastructure. In many parts of Europe, rainfall patterns are changing, in that longer dry spells are followed by periods of intense precipitation. Aged railway embankments are extremely vulnerable to the impact of such events, as drying and cracking of near surface soils during dry periods allows rapid infiltration of water during rainfall, thereby reducing the soil strength and causing sudden failure [1].

The current approach for planning maintenance works on railway infrastructure is mostly reactive [2], since infrastructure managers usually do not have sufficient information and accurate models to assess and predict the condition. Forensic analyses of historic failures often reveal that indicators of distress were ignored due to lack of understanding or the absence of a proper framework for decision-making [3]. The decisions to perform maintenance are based mostly on visual observations, subjective judgments, and choices which are ruled by available budgets, planned schedules, or abrupt failures [4,5]. Decisions based on these drivers often lead to undue maintenance and increased cost. Reactive maintenance should be avoided and railway agencies across the world are trying to move toward proactive maintenance planning, which would ensure safer, cost-effective, and improved network availability and reduce environmental impacts. The optimization of maintenance activities regarding technical and economic requirements is essential for transport infrastructure owners to fulfill societal expectations. Due to the long life time of rail infrastructure, especially engineering structures (often longer than 50 years), the assessment of technical and economic performance is necessary in order to optimize budget expenditure. Life cycle cost (LCC) analysis is a well-established methodology for the identification and assessment of maintenance trade-offs [6–9]. Nevertheless, in order to predict maintenance interventions accurately, it is necessary to assess the current condition and predict the future performance. Through early identification of problems or hot-spot locations, low-cost remediation can be applied and thus costs can be reduced and failures avoided.

The overall aim of the study is to develop a multi-attribute decision-making model to enable categorization of the condition of railway embankments across a network through the development of a ranking list. The steps involved in the categorization methodology are given within the paper. As a first step, several attributes are identified, and these include data from visual inspections and maintenance records available from the railway agency, as well as the information gathered by geophysical testing using a ground-penetrating radar. Ground Penetrating Radar (GPR) is a non-destructive tool that is widely implemented by railway owners, for example, to detect ballast fouling [10], ballast pockets [11], anomalies such as animal burrows [12], and the water content of the soil [13]. It has also been applied in a number of studies of embankment condition [14,15]. The method is affected by some limitations, primarily related to the reliability of results, which is dependent on the set-up of the equipment and the knowledge and experience of the operators and those analyzing and interpreting the data. Despite these limitations, the rapid and non-destructive nature of GPR investigations makes the method ideal for the categorization of embankments. After evaluating the selected attributes on the investigated line, we utilize the multi-attribute utility theory to develop a categorization procedure to be used for proactive decision-making related to the maintenance of railway embankments. The developed Multi Attribute Utility Theory (MAUT) based methodology is applied for the categorization of 181 km of railway embankments in Croatia, located along 18 railway lines. The results presented in the paper clearly demonstrate the potential of the methodology to ensure maximum return for use of the limited financial resources available. By defining the level of safety and potential risks that may arise, an optimized program of maintenance planning was developed, including additional investigation works, monitoring, and/or remedial measures. Further, a secondary advantage of application of complex decision-making processes in infrastructure management is to increase the attraction of traditional engineering disciplines to students with an interest in Information and Communications Technology (ICT) [16,17].

2. Methodology

2.1. The Multi-Attribute Utility Theory (MAUT)

Implicit in any decision-making process is the need to construct either directly or indirectly, the preference order, so that alternatives can be ranked and the best alternative can be selected. For some decision-making problems, this may be easily accomplished. For example, in case of a decision based on a cost-minimization rule (where the lowest cost alternative is chosen), the preference order is adequately represented by the natural order of real numbers, representing costs. Hence, in such a case, the preference order need not be constructed explicitly [18].

Multi-criteria decision-making (MCDM) provides a systematic approach to evaluate multiple conflicting attributes in decision-making. Conflicting attributes usually arise when evaluating options, for example, minimizing costs while maximizing performance. MCDM is used to identify and quantify decision-makers' and stakeholders' considerations about various (mostly) non-monetary factors, in order to compare alternative courses of action [19]. The multiple performance attribute can be combined into a so-called utility function, in which all the attributes are brought into a single scale [20]. One of the decision-making techniques that attempts to construct the preference order by directly eliciting the decision maker's preference and using multiple attributes is known as the multi-attribute utility theory (MAUT). The assumption is that a decision maker, who must select one alternative from a recognized set of decision alternatives, will be governed by preferences. In order to build a model which will represent the decision maker's preference and implement different attributes, a real-valued function called the utility function has to be determined for each attribute [19,20]. Once the functions are constructed, the selection of the appropriate alternative can be done using an optimization method. This technique involves several steps [21], including the definition of objectives and constraints, followed by the definition of attributes and by the development of a single utility function for each of the selected attributes. By assigning relative weights to the multiple attributes, an amalgamation step follows, which includes combining the single criterion utility functions using the relative weights into one measure based on mathematical assumptions about the decision maker's preference structure. Cerić [22] states that MAUT is used in cases when the best alternative solution must be chosen, i.e., for compiling a ranking list of the alternatives offered.

MAUT has been widely used in decision-making processes in the transportation infrastructure domain. Several papers consider the application of theory in the transport sector for the multi-objective optimization of multi-alternative decisions [23], the assessment of quality in bridge construction [24], road bridge management [25], or the development of a rating model that incorporates a wide range of factors affecting flexible pavements [26]. Several publications also address the implementation of MAUT in the railway sector for selecting transportation corridors linked to a traffic simulation model [27], selecting railway lines for reconstruction [28], or railway route planning and design [29]. Additional papers consider civil engineering infrastructure assets, where MAUT is incorporated in maintenance decision-making [21,30,31]. In order to implement MAUT for the categorization of railway embankments, a selection of proper attributes and alternatives shall be conducted.

2.2. Selection of Attributes

The aim of the proposed methodology is to improve the process of prioritization of multiple assets by taking into account multiple attributes. The model makes use of existing data available from railway agencies and improves the quality of information by including the results of non-destructive GPR inspection results. Therefore, the attributes which represent historical performance (maintenance data), experts' judgment on the condition (visual inspection), and GPR inspection results are selected. In total five attributes are selected, and the quantification of each attribute is in the range 0 to 1. For each of the attributes, it is important to define the so called 'quantification starting value' (QSV), since it provides clear quantification procedure, making the decision-making process fully transparent and well followed. All other quantification values (QV) are evaluated in respect to QSV.

2.2.1. Attribute 1: Maintenance History

One of the useful indicators about the actual performance of the infrastructure is information on the frequency and extent of past maintenance. For embankments, the main regular intervention performed is ballast tamping, while ballast cleaning or renewal interventions are performed occasionally [11]. In this work, the frequency of tampering activities is taken into account as the relevant attribute. From the whole data set, a maximum number ($f_{max,t/y,i}$) of tampering per year is assigned a quantification starting value (QSV) of 1, while the no-tampering events per year get the quantification value 0. Intermediate values are calculated as follows:

$$QV_{C1,i} = \frac{f_{t/y,i}}{f_{max,t/y,i}}, \qquad (1)$$

where C1 represents Attribute 1, QVC_1 represents the quantification value of Attribute 1, and $f_{t/y,i}$ represents the tampering frequency per year for an observed section of a railway line, while $f_{max,t/y,i}$ represents the tampering frequency per year for the most tampered section along the investigated railway line.

2.2.2. Attribute 2: Visual Assessment of External Irregularities

Visual inspections of railway embankments are performed by experienced engineers, with the overall aim of recording visible irregularities along the line. The engineers' remarks about the observed irregularities, as well as detailed photo-documentation, are used to define Attribute 2. It has to be noted that it is very challenging to quantify the irregularities due to their different nature and impact on the overall embankment condition as well as their extent. For example, a logical question could be, would a section with two small landslides observed have a higher QV value than a section with a single larger landslide? To overcome these issues, a percentage of length of irregularity extent in comparison to total section length is defined as the relevant quantifiable attribute. The following irregularities should be considered:

- Observable embankment slope slips (material failure in the side slope);
- Observable embankment bulging (evidence of the slide slopes expanding laterally);
- Observable crest settlement (suggesting settlement of the material beneath the embankment);
- Broken sleepers.

From the whole data set, the maximum number of irregularities ($V_{max,RI,i}$) per section is assigned a quantification starting value (QSV) of 1, while 'no visible irregularities' is assigned a value 0. The rest are calculated as follows:

$$QV_{C2,i} = \frac{V_{IR,i}}{V_{max,IR,i}}, \quad (2)$$

where C_2 represents Attribute 2, QV_{C2} represents the quantification value of Attribute 2, and $V_{IR,i}$ represents the number of visually determined irregularities for an observed section of a railway line, while $V_{max,IR,i}$ represents the number of visually determined irregularities for the section with the most irregularities along the investigated railway line.

2.2.3. Attributes 3–5: A GPR Investigation Data

To obtain GPR data relevant for the proposed methodology, a multi-antenna set-up should be implemented. The reason for this is that higher frequency antennas emit electromagnetic waves with a shorter pulse wavelength, therefore enabling the detection of smaller features and providing a high-resolution profile of the ballast bottom and data for analysis of ballast fouling. On the other hand, lower frequency antennas are used to locate potential anomalies in the sub-ballast and embankment fill. The investigation set-up should include both ground-coupled/lower frequency and air-coupled/higher frequency units. A high-frequency antenna (at least 1 GHz) is used to investigate shallow features such as ballast pockets and ballast quality (fouling), which represent Attribute 3 and Attribute 4, respectively. Lower frequency antennas, for example up to 400 MHz (depending on the height of the investigated embankment), should be used to map deep irregularities within the sub-ballast and embankment material, labeled as Attribute 5.

Attribute 3: The Depth of Ballast Layer (Ballast Pockets)

When a ballast penetrates into the lower layers, a depression beneath the ballast layer is formed and referred to as the 'ballast pocket' (Figure 1). Usually, the formation of ballast pockets occurs together with fouling, since during penetration fine grained material intermixes with the clean ballast. When the GPR trace propagates through the clean ballast layer, significant signal scattering occurs, leading to higher signal amplitudes (Figure 1a). Within the sub-ballast layer, the signal is attenuated. By combining the phase and manual layer pick method, the contact between the ballast layer and sub-ballast layer can be determined in a semi-automatic manner. During the determination of the ballast depth, it is important to properly evaluate the value of the dielectric constant of the ballast, since this value affects the depth of the ballast bottom. After the ballast bottom has been marked on the radargram by manual and phase picking (Figure 1b), the depth of the ballast bottom can be easily obtained. Results are extracted in the form of a report containing information on trace number, profile distance, trace amplitude, two-way travel time, and depth to the ballast layer bottom.

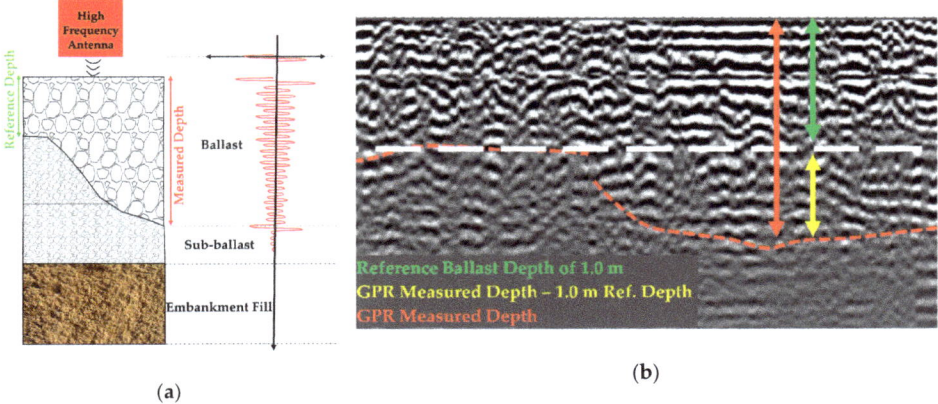

Figure 1. Determination of ballast pockets: (**a**) Ground Penetrating Radar trace in case of ballast pocket formation and (**b**) radagram showing the features for the determination of Attribute 3.

On the Croatian network, the normal ballast depth is 0.5 m (0.30 m below the sleeper bottom, with the addition of sleeper thickness, since the reference investigation surface is top of the sleeper). Therefore, Attribute 3 is defined as the measured thickness of a ballast layer in excess of 0.5 m. If the measured ballast depth is 0.5 m, there is no ballast pocket present. The overall steps for the determination of the ballast depth for a provisional 100 m long section of a railway line are shown in Table 1. It is recommended to utilize signal trace separation of maximum 5 cm, leading to 20 traces/m of the investigated line, in order to have a better insight into the position of the layer bottom. The results can be evaluated on each 1 m of the investigated line, that is for every 20th signal trace. After determination of measured depth for each trace, an averaging procedure for the whole investigated section follows.

Table 1. Methodology steps for the determination of ballast layer depth.

Trace Number	Investigation Distance (m)	Measured Depth (m)	Average Measured Depth (m)	Δd_i (m)	$QV_{C3,i}$ (–)
1 (1st overall)	0.0	$d_{measured,0}$			
2 (21st overall)	1.0	$d_{measured,1}$	$\frac{\sum d_{measured,i}}{100}$	$\frac{\sum d_{measured,i}}{100} - 0.5\text{ m}$	$\frac{\Delta d_i}{\Delta d_{max,i}}$
3 (41st overall)	2.0	$d_{measured,2}$			
...	...	...			
100 (2001st overall)	100.0	$d_{measured,100}$			

$\Delta d_{max,i}$ is determined by the same methodology as given in Table 1, and it represents the line section with the highest average measure depth ($\frac{\sum d_{measured,max,i}}{100} - 0.5$ m). From the data used to formulate the approach, the maximum ballast layer depth measured is assigned a quantification starting value (QSV) of 1, while a ballast layer of 0.5 m is assigned a value 0. The rest are calculated as follows:

$$QV_{C3,i} = \frac{\Delta d_i}{\Delta d_{max,i}}, \tag{3}$$

where C_3 represents Attribute 3, QV_{C3} represents the quantification value of Attribute 3, and Δd_i represents the measured value of ballast depth minus the designed depth of 0.5 m for an observed section of a railway line, while $\Delta d_{max,i}$ represents the maximum measured value of ballast depth minus the designed depth of 0.5 m for the section with measured maximum ballast depth along the investigated railway line.

Attribute 4: Ballast Fouling (Quality of Ballast Layer)

High-frequency antennas are also used for the determination of ballast fouling. Ballast fouling is a direct consequence of the ballast aging process, where fine-grained materials fill the ballast void spaces, leading to track instability with serious implications on track drainage [32]. Even though GPR is commonly used for the inspection of ballast fouling, Panjamani et al. [10] state that there are no robust guidelines to find the degree and type of fouling quantitatively. A good quality, clean ballast is approximately of 30–60 mm size and should be dry. Therefore, GPR signal scattering should be prominent as shown in Figure 1a. However, if the layer is infiltrated with fine-grained materials and consequently with water, more attenuation could be expected (Figure 2a). A fouled ballast does not scatter emitted energy so much, since the air gaps are now filled with smaller grain materials. This principle is widely accepted as a good way to assess the degree of ballast fouling [33]. An example of a clean ballast zone and a fouled ballast zone is given on the radargram in Figure 2b.

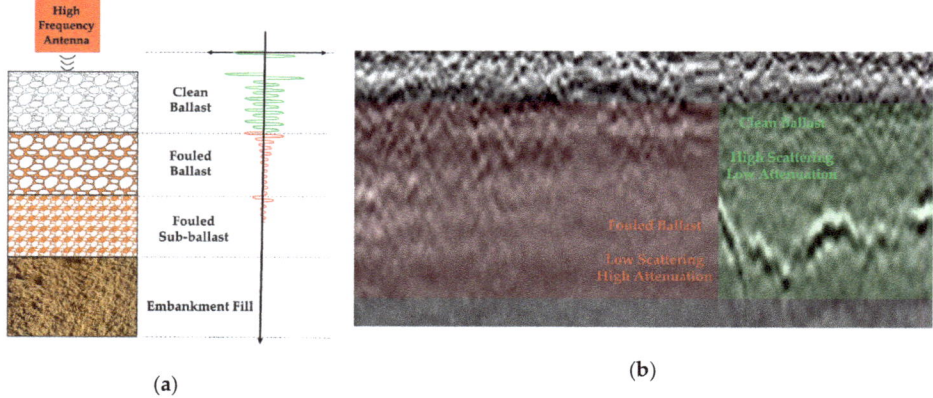

Figure 2. Determination of ballast fouling: (**a**) Ground Penetrating Radar trace in case of ballast fouling and (**b**) radagram showing the difference between clean and fouled ballasts.

After basic processing, each GPR trace is extracted from the radargram and subjected to further analysis by using a MATLAB developed code. Several steps are conducted in order to determine a quantifiable value for Attribute 4, relevant for each section of a line. As a first step, a trace is divided into several depth zones, each with a pre-defined depth of 10 cm. The total number of zones (k_{max}) depends on the overall position of the ballast layer bottom. Next, the maximum amplitude of each depth zone is determined, where an amplitude envelope is constructed by connecting the peaks of the reflections down a trace (Figure 3).

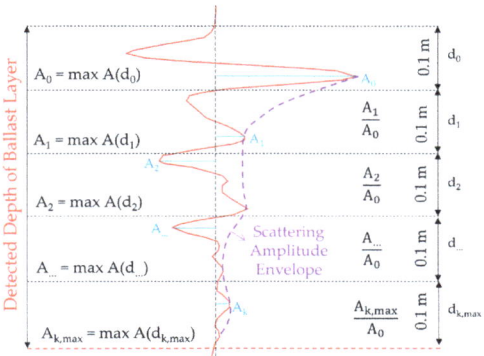

Figure 3. Elements of a single Ground Penetrating Radar trace for the determination of ballast fouling.

A lower amplitude value could be associated with a fouled ballast. After determining the amplitude ratio A_i/A_0 for each depth zone, an averaging procedure for the whole trace is conducted with the determination of an average amplitude decrease (AAD) ratio for a single trace. From there, ballast fouling (%) is determined for a single trace, followed by the calculation of average ballast fouling for the whole investigated section. The overall steps for the determination of a ballast fouling percentage for a provisional 100 m long section of railway line are shown in Table 2. It is recommended to utilize signal trace separation of maximum 5 cm (horizontal resolution), leading to 20 traces/m of the investigated line, with each trace divided into at least 512 samples (vertical resolution). For the ballast fouling evaluation, each signal trace should be analyzed.

Table 2. Methodology steps for the determination of ballast fouling.

Trace Number	Investig. Distance (m)	Average Amplitude Decrease AAD (−)	Signal Attenuation, i.e., Ballast Fouling (BQ) (−)	BQ Average × 100 (%)	$QV_{C4,i}$ (−)
1 (1st overall)	0.0	$AAD_n = \left(\frac{\sum_{k=0}^{k_{max}} \left(\frac{A_k}{A_0} \right)}{k_{max}} \right)_1$	$(BQ)_1 = 1 - AAD_1$		
1 (2nd overall)	0.05	$AAD_{n+1} = \left(\frac{\sum_{k=0}^{k_{max}} \left(\frac{A_k}{A_0} \right)}{k_{max}} \right)_2$	$(BQ)_2 = 1 - AAD_2$	$BQ_i = \frac{\sum BQ}{2001}$	$\frac{BQ_i}{BQ_{max,i}}$
2 (3rd overall)	0.10	$AAD_{n+2} = \left(\frac{\sum_{k=0}^{k_{max}} \left(\frac{A_k}{A_0} \right)}{k_{max}} \right)_3$	$(BQ)_3 = 1 - AAD_3$		
3 (4th overall)	0.15	$AAD_{n+3} = \left(\frac{\sum_{k=0}^{k_{max}} \left(\frac{A_k}{A_0} \right)}{k_{max}} \right)_4$	$(BQ)_4 = 1 - AAD_4$		
...	...	...	...		
2001 (2001st overall)	100.0	$AAD_{2001} = \left(\frac{\sum_{k=0}^{k_{max}} \left(\frac{A_k}{A_0} \right)}{k_{max}} \right)_{2001}$	$(BQ)_{2001} = 1 - AAD_{2001}$		

$BQ_{max,i}$ is determined by the same methodology as given in Table 2, and it represents the line section with the highest average fouling percentage. A non-fouled ballast (clean-ballast) has a 0% fouling degree (QSV of 0). A QSV of 1 is attributed to the section with highest fouling degree, while the rest are calculated as follows:

$$QV_{C4,i} = \frac{BQ_i}{BQ_{max,i}}, \qquad (4)$$

where C_4 represents Attribute 4, QV_{C4} represents the quantification value of Attribute 4, and BQ_i represents the value of ballast fouling (in %) for an observed section of a railway line, while $BQ_{max,i}$

represents the value of ballast fouling for the section with the lowest quality/maximum fouling of ballast along the investigated railway line.

Attribute 5: Irregularities in Sub-Ballast and Embankment Fill

Although Selig and Waters [34] note that degradation mostly affects the ballast layer, it is important to map and detect defects in the sub-ballast and subgrade layers. In this case, lower frequency antennas are used to locate anomalies in the sub-ballast and embankment fill (voids, animal burrows, water erosion channels, etc.). Reflections of the emitted waves occur when the signal reaches boundaries and/or anomalies at larger depths. The presence of irregularities results in specific signal trace peaks (Figure 4a). If the anomaly is in the form of a void within the investigated layers, for example, animal burrows, it shows in the radargram in the form of a hyperbola (Figure 4b). The parameters of these hyperbolas are commonly determined by utilization of generalized Hough transform or, recently, by utilization of neural network tools to reduce the analysis time [35].

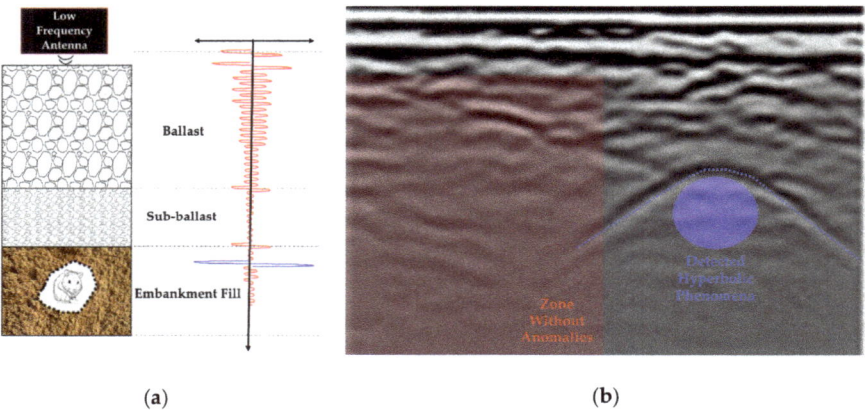

Figure 4. Determination of deeper irregularities: (**a**) Ground Penetrating Radar trace in case of deep irregularity presence and (**b**) radargram showing the detected hyperbolic phenomena.

From the whole data set, the maximum number of deep irregularities ($D_{max,IR,i}$) per section is assigned a quantification starting value (QSV) of 1, while 'no deep irregularities' is assigned a value 0. The rest are calculated as follows:

$$QV_{C5,i} = \frac{D_{IR,i}}{D_{max,IR,i}}, \quad (5)$$

where C_5 represents Attribute 5, QV_{C5} represents the quantification value of Attribute 5, and $D_{IR,i}$ represents the number of deep irregularities for an observed section of a railway line, while $D_{max,IR,i}$ represents the number of deep irregularities for the section with the maximum number of irregularities along the investigated railway line. These values are determined based on a visual assessment of each radargram, by counting the number of hyperbolic phenomena for each investigated section.

2.3. Implementation of Multi-Attribute Utility Theory

To effectively implement MAUT into a methodology for railway embankment condition assessment, several steps need to be considered. A detailed description of the mathematical basis of the MAUT approach is presented in [36–38]. After the selection and determination of the quantification values of the attributes (QV_C) using Equations (1)–(5), the following step includes calculation of the utility function values for the selected 'n' number of attributes (five attributes are proposed in the paper) and 'm' number of alternatives. An alternative (S) is defined as 'a sub-section of pre-defined length'. In this case, 100 m long sub-sections were considered, which determine the resolution for categorization. Any other resolution length, adapted to the character of specific problem and to the needs of infrastructure managers, can also be considered.

The overall MAUT problem can be expressed by an $m \times n$ decisional matrix, having the form as shown in Table 3.

Table 3. The Multi Attribute Utility Theory matrix for railway embankment condition assessment.

Alternatives (m)	Attribute QV (n)	QV_{C1}	QV_{C2}	QV_{C3}	QV_{C4}	QV_{C5}
S_1		$\overline{U_1}(S_1)$	$\overline{U_2}(S_1)$	$\overline{U_3}(S_1)$	$\overline{U_4}(S_1)$	$\overline{U_5}(S_1)$
S_2		$\overline{U_1}(S_2)$	$\overline{U_2}(S_2)$	$\overline{U_3}(S_2)$	$\overline{U_4}(S_2)$	$\overline{U_5}(S_2)$
...		...	...	...	...	...
S_m		$\overline{U_1}(S_m)$	$\overline{U_2}(S_m)$	$\overline{U_3}(S_m)$	$\overline{U_4}(S_m)$	$\overline{U_5}(S_m)$

The normalized utility function values can be determined using the following equation:

$$\overline{U_i}(S_j) = \frac{QV_{C,i}(S_j)}{\sum_{j=1}^{m} QV_{C,i}(S_j)}, \qquad (6)$$

where

$\overline{U_i}(S_j)$—normalized utility function value for Attribute i and Alternative j
S_j—Alternative j, subsection of 100 m in length; $j = 1, 2, \ldots, m$
m—number of alternatives
$QV_{C,i}$—quantification value of Attribute i; $i = 1, 2, \ldots, n$
n—number of attributes

The sum of all utility function values for a specific alternative is equal to 1:

$$\sum_{i=1}^{n} \overline{U_i}(S_j) = 1. \qquad (7)$$

The next step in the implementation of MAUT includes the evaluation of the importance of each selected attribute. To do this, a grading scheme is established, where the grading of importance of each attribute for the embankment condition assessment is conducted by attributing a grade ranging from 1 (negligible importance) to 10 (extremely important). This information was collected using a questionnaire completed by experienced engineers working in the field of infrastructure management. The determination of each attribute's importance is thus by nature subjective, reflecting the experience and risk acceptance of the engineers in question. After determining mean values and standard deviations of the assigned grades, a weight of importance for a specific attribute can be calculated by:

$$w_i = \frac{QV_{C,i}}{\sum_{i=1}^{n} QV_{C,i}}, \qquad (8)$$

where

w_i—weight of importance for Attribute i

The sum of all weight values for all attributes is equal to 1:

$$\sum_{i=1}^{n} w_i = 1. \qquad (9)$$

Finally, the overall utility function values $U(S_j)$ for each alternative are calculated by combining the calculated utility functions (Equation (6)) and the weight of importance (Equation (8)):

$$U(S_j) = \sum_{i=1}^{n} w_i \cdot \overline{U_i}(S_j), \qquad (10)$$

where

$U(S_j)$—the overall utility function value for Alternative j

After the overall utility function is calculated for each alternative, the development of a final ranking list is completed as follows. A classification list is developed for five (5) categories ranging from an embankment in a very poor to very good condition. The appropriate MAUT condition for each category is shown in Table 4. The list is color-coded in order to be easily comprehensible for railway infrastructure managers.

Table 4. Categorization representation with Multi Attribute Utility Theory conditions.

Category	Condition	Graphical Representation	MAUT Condition
1	Very poor		$0.8 < U(S_j) \leq 1.0$
2	Poor		$0.6 < U(S_j) \leq 0.8$
3	Adequate		$0.4 < U(S_j) \leq 0.6$
4	Good		$0.2 < U(S_j) \leq 0.4$
5	Very good		$0.0 < U(S_j) \leq 0.2$

A higher value of overall utility function for a specific alternative reflects a larger number of visible as well as deeper (GPR) anomalies, larger ballast depth, lower ballast quality, and more tampering activities per year.

3. Case Study Example

3.1. Description of the Case Study Area

The railway network in Croatia consists of more than 2600 km of lines, with the majority forming parts of European railway transport corridors (Figure 5). Given the general age of the railway infrastructure and deterioration including the impact of war conflicts [39], a number of sections have traffic speed restrictions with maximum speeds of only 20 km/h on some sections.

To validate the categorization procedure proposed in this paper, a case study was chosen which included 181 km of railway embankments in Croatia along 18 different railway lines. For each investigated line, a categorization procedure was conducted. The investigated lines were chosen by railway managers from eight Supervision Centers (SC) of the Croatian Railway Infrastructure: SC Ogulin (three investigated lines), SC Osijek (three investigated lines), SC Pula (one investigated line), SC Rijeka (one investigated line), SC Slavonski Brod (two investigated lines), SC Varaždin (four investigated lines), SC Vinkovci (one investigated line), and SC Zagreb (three investigated lines).

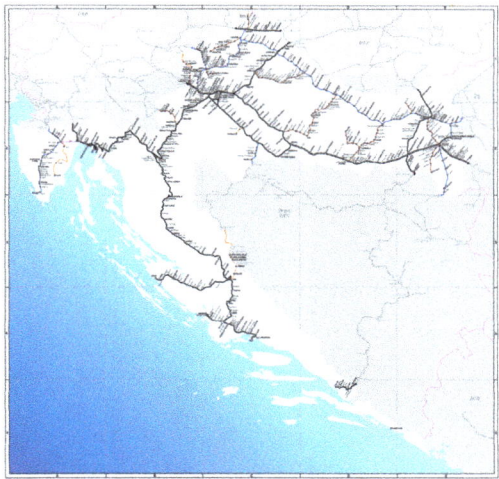

Figure 5. Railway network in Croatia, scale 1:6,000,000, from [40].

3.2. Data Acquisition Procedure

As a starting point in the categorization procedure, each SC provided information on the tamping frequency for the railway lines under their supervision. The tamping activities are usually determined solely on the basis of visual assessment and periodical measurements of the track level. In the study area, the tamping frequency was provided for the year prior to the in-situ assessment which followed.

The next step included expert visual assessment of the sections and the performance of ground-penetrating radar (GPR) investigations. The visual assessment included walking along the selected line sections and recording of detailed photos and video-documentation along with textual description of each phenomena, including its shape, extent, and the potential nature of the irregularity. The assessment was done by a multidisciplinary team consisting of railway inspectors, geotechnical engineers, and geologists.

The GPR investigation methodology included a multi-channel approach in order to detect the phenomena which are defined as attributes for methodology of embankment categorization (Figure 6). For the purpose of distance measurements during data acquisition, a distance measuring instrument (DMI) was attached to the survey wheel of a specially constructed bogey which housed the GPR equipment. In this study, the signal trace separation was 5 cm, leading to 20 traces per m of investigated track. The rapid nature of GPR testing was essential, since the investigations were conducted during the day with minimal line closure as requested by infrastructure managers.

Figure 6. Conduction of GPR investigations.

3.3. Results and Discussion

The terminology used in this section includes the following terms: 'line' as the railway line, 'section' as part of the line under investigation, and 'subsection' is a 100 m length of the investigated line. Therefore, for the total of 181.1 km of 18 investigated sections on 18 different railway lines, we had 1811 subsections, which were analyzed and represented the alternatives for the implemented multi-attribute utility theory model. Taking into consideration that sections selected for investigation were part of different lines, the analysis for 18 sections were done independently, bearing in mind that maintenance on those sections would also be performed independently. Sections differed from 1.0 km of length, which had 10 alternatives (10 subsections of 100 m length) to the largest one, which was 31.3 km long and consisted of 313 alternatives. This division was done in agreement with the Supervision Centers, since it was rational to separately evaluate these groups, rather than implementing categorization procedure for all the lines together. The division will enable infrastructure managers' better insight into the condition of each section within the relevant line, where a subsequent resource optimization can be conducted.

As a first step in implementing the developed methodology in the case study, attributes were evaluated for all alternatives within each section. The quantification values were determined as given by Equations (1)–(5). An example of attribute evaluation is given for a 1000 m long section, divided into 100 m subsections, in Table 5.

Next, a multi-attribute decision-making problem was expressed by an $n \times m$ decisional matrix, n being the number of attributes (5 in our case) and m being the number of alternatives within one section. Considering the 18 investigated lines, the alternatives were divided into 18 groups ($k = 1, 2, \ldots, 18$), leading to the formation of 18 $n \times m$ decisional matrices, the smallest one being 5×10 and the largest one being 5×313. As a first step in calculating the overall utility function of each alternative $U_i(S_j^k)$, the normalized utility functions were determined using Equation (6). For example, the normalized utility function value for the GPR fouling Attribute (QV_{C4}) and Alternative (S_5^1) which stands for 'fifth alternative within the first section' was calculated as:

$$\overline{U_4}(S_5^1) = \frac{QV_{C4}(S_5^1)}{\sum_{j=1}^{m} QV_{C4}(S_j^1)} \qquad (11)$$

with

$$\sum_{i=1}^{5} \overline{U_i}(S_5^1) = 1. \tag{12}$$

Further, to define the importance of each of the five attributes in the overall embankment categorization procedure, a developed questionnaire was used to obtain information from experienced experts working in the railway maintenance sector. To overcome the mentioned subjectivity aspect, a questionnaire was delivered to 12 experts, and the results are given in Table 6.

Table 5. An example of attribute evaluation from one of the sections.

Attribute	Alternative	1 km 0.0–0.1	2 km 0.1–0.2	3 km 0.2–0.3	4 km 0.3–0.4	5 km 0.4–0.5	6 km 0.5–0.6	7 km 0.6–0.7	8 km 0.7–0.8	9 km 0.8–0.9	10 km 0.9–1.0
Tampering per year		3	0	0	5	2	3	3	4	2	2
QV_{C1}		0.60	0.00	0.00	1.00 (QSV_{C1})	0.40	0.60	0.60	0.80	0.40	0.40
% of visual irregularities per section		10	0	5	70	90	40	40	50	50	40
QV_{C2}		0.11	0.00	0.56	0.77	1.00 (QSV_{C2})	0.44	0.44	0.56	0.56	0.44
average depth (m)		1.32	1.25	1.29	1.34	1.36	1.17	1.23	1.09	1.05	1.06
QV_{C3}		0.95	0.87	0.92	0.98	1.00 (QSV_{C3})	0.78	0.84	0.69	0.64	0.65
fouled % per section		40	10	50	80	60	20	20	30	30	40
QV_{C4}		0.50	0.13	0.63	1.00 (QSV_{C4})	0.75	0.25	0.25	0.38	0.38	0.50
GPR irregularities per section		0	1	2	4	0	0	2	4	0	0
QV_{C5}		0.00	0.25	0.50	1.00 (QSV_{C5})	0.00	0.00	0.50	1.00 (QSV_{C5})	0.00	0.00

Table 6. Mean values, standard deviations, and calculated weight of importance for attributes.

Attribute	Label	Mean	SD	Calculated Weight of Importance (w)
Supervision Center Information	QV_{C1}	9.30	0.78	0.243
Visual Assessment	QV_{C2}	5.60	1.19	0.146
Ballast depth (ballast pockets)	QV_{C3}	8.40	0.91	0.219
Ballast fouling (ballast quality)	QV_{C4}	8.70	0.94	0.227
Irregularities in sub-ballast and embankment fill	QV_{C5}	6.30	1.21	0.164

Taking into consideration the mean and standard deviation (SD) data from Table 6, the weight of importance for each attribute could be calculated. For example, the importance of ballast fouling Attribute (C_4) was calculated using Equation (8):

$$w_4 = \frac{QV_{C4}}{\sum_{i=1}^{5} QV_{C,i}} = \frac{8.70}{9.30 + 5.60 + 8.40 + 8.70 + 6.30} = 0.227. \tag{13}$$

The sum of all weights of importance concerning attribute parameters, based on Equation (9), equals 1:

$$\sum_{i=1}^{5} w_i = 0.243 + 0.146 + 0.219 + 0.227 + 0.164 = 1. \tag{14}$$

Finally, the overall utility function values $U(S_j^k)$ for each alternative was calculated by combining five weights of attribute importance with the normalized utility functions for the five (n) attributes and the (m) alternatives of specific section using Equation (10). The overall utility function values give the ranking for each alternative. For example, it is calculated for the fifth subsection within the first section:

$$U(S_5^1) = \sum_{i=1}^{5} w_i \cdot \overline{U_i}(S_5^1). \tag{15}$$

After the overall utility function was calculated for each alternative within each section (railway line), the development of the final ranking list followed. A classification scheme, as shown in Table 4, was assigned to each subsection and the results are presented in the form, shown in Figure 7, for a 500 m length where the overall condition of the embankment was in the range from very poor to adequate and a second 500 m section where the overall condition of embankment was in the range from adequate to very good.

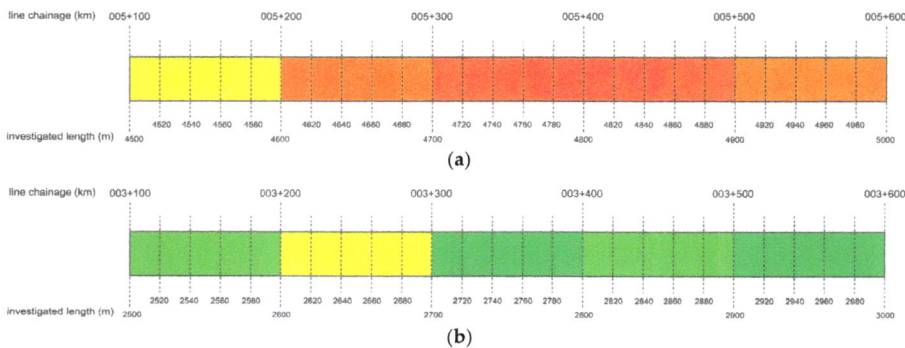

Figure 7. A graphical representation of: (**a**) a 500 m long line section with very poor to adequate condition and (**b**) a 500 m long line section with adequate to very good condition.

The implementation of the multi attribute utility theory thus provided classification of the condition for the investigated railway embankments. The overall classification results, given in Figure 8, represent the summation of classification results for lines selected by each Supervision Center (SC), as well as averaged values, in the form of an overall utility function—an overall percentage graph.

As it can be seen from Figure 8, the categorization results show that most of the investigated embankments, 41.8%, are in an adequate condition and that 33.7% of the embankments are in poor condition, while 12.6% are in a very poor condition. On the other hand, 10% of the investigated embankments are in good condition, with only 1.9% is in a very good condition. The categorization results are consistent for line sections within each Supervision Center. The described procedure of implementing MAUT for railway embankment categorization gave the solid basis for decision-makers to plan further detailed investigation works and monitoring programs as well as remediation measures.

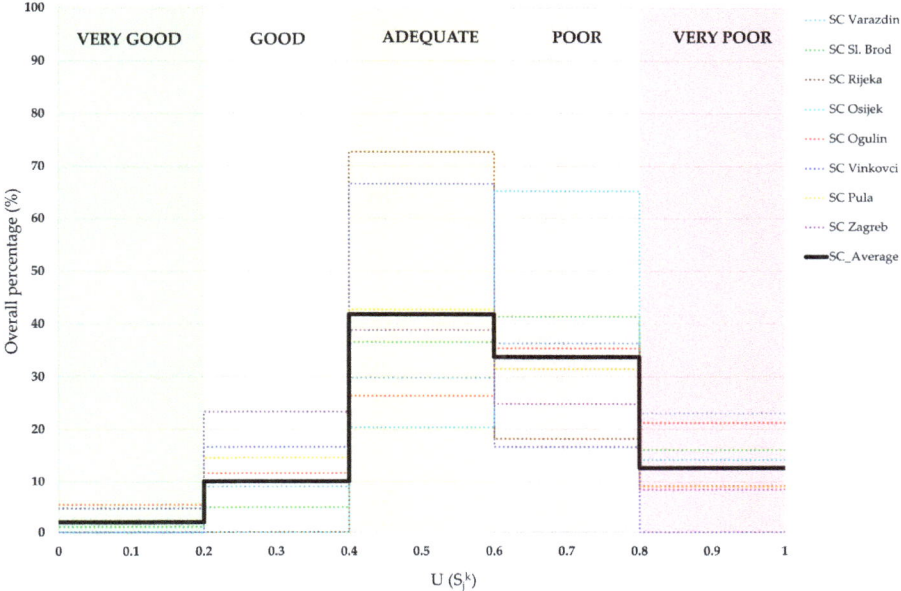

Figure 8. Overall utility function—overall percentage categorization summation results.

4. Conclusions

The paper presents a framework to apply the multi-attribute utility theory (MAUT) for the categorization of the condition of railway embankments using multiple sources of data available to the infrastructure managers. The overall methodology proposes the use of five attributes. These include frequency of tamping activities, extent of external irregularities obtained from visual inspections, frequency of internal irregularities obtained from GPR lower frequency data, ballast quality (fouling), and ballast depth (pockets). Ballast quality and depth data were obtained from higher frequency GPR antennas. The selection of these attributes is made because of their relevance to the problem and that they are readily available. By weighting the importance of each of the attributes and by evaluating them on the investigated railway lines, the MAUT-based methodology provides calculation of the overall utility function values used to form a ranking list of the condition for the investigated railway embankments. To verify the usefulness of the methodology, the paper presents its application on 181 km of railway embankments in Croatia. The calculated overall utility function for each section provided the categorization of embankments in five categories, ranging from very poor to very good. The developed MAUT model provides a transparent and comprehensive procedure that can support decision-makers to plan maintenance works and further detailed investigation works and monitor programs and remediation measures. The framework is flexible and methodology could be expanded to consider additional attributes.

Author Contributions: Conceptualization, M.S.K. and M.B.; Data curation, M.S.K.; Formal analysis, M.S.K. and M.B.; Investigation, M.S.K. and M.B.; Methodology, M.S.K. and I.S.; Project administration, M.S.K.; Validation, M.B., I.S. and K.G.; Visualization, M.B. and K.G.; Writing—original draft, M.B. and I.S.; Writing—review & editing, M.S.K. and K.G.

Funding: The authors gratefully acknowledge the support from the H2020 Programme for DESTination RAIL project, funded under MG-2.1-2014 I^2I Intelligent Infrastructure call, grant agreement No 636285.

Conflicts of Interest: The authors declare no conflicts of interest.

References

1. Gavin, K.G.; Xue, J.F. Use of a genetic algorithm to perform reliability analysis of unsaturated soil slopes. *Geotechnique* **2009**, *59*, 545–549. [CrossRef]
2. Garramiola, F.; Poza, J.; Madina, P.; Del Olmo, J.; Almandoz, G. A Review in Fault Diagnosis and Health Assessment for Railway Traction Drives. *Appl. Sci.* **2018**, *8*, 2475. [CrossRef]
3. Kovacevic, M.S.; Gavin, K.; Stipanovic Oslakovic, I.; Bacic, M. A New Methodology for Assessment of Railway Infrastructure Condition. *Transp. Res. Procedia* **2016**, *14*, 1930–1939. [CrossRef]
4. Dhillon, B.S. *Engineering Maintenance: A Modern Approach*; CRC Press: Boca Raton, FL, USA, 2002.
5. Wu, Y.; Qin, Y.; Wang, Z.; Jia, L. A UAV-Based Visual Inspection Method for Rail Surface Defects. *Appl. Sci.* **2018**, *8*, 1028. [CrossRef]
6. Woodward, D.G. Life cycle costing—Theory, information acquisition and application. *Int. J. Proj. Manag.* **1997**, *15*, 335–344. [CrossRef]
7. Kendall, A.; Keoleian, G.; Helfand, G. Integrated Life-Cycle Assessment and Life-Cycle Cost Analysis Model for Concrete Bridge Deck Applications. *J. Infrastruct. Syst.* **2008**, *14*, 214–222. [CrossRef]
8. Kabir, G.; Sadiq, R.; Tesfamariam, S. A review of multi-Attribute decision-making methods for infrastructure management. *Struct. Infrastruct. Eng.* **2014**, *10*, 1176–1210. [CrossRef]
9. Breemer, J.; Al-Jibouri, S.H.S.; Veenvliet, K.T.; Heijmans, H.W.N. *RAMS and LCC in the Design Process of Infrastructural Construction Projects: An Implementation Case*; University of Twente: Enschede, The Netherlands, 2010; pp. 1–18.
10. Panjamani, A.; Bharatha, T.P.; Amarajeevi, G. Study of Ballast Fouling in Railway Track Formations. *Indian Geotech. J.* **2012**, *42*, 87–99. [CrossRef]
11. Funtul, S.; Fortunato, E.; De Chiara, F.; Burrinha, R.; Baldeiras, M. Railways Track Characterization Using Ground Penetrating Radar. *Procedia Eng.* **2016**, *143*, 1193–1200. [CrossRef]
12. Di Prinzio, M.; Bittelli, M.; Castellarin, A.; Rossi Pisa, P. Application of GPR to the monitoring of river embankments. *J. Appl. Geophys.* **2010**, *71*, 53–61. [CrossRef]
13. Cai, J.Q.; Liu, S.X.; Fu, L.; Feng, Y.Q. Detection of railway subgrade moisture content by GPR. In Proceedings of the 16th International Conference on Ground Penetrating Radar (GPR), Hong Kong, China, 13–16 June 2016; pp. 1–5. [CrossRef]
14. Sussmann, T.; Selig, E.T.; Hyslip, J.P. Railway track condition indicators from ground penetrating radar. *NDT E Int.* **2003**, *36*, 157–167. [CrossRef]
15. Donohue, S.; Gavin, K.; Tolooiyan, A. Geophysical and geotechnical assessment of a railway embankment failure. *Near Surf. Geophys.* **2011**, *9*, 33–44. [CrossRef]
16. Burrows, A.; Lockwood, M.; Borowczak, M.; Janak, E.; Barber, B. Integrated STEM: Focus on Informal Education and Community Collaboration through Engineering. *Educ. Sci.* **2018**, *8*, 4. [CrossRef]
17. Grout, I. Remote Laboratories as a Means to Widen Participation in STEM Education. *Educ. Sci.* **2017**, *7*, 85. [CrossRef]
18. Von Neumann, J.; Morgenstern, O. *Theory of Games and Economic Behavior*; Princeton University Press: Princeton, NJ, USA, 1945.
19. Patidar, V.; Labi, S.; Sinha, K.; Thompson, P.D. *Multi-Objective Optimization for Bridge Management Systems*; National Cooperative Highway Research Program Report 590; Transportation Research Board of the National Academies: Washington, DC, USA, 2007.
20. Fishburn, P.C. *Utility Theory for Decision Making*; Research Analysis Corporation: McLean, VA, USA; John Wiley & Sons: Hoboken, NJ, USA, 1970.
21. Allah Bukhsh, Z.; Stipanovic Oslakovic, I.; Klanker, G.; O' Connor, A.; Doree, A.G. Network level bridges maintenance planning using Multi-Attribute Utility Theory. *Struct. Infrastruct. Eng.* **2018**, 872–885. [CrossRef]
22. Cerić, A. *Trust in Construction Projects*; Routledge: Abingdon, UK; Taylor & Francis Group: Didcot, UK, 2015.
23. Brauers, W.K.M.; Zavadskas, E.K.; Peldschus, F.; Turskis, Z. Multi-objective decision-making for road design. *Transport* **2008**, *23*, 183–193. [CrossRef]
24. Zavadskas, E.K.; Liias, R.; Turskis, Z. Multi-Attribute Decision-Making Methods for Assessment of Quality in Bridges and Road Construction: State-Of-The-Art Survey. *Balt. J. Road Bridge Eng.* **2008**, *3*, 152–160. [CrossRef]

25. Dabous, S.A.; Alkass, S. A multi-attribute ranking method for bridge management. *Eng. Constr. Archit. Manag.* **2010**, *17*, 282–291. [CrossRef]
26. Abu-Samra, S.; Zayed, T.; Tabra, W. Pavement Condition Rating Using Multiattribute Utility Theory. *J. Transp. Eng. Part B Pavements* **2017**, *143*. [CrossRef]
27. Zietsman, J.; Rilett, R.L.; Kim, S.J. Transportation corridor decision-making with multi-attribute utility theory. *Int. J. Manag. Decis. Mak.* **2006**, *73*, 254–266. [CrossRef]
28. Barić, D.; Radačić, Ž.; Čurepić, D. Implementation of multi-attribute decision-making method in selecting the railway line for reconstruction. In Proceedings of the International Conference on Traffic Science ICTS 2006 "Transportation Logistics in Science and Practice", Portorož, Slovenia, 5–7 December 2006; Zanne, M., Fabjan, D., Janček, P., Eds.; Fakultet za Pomorstvo in promet: Portorož, Slovenia, 2006; pp. 1–5.
29. Ivic, M.; Markovic, M.; Belosevic, I.; Kosijer, M. Multi Attribute decision-making in railway route planning and design. *Građevinar* **2012**, *64*, 195–205. [CrossRef]
30. Garmabaki, A.H.S.; Ahmadi, A.; Ahmadi, M. *Maintenance Optimisation Using Multi-Attribute Utility Theory. Current Trends in Reliability, Availability, Maintainability and Safety: An Industry Perspective*; Kumar, U., Ahmadi, A., Verma, A.K., Varde, P., Eds.; Springer International Publishing: Cham, Switzerland, 2016; pp. 13–25. [CrossRef]
31. Jajac, N.; Knezić, S.; Marovic, I. Decision support system to urban infrastructure maintenance management. *Organ. Technol. Manag. Constr.* **2009**, *1*, 72–79.
32. Tennakoon, N.; Indraratna, B.; Rujikiatkamjorn, C.; Neville, T. The Role of Ballast Fouling Characteristics on the Drainage Capacity of Rail Substructure. *Geotech. Test. J.* **2012**, *35*, 1–12. [CrossRef]
33. Al-Qadi, I.L.; Xie, W.; Roberts, R. Scattering analysis of ground-penetrating radar data to quantify railroad ballast contamination. *NDT E Int.* **2008**, *41*, 441–447. [CrossRef]
34. Selig, E.T.; Waters, J.M. *Track Geotechnology and Substructure Management*; Thomas Telford: London, UK, 1994.
35. Long, Z.J.; Xing, B.; Liu, H.; Liu, Q. Hyperbola Recognition from Ground Penetrating Radar Using Deep Convolutional Neural Networks. In Proceedings of the 2nd International Conference on Artificial Intelligence: Techniques and Applications (AITA 2017), Shenzen, China, 17–18 September 2017; pp. 11–13. [CrossRef]
36. Yakowitz, D.S.; Lane, L.J.; Szidarovszky, F. Multi-attribute decision making: Dominance with respect to an importance order of the attributes. *Appl. Math. Comput.* **1993**, *54*, 167–181. [CrossRef]
37. Keeney, R.; Raiffa, H. *Decisions with Multiple Objectives: Preferences and Value Tradeoffs*; Cambridge University Press: Cambridge, UK, 1993.
38. Dyer, J.S. Maut—Multiattribute Utility Theory. In *Multiple Criteria Decision Analysis: State of the Art Surveys*; International Series in Operations Research & Management Science; Springer: New York, NY, USA, 2005; Volume 78.
39. Lajnert, S. An overview of the organization and operation of Croatian Railways in the Homeland War. *Arh. Vjesn.* **2007**, *50*, 131–174. (In Croatian)
40. HZ Infrastruktura. Available online: www.hzinfra.hr (accessed on 22 September 2019).

© 2019 by the authors. Licensee MDPI, Basel, Switzerland. This article is an open access article distributed under the terms and conditions of the Creative Commons Attribution (CC BY) license (http://creativecommons.org/licenses/by/4.0/).

Article

Evaluation of Long-Range Mobile Mapping System (MMS) and Close-Range Photogrammetry for Deformation Monitoring. A Case Study of Cortes de Pallás in Valencia (Spain)

Francesco Di Stefano [1], Miriam Cabrelles [2], Luis García-Asenjo [2], José Luis Lerma [2], Eva Savina Malinverni [1], Sergio Baselga [2], Pascual Garrigues [2] and Roberto Pierdicca [1],*

[1] Dipartimento di Ingegneria Civile, Edile, Architettura (DICEA), Università Politecnica delle Marche, Via Brecce Bianche, 12, 60131 Ancona, Italy; f.distefano@pm.univpm.it (F.D.S.); e.s.malinverni@staff.univpm.it (E.S.M.)
[2] Department of Cartographic Engineering, Geodesy and Photogrammetry, Universitat Politècnica de València, Building 7i. Camino de Vera s/n, 46022 Valencia, Spain; micablo@upvnet.upv.es (M.C.); lugarcia@cgf.upv.es (L.G.-A.); jllerma@cgf.upv.es (J.L.L.); serbamo@cgf.upv.es (S.B.); pasgarta@cgf.upv.es (P.G.)
* Correspondence: r.pierdicca@staff.univpm.it

Received: 27 August 2020; Accepted: 24 September 2020; Published: 29 September 2020

Abstract: This contribution describes the methodology applied to evaluate the suitability of a Long-Range Mobile Mapping System to be integrated with other techniques that are currently used in a large and complex landslide deformation monitoring project carried out in Cortes de Pallás, in Valencia (Spain). Periodical geodetic surveys provide a reference frame realized by 10 pillars and 15 additional check points placed in specific points of interest, all with millimetric accuracy. The combined use of Close-Range Photogrammetry provides a well-controlled 3D model with 1–3 cm accuracy, making the area ideal for testing new technologies. Since some zones of interest are usually obstructed by construction, trees, or lamp posts, a possible solution might be the supplementary use of dynamic scanning instruments with the mobile mapping solution Kaarta Stencil 2 to collect the missing data. However, the reliability of this technology has to be assessed and validated before being integrated into the existing 3D models in the well-controlled area of Cortes de Pallás. The results of the experiment show that the accuracy achieved are compatible with those obtained from Close-Range Photogrammetry and can also be safely used to supplement image-based information for monitoring with 3–8 cm overall accuracy.

Keywords: deformation; environmental monitoring; long-range mapping; MMS; close-range photogrammetry; sub-millimetric EDM geodetic techniques

1. Introduction

The evaluation of deformations in civil infrastructures or natural environments is normally assessed by the realization of very accurate three-dimensional (3D) models by means of a range of complementary geomatics techniques such as total stations, global navigation satellite systems (GNSS), photogrammetry, or laser scanning [1]. Nonetheless, each geomatic technique is affected by its own instrumental and physical limitations, especially when the monitored area is large and has a complex topography [2]. For instance, atmospheric refraction can severely limit the achievable accuracy in the distance and angular measurements, also in laser and image scanning [3]. Therefore, the integration of data, collected by different techniques, in a unique coordinate reference frame becomes crucial to producing consistent 3D models able to be used for overtime deformation monitoring purposes [4].

In recent decades, advanced monitoring technologies have spread and increasingly been used for the study and management of geological hazard and risk, which may compromise the state of preservation of civil infrastructure [5]. Monitoring actions are necessary to guarantee health and safety conditions by controlling the evolution of deformation patterns or detecting significant instabilities. In terms of spatial and temporal resolution, the improvement of geomatics techniques represents a significant achievement. These methods provide innovative tools in supporting mapping products and geological analysis required for assessment and evaluation. Accurate and fully geo-referenced 3D datasets can be used to characterize in detail structural and geological settings, as well as the geomorphology of a studied area. Geological applications, for example, geo-hydrological risk assessment, rockfall runout modellings, or slope stability analysis, can have a great benefit through non-destructive investigation.

Where a topographic survey is based on limited distances (e.g., tens to hundreds of meters), it has historically been carried out with total stations. Although such method provides high accuracy and precision for the measurement of individual points, significant time is required to collect a sufficient density of data to produce rough landscape Digital Elevation Models (DEMs) [6]. Laser Scanning (LS) and Close-Range Photogrammetry (CRP) are state of the art techniques for acquiring dense and precise topographic data at the output detail, for accurate volume measurements or modeling.

Accurate mapping and monitoring of lakeside reservoirs, as well as coastal areas and fluvial processes, are critical tasks to which several techniques have been used, from aerial photographs, remote sensing, land surveying, CRP and, more recently Terrestrial Laser Scanning (TLS) and Mobile Laser Scanning (MLS). These latter technologies are in principle advantageous because of their good accuracy, easiness of use and lower time of response [7].

Most of the existing geomatics techniques are sometimes unaffordable, and there is not an all-in-one solution able to provide spatial information with suitable accuracy and temporal frequency. For the completion of existing surveys in particular, MLS was considered the most suitable alternative for the challenges established in the project requirements, concerning productivity, sample density and final costs. In particular, Mobile Mapping System (MMS) technology enables users to reach complex and enclosed spaces, either scanning by hand or by attaching a scanner to a trolley, drone, or mounting on a pole. As a result, the variety of difficult-to-survey environments becomes wider. This solves the problem linked to GNNS-based systems where it does not work well in complex contexts, for example, woods, where tree canopies block signals, as occurred in this case study. With no reliance on remote data, MMS are a priori a truly go-anywhere technology.

In this paper, a comparison of a long-range handheld MMS with the CRP has been evaluated that offers particular promise for site-scale topographic surveys. Experiments were conducted exploiting the case study of Cortes de Pallas [2], where three-year period monitoring surveys have been undertaken with sub-millimetric electronic distance meter (EDM) techniques and CRP. The result of the survey consists of a 3D model where the combination of these techniques ensures a complete mapping of the site, avoiding the creation of gaps in the point clouds thanks to the compensation of one technique on the other and vice versa. This 3D model provides a basis for the analysis of the rocky landslide deformation monitoring. The main contribution of this manuscript is to demonstrate how MMS can complement the 3D mapping of a challenging environment, by integrating multi-source data in a unique reference frame and with an accuracy comparable with other state of art methods. Several tests are presented, providing the research community with guidelines that will be useful for other similar settings, presenting the MMS as an alternative approach when other geomatic methods fail.

This article is structured as follows: after a first state-of-the-art review in using MLS for mapping and monitoring purposes, Section 3 describes the geomatics techniques used for the data acquisition in the selected case study. Section 4 focuses on the data processing and the integration of georeferenced point clouds from MMS with the CRP. In Section 5 the results of this comparison are evaluated to assess the accuracy achieved in the combination of these data. The discussion and conclusion are finally presented in Sections 6 and 7.

2. Related Work

Mobile mapping systems are becoming popular as they can build 3D point clouds of any type of environment rapidly by using a laser scanner that is integrated with a navigation system. The laser scanner is able to record millions of points. The files with the point clouds can be viewed, navigated, measured and analyzed as discrete 3D models. The evolution of the laser scanner as a surveying technique has resulted in this tool being used not only to obtain an optimal geometric reconstruction of the scene, but also to assess changes in a particular state.

Mobile LiDAR (Light Detection And Ranging) technology presents some advantages: high-speed data capture through reduction of time and cost, remote acquisition and measurement increasing survey in efficiency and safety, high point density data ensuring a comprehensive representation of the detected scene, an abundance of data acquired in movement. The light weight of these MLSs makes them flexible and versatile instruments, so they can be mounted on any mobile platform. In places with complex topography, the use of MMS, following a continuous path, is more advantageous than a tripod-based laser scanner that requires multiple scan positions to cover all the areas of the survey [6]. Moreover, integrating MMS with other geomatics techniques, such as Digital Photogrammetry (DP), allows giving an added value and greater richness of the acquired data providing a high detailed DEM or DTM (Digital Terrain Model) of the selected area [8].

Alternative purposes to use mobile LiDAR technologies, in addition to the assessment of the geometrical state, concern change detection, deformation analysis [9], hazard assessment and structural and infrastructural health monitoring [4] in different types of natural environment. In technical terms, mobile mapping solutions contemporaneously allow users to acquire geometrical aspects for geological studies and geomorphological analysis, to operate mapping of all the elements present in the detected area (e.g., vegetation, road, etc.), and to define basic modeling for monitoring operations (e.g., rockfalls, coastal erosion, river dynamics, etc.). Examples described by literature are various, depending on natural effects as geological and atmospheric actions or anthropogenic consequences of the built environment which compromise the stability of the natural landscape.

The major case studies that include the use of MLS for mapping and monitoring purposes are those related to geomorphology detection and landscape dynamics such as landslides and rockfall displacements. The combination of MLS and DP can greatly improve the realization of detailed modeling of the geometrical discontinuity of a slope to analyze a landslide susceptibility and potential rockfall mechanisms [9]. Examples of slope stability assessment through the kinematic method are applied in rock outcrops in British Columbia, Canada, and Carrara marble quarry, Italy [10]. The use of MLS was presented by Francioni et al. [11] in a geological study of landslides applied to an interesting case study in Normandy. They used a boat-based MMS to scan 3D point clouds of unstable coastal cliffs to detect rockfalls and erosional deposits. A similar case is proposed by [12] who described a new approach of coastal cliff monitoring, in Poland, which is based on a combination of MLS from the sea with the geotechnical stability calculations. A 3D mapping was carried out by an airborne MMS, installed on a helicopter, to monitor a small landslide in the North Yorkshire coast in the UK. As a result, more accurate modeling of the terrain was obtained, especially areas covered by vegetation [13]. A ground-based approach was evaluated by James et al. [5] using a handheld MLS to collect topographic data in complex terrain as a gully site at the coastal cliff in Sunderland, UK. To carry out the survey, the hand-held mobile device is walked around the site following a close loop path to facilitate accurate 3D reconstruction and avoiding problems associated with drift.

Landslides happen also inside residential areas causing cracks to buildings and drainage, for example, the study area located in a Malaysian city. Here, to monitor an active landslide, researchers used MLS to assess the movements of the land, in both vehicle-based and human-based mode in a complementary way to acquire completely the surface of the study area [14].

Mobile mapping solutions have been also applied in change detection and deformation analysis for sandy dunes and river courses. The adoption of MLS was illustrated by Nahon et al. [15] that provides an accurate and robust method to obtain high resolution space-time datasets along a Dutch beach useful to understand the changes of dunes volume, under the influence of both marine and aeolian processes. The MLS was carried out mounting the device on a car and using airborne LiDAR to process several DEMs of the sandy dunes. The monitoring of beach dunes is needed to improve the scientific observation of their dynamics. A test was realized in Cap Ferret, France, where researchers combined MLS with aerial photogrammetry to deliver accurate 3D reconstruction of the dunes. They confirmed the good results of using mobile mapping devices for their high level of detail and greater spatial coverage [16]. The best results were also achieved in monitoring the lagoon area of Padre Island National Seashore, located in Texas, where mobile scanning devices may be preferred for the detailed and comprehensive final DEMs useful to detect 3D terrain features and so to monitor geomorphic changes [17].

These challenging trials prove that a mobile LiDAR system, in recent years, has emerged as a viable alternative for surveying coastal beaches and foredunes, but also for riverine topography. Measurements were conducted by Vaaja et al. [18] using an MMS mounted on a boat and mounted on a manually-operated cart to define the riverine topography and modeling in 3D of the coarse fluvial sediment along the river Tenojoki, in Finland. The application of a MMS was tested by Williams et al. [19] for a complex relief and the following reconstruction of fluvial surface sedimentology and topography of river Feshie, in Scotland.

Monitoring actions also occur to control the actual state of civil infrastructures, for example, the maintenance of pavement condition of roads to detect instabilities and to assure safe conditions. It is possible to detect the road surface on 3D models derived from a dense point cloud acquired by MLS. LiDAR instruments are suitable to collect data with adequate accuracy and high resolution for mapping and inventory purposes, and, also, the surveyors can make surveys safely with minimal interruptions of the traffic flow [9]. The use of accurate and dense point cloud data along a route corridor enables the detection of surface distortion, joints, cracks and other roughness conditions [20]. Also retaining walls along roads needs to be monitored to assess their stability, especially in mountainous regions to support either roads or slopes adjacent to roads. An efficient method was proposed by Lienhart et al. [21] based on a mobile mapping system to detect a retaining wall used to construct a highway, in Austria. The MMS, mounted on a car, generated a high-density point cloud where tilt changes of the structure can be calculated to define the current state of the structure itself. The 3D model helped to recognize forms of damage and their distribution on the surface of the wall, thus compensating for the typical punctual analysis carried out with the total station. Moreover, referring to the case study presented in that article, among the monitored infrastructures they also include water reservoirs and hydroelectric power plants. A boat-based MMS was selected by Brazilian researchers to scan the progression of marginal erosion in different reservoirs of hydroelectric plants. The processed point clouds and rendered meshes and the creation of cross-sections were used to compute the rate and the dynamics of the erosion phenomenon [6,22].

3. Data Acquisition

Since 2017, a monitoring plan to the rock-wall of the lakeside reservoir has been commissioned by the Infrastructures Area of the Diputació de València. This change detection operation is performed yearly. After a first geodetic survey to detect fixed targets and select some check points, the data acquisition was carried out with a reflex camera for a CRP from different points of view to cover the whole area of interest. In 2019, in addition to the two techniques mentioned, a third survey was added: the MMS was tested to make this type of survey more complete, and verified through a comparison operation between point clouds. The combination of the two surveys had the objective of completing the 3D mapping of the site, needed to support the monitoring plan of the rock-wall.

3.1. Cortes de Pallás Test Site

Cortes de Pallás is a Cretaceous limestone area with historical geotechnical issues. On 6 April 2015, a cliff called La Muela partially collapsed, and some facilities of the electricity power plant and the main access road to the village were seriously damaged. At the end of 2017, once the consolidation works were finished, the Infrastructures Area of the Diputació de València commissioned a deformation monitoring plan to detect possible displacements of huge boulders or potential malfunction of the installed anchoring systems [2]. However, the detection of possible displacements of some centimeters with the required level of significance in a short period, e.g., two or three years, is a quite challenging task that can only be approached by means of high-precision geodetic techniques due to the peculiar topography of the zone. The whole area involves distances from 500 to 2000 m with height differences reaching 500 m. Moreover, 10 geodetic pillars mounted on presumably stable locations at 15 target points (referred to as check points) were installed in the rock wall by professional climbers using abseiling techniques, because they are not directly accessible. Furthermore, the measurements have to be undertaken by necessity from the opposite shoreline which is about 600 m away because the cliff of interest is facing a water reservoir (Figure 1).

Figure 1. Perspective view of Cortes de Pallás site area.

Therefore, the Diputació de València opted for periodical geodetic surveys along with image-based techniques like CRP and long-range TLS (since 3D models derived from CRP and TLS techniques proved compatible, only the former will be use for the experiment) as the most feasible join solution for the deformation monitoring plan. The periodical geodetic surveys, which are based on sub-millimetric EDM techniques and performed annually, have a triple objective: (i) to establish and monitor a high-precision 3D reference frame realized by 10 pillars, coded 8000+ (Figure 2a), (ii) to determine the 3D coordinates of 15 additional check points, represented by 360° prisms coded 1000+ (Figures 2b and 3) placed in specific points of interest on the monitored rock wall, and (iii) to provide a ground control network for image-based techniques.

As a demonstration of the potentiality of the high-precision geodetic techniques used [23,24], the displacements found for the reference frame pillars between the years 2018 and 2019 are shown in Table 1. Except for pillar number 8005 (Figure 2a), where a significative vertical displacement of −6.12 mm (see Table 1) was detected, the reference frame can be considered stable and well-controlled. The resulting reference frame has an overall accuracy of 1 mm and its scale is metrologically consistent with the unit of length of the International System (SI). Further information about the geodetic surveys is given in Section 3.2.

This high-precision geodetic method, which is very demanding and time-consuming, can be only applied to a limited number of relevant points of interest, which are 15 targets in the case at hand.

Complementary geomatic techniques like long-range TLS or CRP are needed to massively collect information with a density of around 300 points/m^2. However, the accuracy of this type of points is reduced up to 1 to 3 cm. The main problem with both geomatic techniques is that photographs and cloud points provided by respectively CRP and TLS need to be registered in the same reference frame. In the case at hand, the reference frame as well as the coordinates of the 15 check points of interest, which are periodically provided by the sub-millimetric geodetic techniques, are considered the ground truth for the integration of the three techniques so that the combined numeric models are fully consistent.

Figure 2. Example of target devices used on pillars of the reference frame and check points used for monitoring the site: (**a**) 50 cm white spherical point on the pillar (number 8005); (**b**) check point (in the center of the 360° prism) with a 15 cm white sphere on top. Please note both images are not to the same scale.

Figure 3. Example of check point (center of the 360° prism) on site: (**a**) number 1009; (**b**) number 1010.

Table 1. Displacement of the reference frame pillars obtained by using sub-millimetric electronic distance meter (EDM) techniques between years 2018 and 2019. Points with statistical T above the cut-off value 5.739 are considered to have a significative displacement with a 99% probability. Pillar 8006 was not measured in the 2018 campaign.

Pillar	Displacements and Their Corresponding Errors			Deformation	
No.	Dx (mm)	Dy (mm)	Dz (mm)	T	Conclusion
8001	1.23 ± 0.70	0.84 ± 0.48	−2.45 ± 1.74	1.253	No
8002	0.42 ± 0.37	0.32 ± 0.34	3.31 ± 1.80	3.340	No
8003	0.87 ± 0.27	0.72 ± 0.37	0.48 ± 1.23	3.457	No
8004	0.28 ± 0.70	−1.11 ± 0.52	−3.51 ± 3.37	1.952	No
8005	0.58 ± 0.61	−1.48 ± 0.67	−6.12 ± 2.29	11.609	Yes
8007	−0.43 ± 0.65	−1.35 ± 0.97	−0.80 ± 1.95	1.272	No
8008	0.15 ± 0.59	0.57 ± 0.50	1.87 ± 1.75	1.024	No
8009	0.21 ± 0.47	2.11 ± 0.66	−5.55 ± 2.59	4.858	No
8010	−0.00 ± 0.99	−1.51 ± 0.58	1.10 ± 1.49	3.477	No

An additional problem with both CRP and TLS techniques, especially when they are performed statically, like in the case at Cortes de Pallás, is the presence of occlusions. Since the CRP images can easily miss some shadowed areas, the 3D model obtained in each campaign usually has small patches with no information. A possible and efficient solution could be the use of mobile mapping solutions like the Kaarta Stencil 2 (https://www.kaarta.com/products/stencil-2-for-rapid-long-range-mobile-mapping/). Being dynamic, this system can provide a continuous 3D survey of the problematic area in Cortes de Pallás in less than one hour.

However, prior to being accepted as part of the combined 3D numeric models, the consistency of the points cloud obtained with the Kaarta system has to prove that it is consistent with the data provided by the EDM and CRP techniques. Since EDM and CRP are compatible at the 1–3 cm level (see Table 2), the point cloud obtained with the Kaarta system can be compared with the CRP solution in two ways to facilitate the analysis: first, using the dense CRP point cloud; second, employing manually measured natural photogrammetric check points.

Table 2. Differences between the close-range photogrammetry (CRP) and the EDM coordinates obtained for the check points.

Target	Differences CRP—EDM (mm)			
	Δx	Δy	Δz	Total
1001	13.1	23.9	21.9	35.0
1002	9.7	21.2	8.3	24.7
1003	4.9	18.7	16.2	25.2
1004	−0.6	6.5	12.3	13.9
1005	5.0	11.3	1.5	12.5
1006	3.2	−9.4	7.9	12.7
1007	−11.8	−11.8	2.1	16.8
1008	11.3	23.0	−3.9	25.9
1009	12.9	13.7	3.8	19.2
1010	9.8	−7.7	10.5	16.3
1011	16.8	−13.2	12.5	24.7
1012	0.5	−19.0	19.9	27.5
1013	3.5	−17.8	−0.1	18.1
1014	12.4	−10.5	3.3	16.6
1015	13.7	14.9	9.4	22.3

3.2. Geodetic Survey

The reference frame is a geodetic network with 10 concrete pillars that surround the 15 check points. Pillars and target points (Figure 2) were set up in 2017. Each target point consists of one Leica 360 reflector (RFL) and one standard target sphere (Ø145 mm) which are rigidly mounted and firmly attached to the rock.

Since their installation had to be undertaken by abseiling, their verticality cannot be taken for granted. The determination of their attitude, which is crucial to transfer coordinates between the center for distances and the center for images, was obtained by photogrammetric methods. Figure 3 shows the photogrammetric data acquisition followed to obtain the vector and thus their attitude between centers.

So far, two geodetic campaigns have been carried out by using a sub-millimetric EDM Kern ME5000 Mekometer (ME5000) [25,26]. For each distance measurement, the meteorological parameters were measured and double-checked by using a traditional Thies Clima Assmann-Type psychrometer (±0.2 K) and a Thommen 3B4.01.1 aneroid barometer (±0.3 hPa) along with a network of 10 Testo 176P1 data-loggers. All the meteorological sensors were previously calibrated at the Universitat Politècnica de Valencia (UPV) calibration laboratory.

Prior to the EDM adjustment, the following corrections were applied: refraction correction [27–29], EDM frequency drift correction and geometric correction [25]. Once these corrections were applied and their corresponding errors computed in order to contribute to the stochastic model, the resulting slope distances were 3D adjusted in the local coordinate system CP2017 (x,y,z) in two steps. In the first step, only distances between pillars were adjusted to provide a solution for the frame. In the second step, distances to check points are included in order to obtain their 3D coordinates. The applied method not only provides coordinates with an overall accuracy of 0.5–1 mm for pillars and 1–3 mm for check points, respectively, but also allows us to monitor the possible displacement of pillars (see Table 1). Finally, the EDM coordinates and their precisions were also converted into geodetic coordinates with ellipsoidal height (φ,λ,h) and TM30 with orthometric height (E,N,H) [2].

As it can be seen in Table 2, the differences between the CRP and the EDM coordinates that were obtained in November 2019 are in the range of 1 to 3 cm, which are the expected values and demonstrates that the sub-millimetric EDM-based coordinates are consistent with the CRP-based 3D numeric model that is used as ground truth to validate the accuracy of the CRP survey.

3.3. Close-Range Photogrammetry

The geometry of the site is rather peculiar, with a huge reservoir in front of the hillside (Figure 1), with many crossing hydroelectric power lines from one side to the other, which hamper the exploitation of close-range imagery with RPAS–UAVs (remotely piloted aircraft system–unmanned aerial vehicles). The approximate dimension of the interest hillside displayed in Figure 1 (yellow box) is 700 m wide and 200 m high. In addition, pictures were taken not from ideally planned positions but realistically ideal positions on-site, after defining some minimum constraints such as ideal ground sampling distance (GSD) e.g., following roads parallel to the hillside, and from surrounding hills.

For the CRP campaign, two 21 Mpix full-frame cameras were used mounted on stable tripods: Canon EOS D5 Mark II with a normal Canon EF 50 mm f/1.4 USM lens, and a Canon 1Ds Mark III with a telephoto Canon EF 200 mm f/2,8 L II USMlens (Figure 4). The first camera was used to get the whole area, summing 280 normal and convergent imagery; the second camera was used to measure only three areas with the fixed on-the-rock 15 check points, summing 404 imagery. The data acquisition was planned to achieve a GSD of 3.5 cm. The mean camera-object distance was 512 m. A total of 6 h were devoted to taking images on site.

Figure 4. Images acquired from the same spatial position with two fixed lenses: (**a**) normal 50 mm; (**b**) telephoto 200 mm.

The three-step workflow followed to achieve accuracy better than 3 cm is presented next. First step, each camera was calibrated on the job selecting a subsampled ideal network with convergent and rotated imagery, summing up to 57 images for the 50 mm lens and 47 images for the 200 mm lens. For the normal lens, 8 interior orientation parameters (f, cx, cy, K1, K2, K3, P1, P2) were computed to achieve 0.34 pixel image reprojection error; for the telephoto lens, 4 additional parameters (f, cx, cy, K1) were computed to achieve a 0.50 pixel image reprojection error. Second step, a free bundle block adjustment with 230 imagery (142 normal + 88 tele) was undertaken. This step included the manual measurement of the white spherical targets (Figures 2b and 3) on the imagery, including 4 (visible) control points on pillars of the reference frame and 15 check points (Figure 5). A mean reprojection error of 0.50 pixels was achieved in the photogrammetric adjustment. Third step, absolute orientation by means of a 3D similarity transformation yielding an RMSE (root mean square error) of 1.55 cm.

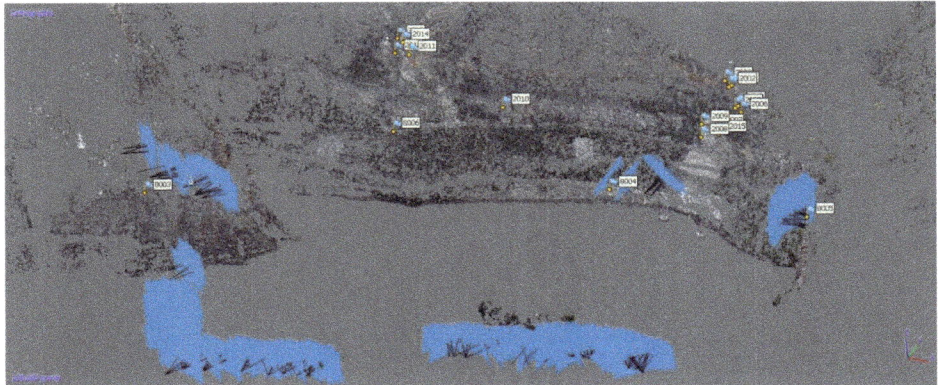

Figure 5. Set of imagery used for the bundle adjustment of the normal and tele cameras and visualization of the measured points on geodetic pillars and check points.

3.4. Mobile Mapping System

After the acquisition phases performed by EDM and CRP campaigns, a survey with the MMS was conducted. As previously stated, despite the high accuracy reached with the aforementioned methods, visual occlusions and vegetation prevented the whole mapping; mainly the streets and the lower parts of the rock walls could not be surveyed. Kaarta Stencil 2 is a stand-alone, lightweight instrument, with an integrated system of mapping and real-time position estimation. An MMS consists mainly of three components: mapping sensors for data acquisition, a positioning and navigation unit for data localization and a time referencing unit for data recording [30]. Kaarta Stencil 2 depends on LiDAR consisting of a Velodyne VLP-16, connected to a low-cost MEMS (micro electro-mechanical system) IMU (inertial measurement unit) and a processing computer for localization and real-time mapping. VLP-16 has a maximum laser range of 100 m and 360° horizontal FOV (field of view) with a 30° azimuthal opening with a band of 16 laser beams. The LiDAR has a centimetric accuracy with a value variation of ±30 mm and a speed of 300,000 points per second. The acquisition phase was carried out using the appropriate configuration parameters, set for outdoor environments, that include values for the voxelSize, namely the resolution of the point cloud in the map file, for cornerVoxelSize, surfVoxelSize, sorroundVoxelSize, those indicate the resolution of the point cloud for scan matching and display, and for blindRadius, that is the minimum distance of the points to be used for the mapping (Table 3).

Table 3. Parameters setting of Kaarta Stencil 2, optimized for outdoor scenario.

Parameters	Default Value [m]
voxelSize	0.4
cornerVoxelSize	0.2
surfVoxelSize	0.4
surroundVoxelSize	0.6
blindRadius	2.0

In realizing this ground-based survey characterized by a long path on foot, walking along the edge of the road, the laser scanner was mounted on a small pole held by hand (Figure 6). Human-based LS has not been widely used in the monitoring of rock slopes, but in this case study was useful to detect areas of the surveyed region not clearly visible through the CRP. The MMS needed to fill the gaps of the photogrammetric 3D model, such as roads or parts obstructed by the crown of trees, that the photogrammetric camera cannot reach from the different points of view to detect all the features.

(a) (b)

Figure 6. Two sequences of the MMS (Mobile Mapping System) survey with Kaarta Stencil 2. (**a**) Lower part and (**b**) upper part of the rockwall.

The data acquisition operation was performed following a closed path (close-loop) [31,32], to facilitate accurate reconstruction of the surveyed region and avoid problems associated with drift, where the beginning and end of the route coincide [3]. In Cortes de Pallás test site the surveyor completed a trajectory (backwards and forwards) of 2.9 km long in total, at an average distance of 3–5 m from the rock wall, acquired in 1 h and 43 min, collecting over 480 billion points (Table 4).

Table 4. Scan info of the estimated trajectory executed with mobile mapping system (MMS) Kaarta Stencil 2.

Scan Information	
Filename	/2019-11-27-10-58-50/scan_info_2019-11-27-10-58.yaml
Launch mode	Mapping with Camera
Time Stats	
Start Time (y-m-d; h:m:s)	2019-11-27; 10:59:02
Stop Time (y-m-d; h:m:s)	2019-11-27; 12:42:27
Scan Time (h:m:s)	01:43:25
Point Cloud Stats	
Average Confidence	218,943.715342
Number of Points	482,799,466
Trajectory Length	2905.259698 m
Trajectory Points	25,414
File Size	11,587,187,366 Byte

A tracker camera, integrated to the mobile mapping device, needs to show and save the trajectory made during the acquisition operations. The progress of the scanning can be monitored in real-time via an external monitor attached with a USB cable. At the end of the acquisition phase with Kaarta Stencil 2, information about the configuration setting, the 3D point cloud characteristics, the estimated trajectory is stored in a folder automatically created by the MMS processer at every operation of the survey.

4. Data Processing

4.1. Close-Range Photogrammetry (CRP) Point Cloud Processing

The next steps in the photogrammetric processing were devoted to yield as accurately as possible the 3D model of the hillside. 43.58 million points were obtained for the site area, reaching a mean density of 413 points/m^2. The mean distance among points in the 3D point cloud is 7.6 cm. All the steps presented up to this point were processed with Agisoft PhotoScan Professional v. 1.4.4 build 6848. However, for filtering the point cloud, 3DReshaper was used. Noisy points were removed with the option delete points (over 20 cm). Afterward, from the segmented point cloud, tiny sets of points (below 11) were automatically deleted. Next, extensive manual filtering was undertaken to remove points identified as vegetation (trees and shrubs) or isolated points outside of the ground such as power poles or road signs. The result of this phase is visible in Figure 7, which presents the difference in a sector without and after filtering. The final number of points was 42.75 million points.

Figure 7. Cortes de Pallás sector: (**a**) point cloud without filtering; (**b**) cleaned and filtered point cloud.

4.2. Mobile Mapping System (MMS) Point Cloud Processing

The open-source software CloudCompare has been used to process the 3D point cloud by MMS, with the purpose of analysing the raw data acquired (Figure 8). As a default parameter, the number of points composing the sharpened MMS point cloud is 94.30 million.

In order to obtain ground and non-ground points from MLS point cloud, the data filtering algorithm Cloth Simulation Filter (CSF) was used [33]. This method allows users to obtain the "steep slope" model setting some advanced parameters such as cloth resolution, maximum iterations and classification threshold (Table 5). The cloth resolution refers to the grid size of cloth composed by particles interconnected through virtual springs, which is used to cover the terrain. The positions of the particles in three-dimensional space determine the position and shape of the cloth. The grid size has the same unit of the point cloud. The lower the value of cloth resolution, the softer the resulting mesh from the filtered point cloud is. The number of iterations is linked to the maximum iteration times of terrain simulation. Classification threshold is used to classify ground and non-ground points, based on the distances between points and the simulated terrain.

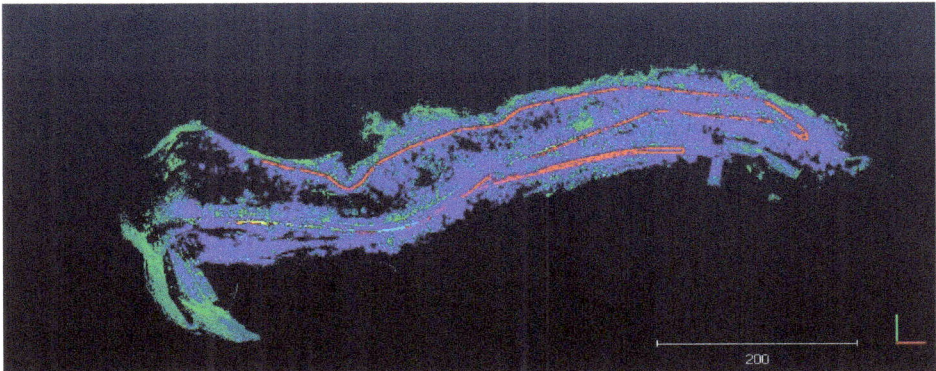

Figure 8. Cortes de Pallás MMS (Mobile Mapping System) point cloud: 3D point cloud (intensity scale) and the closed-loop trajectory (red line) carried out with the MMS.

Table 5. Advanced parameters applied to cloth simulation filter (CSF) operation to perform the segmentation between ground and non-ground points.

Cloth Resolution	Maximum Iterations	Classification Threshold
0.5	n. 150	0.5

The final output is a segmentation of the MMS cloud, where all the vertical features, like trees, pylons and road signs, were automatically removed (Figure 9), reducing the number of points to 47.86 million.

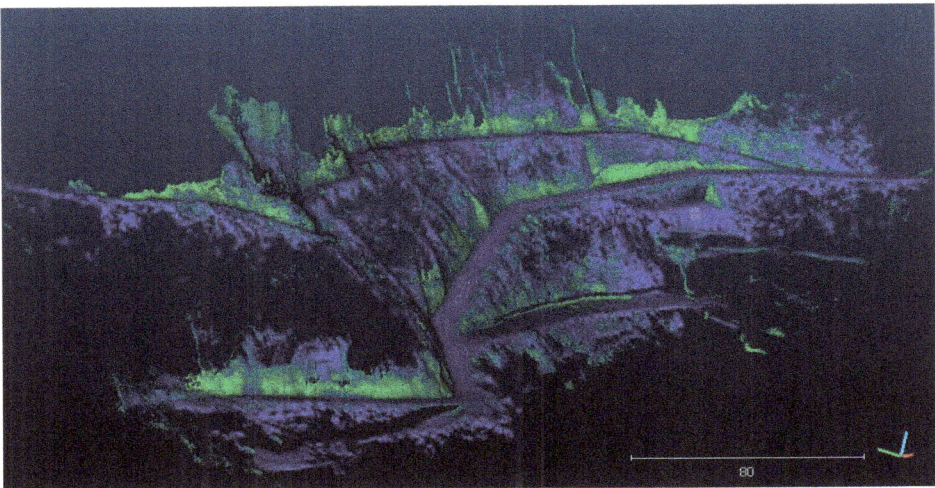

Figure 9. Cortes de Pallás MMS point cloud: the result of filtering algorithm CSF (Cloth Simulation Filter), applying the shader filter EDL (Eye Dome Lighting) which help to show better the topography of the study area.

4.3. MMS Point Cloud Georeferencing

Since a GNSS system was not integrated with Kaarta Stencil 2, the georeferencing of the point cloud could be realized in two ways: using the control points captured by the EDM or aligning the point cloud to the CRP model, already georeferenced. As only EDM points fell within the path trajectory, the second option has been chosen and represents a completely new experiment tested by the Kaarta manufacturing experts. The filtered CRP point cloud presented in Figure 10 was thus used next as the control system for the MMS point cloud.

After the filtering operation of the MMS point cloud, the next step consisted of roughly aligning the trajectory of Kaarta Stencil 2 with the CRP point cloud. First, the yaw angle of the starting point of the MMS close-loop trajectory in the CRP point cloud was identified and calculated. Then, the parameters of the MMS point cloud could be set up iteratively to obtain a better result in the replay operation of the trajectory line, evaluating every time the correspondence with the photogrammetry point cloud. Defined as the right alignment of the trajectory, the MMS point cloud was run following this adjusted close-loop line and automatically adapted with the same roto-translation transformation. At the end of this procedure, the two point clouds were overlapped and so they could be analyzed by executing a comparison operation between the point clouds themselves (Figure 11).

Figure 10. Top view of the CRP (Close-Range Photogrammetry) point cloud. In this image are visible the holes due to occlusions and vegetation.

Figure 11. MMS point cloud (false color scale) aligned to the CRP point cloud (red, green and blue (RGB) scale).

5. Result Comparison

The resulting combination (Figure 12) of the point clouds from MMS and CRP surveys was assessed using the cloud-to-cloud (C2C) distance tool. The C2C tool exploits the nearest neighbour algorithm to compute the Euclidean distance between each point of the compared cloud and the nearest point of the reference cloud [32].

Figure 12. "Closest points set" tool applied to CRP point cloud, taking the MMS point cloud as a reference.

First, we computed the "closest point set" [34] applied to the CRP point cloud. This tool is useful for understanding which points of the CRP point cloud are closest to each point of the MMS cloud.

Taking the MMS point cloud as a reference, a new CRP point cloud was generated, preserving only the closest points. This new point cloud represents a new base for the C2C distance calculation.

Then, the C2C distance was calculated by taking the resampled CRP point cloud as a reference and setting the point distances in different ranges, from 1.00 to 0.10 m to evaluate different computations in term of RMSE, standard deviation and number of points (Figure 13).

Analyzing Figure 13, it can be noted that setting the distance computation at 1 m between the two point clouds, we have lost almost 8.8 million points with respect to the original one. The former value represents the points that have any correspondence in the CRP point cloud, excluding then the parts of the roads that cannot be seen from the camera due to the occlusions. The threshold of 0.50 m between the two point clouds was computed to understand better the distribution fitting of C2C absolute distance. In the end, we set the C2C distance computation at 0.10 m. The choice of this threshold value is motivated by the fact that long-range MMS and CRP are techniques characterized by sub-decimeter accuracy. The number of points has become low: 17 million points of MMS have a higher correspondence with the CRP point cloud. From the results of the point clouds assessment, it emerges that, at 0.10 m distance, a mean difference of 5.6 cm between laser scanner and photogrammetry exits, with a standard deviation of 2.3 cm (Figure 13).

The reduction of the number of points during the different distance computations can be explained as follows: first, the MMS point cloud is denser than the CRP "closest point set" point cloud, so only a low number of points between the two point clouds has the nearest correspondence; second, the MMS point cloud is noisier. The decrease of the number of points in MMS point clouds during the phases of the C2C evaluation does not imply an information waste, since the remaining 30 million points of MMS point clouds are partially useful to fill the "holes" of the CRP point cloud.

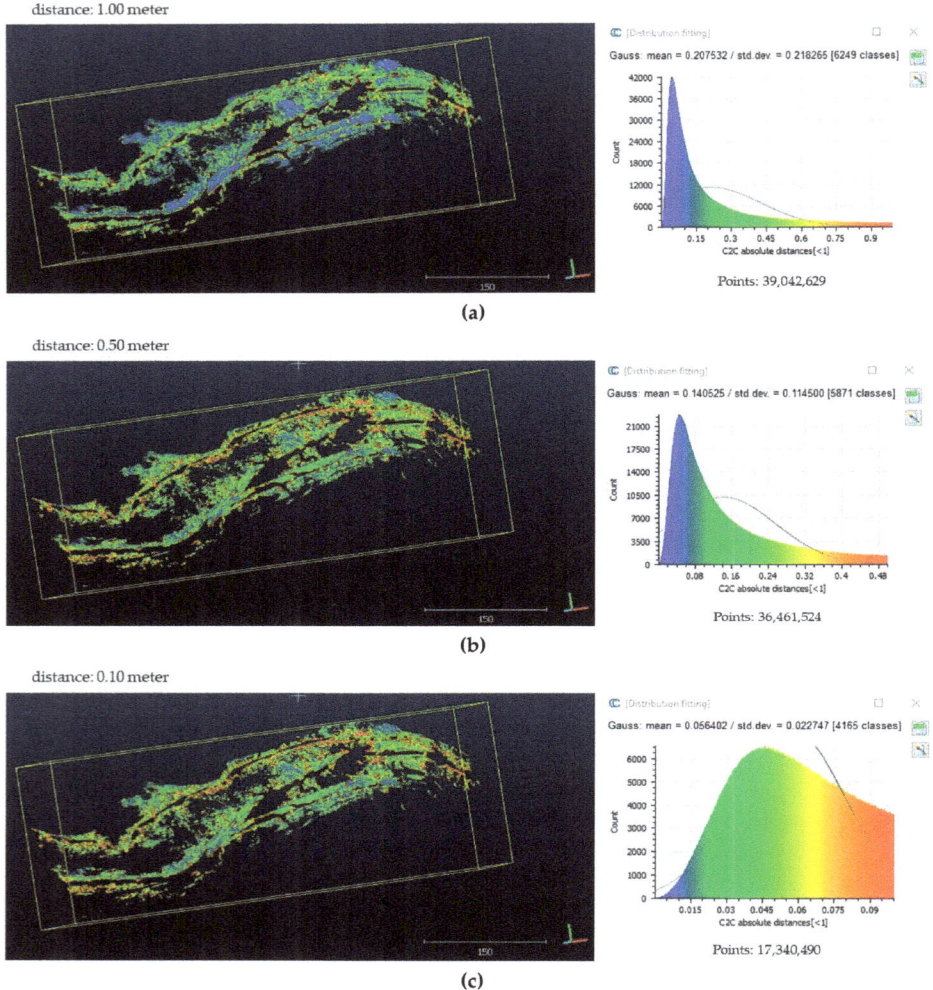

Figure 13. Cloud-to-cloud (C2C) processing between CRP and MMS point clouds at different values of distances: (**a**) 1 m, (**b**) 0.50 m, (**c**) 0.10 m.

A first observation that can be made by seeing the two point clouds, is that the final 3D point cloud (Figure 14) does not present so many holes. The MMS allowed to acquire the elements close to the path followed along the road and to compensate for the gaps that had been created in the CRP point cloud.

This explains that a possible combination of different surveying techniques, their different use and the different data acquired can be complementary and guarantee a complete mapping of the same site of interest. The previous statement is true as far as both the RMSE and the standard deviation are small. Besides the integration between CRP and MMS, we also computed the accuracy evaluation between the aligned MMS point cloud and the photogrammetric control points on natural features. As explained in Section 3.1, the EDM technology system is based on pillars for control and spherical targets for checking the photogrammetric accuracy. Once the quality of the photogrammetric survey has been confirmed, photogrammetry can also be used for checking the accuracy of the MMS based on natural features. Taking the 22 check points of the CRP solution as a reference, we computed the distance on three natural check points (randomly combined) doing a point-pair registration. As shown in Figure 15, the error value is subdecimeter and comparable to the distance evaluation between CRP and MMS point clouds.

Figure 14. Integration of filtered point clouds from MMS (Mobile Mapping System) in intensity scale and CRP (Close-Range Photogrammetry) in RGB scale.

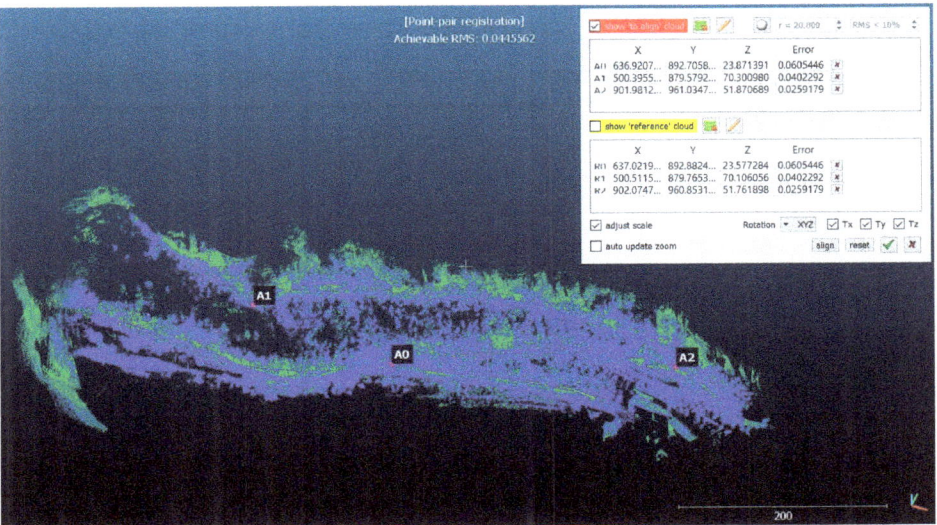

Figure 15. Point-pair registration between MMS (Mobile Mapping System) point cloud and natural photogrammetric check points.

6. Discussion

The experiments presented above involved point clouds produced with CRP and MMS solutions. The experiment was designed to investigate whether MLS can reach a degree of accuracy to complement CRP for mapping monitored areas; it was motivated for understanding the potential of an innovative acquisition method using MMS in challenging environments.

The MMS proved to be very useful due to the rapidity of acquisition and the ease of use. In the presence of an existing reliable photogrammetric survey, the mobile mapping can be easily constrained, reducing the post-processing actions required; the developed methodology made the combination with existing surveys very easy for expert operators. Kaarta Stencil 2 has a strong potentiality to generate a higher number of points when composing the 3D point clouds. The advantage of using the MMS at the ground level allows users to enrich the point cloud in areas difficult to acquire with the CRP from the other side, for example, the roads surface and objects hidden or covered by the crown of trees. It is clear that these geomatics technologies may be complementary to one another in creating complete high-quality fully 3D representations. Nevertheless, the noisy data invalidate their usefulness to create high accuracy 3D models of the area, complementary to the photogrammetric one undertaken in the previous campaigns.

This work demonstrates its usefulness for data acquisition and mapping purposes, presenting similarities with the case studies described in [3,13], referring to the hand-held use of MMS. Indeed, we have been able to manage the occlusions, which are unavoidable from the photogrammetric model. Therefore, the information can be considered complementary to CRP for mapping the entire area, i.e., for a comprehensive survey. However, we have corroborated that the MMS is not robust enough to be used for monitoring complex sites such as this Corte de Pallás site. One promising approach is to integrate an accurate differential GNSS to the MMS solution.

It is fair, however, to highlight some drawbacks that emerged from this research. First of all, the point cloud cannot be exploited from scratch; in fact, further work of alignment was undertaken (as described in Section 4). The more the traveled survey distance extends, the more the drift error increases. The use of some constraints along the path should reduce the drift error. Despite MMS being designed with the main purpose of collecting 3D data without the need to register the point cloud further, this work proved that, the achievement of a result exploitable for mapping purposes relies upon the co-registration with an existing, and more accurate, point cloud. As the survey needs to be integrated with other data, accurate planning of the survey is required; in other words, it is not enough to just walk (see loop closure issue in Section 3.4) in order to improve the accuracy and facilitate the integration. Monitoring data need a submillimeter accuracy; as we reached an accuracy that is not comparable to the photogrammetric one (5.6 ± 2.3 cm), hand-held MMS as it is now implemented (without an accurate implementation of GNNS) cannot be considered suitable for such monitoring purposes. Nevertheless, it is very useful to complete the mapping of challenging areas in a fast, agile, and affordable way.

Another aspect that is worth discussing is the lack of RGB information. Indeed, the Kaarta MMS provides output 3D point clouds without the color information. This is an aspect that cannot be neglected, as environmental monitoring applications require in depth knowledge of a site. To overcome this limitation, the MMS cloud has been colorized, assigning the RGB values of corresponding neighbors from the CRP cloud. The colorized MMS cloud is depicted in Figure 16.

A final note is required with regard to the number of points that can be exploited with respect to the original raw data. As visible in Figure 13, to achieve a satisfactory accuracy comparable with the CRP one, the filtering steps brought to reduce the threshold up to values lower than 10 cm. This step increased the accuracy of the remaining points, at the expense of the number that was significantly reduced. In other words, due to the noise that is introduced by the tool, there is a waste of 3D information.

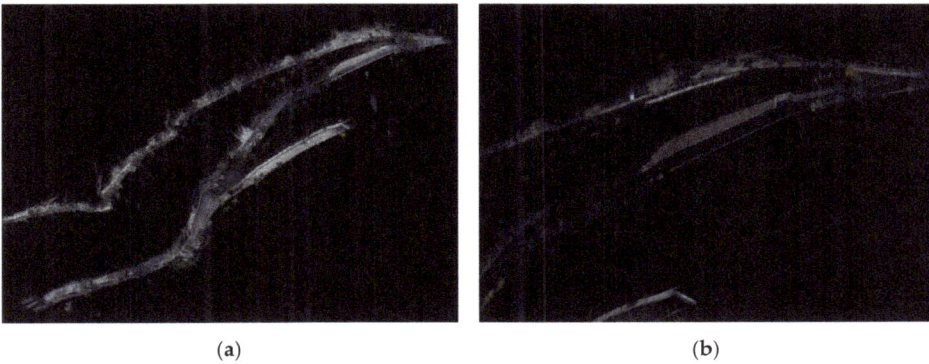

(a) (b)

Figure 16. Colorized MMS (Mobile Mapping System) point cloud: (**a**) overview of the entire survey, (**b**) close-up view of a portion of the surveyed area.

7. Conclusions

This research was aimed at assessing the use of hand-held MMS for environmental applications and, more in depth, for monitoring purposes. The case study investigated in Cortes de Pallas involved a very challenging area. The evidence that emerged from this study is that, as expected, CRP accuracy is strictly dependent on the supporting topographic network. Consequently, the MMS survey can only rely on it if combined in the same reference system. Once aligned, and in the case of the absence of UAV surveys, MMS was revealed to be an alternative to complete the 3D mapping of the area in a fast and agile way. As the management of occlusion is a well-known issue, often irresolvable even with a more sophisticated TLS system, the geomatic community should put more effort into improving the MMS accuracy and reliability.

As the Kaarta Stencil 2 is still under development and a growing number of new tools will enter the surveying market, this research is the occasion to set a useful baseline for further implementations demanded by the market, not only in civil engineering but also in building construction.

Author Contributions: Conceptualization, S.B.; Data curation, F.D.S.; Formal analysis, M.C. and R.P.; Funding acquisition, E.S.M.; Methodology, R.P.; Project administration, J.L.L. and P.G.; Supervision, L.G.-A. and R.P.; Writing—original draft, F.D.S.; Writing—review & editing, J.L.L. and E.S.M. All authors have read and agreed to the published version of the manuscript.

Funding: This research received no external funding.

Acknowledgments: The authors are grateful to the Infrastructures Area of the Diputació de València for his support and interest to test new technologies in Cortes de Pallás. We also acknowledge UCM which granted the use of its ME5000 Mekometer through a co-operation agreement.

Conflicts of Interest: The authors declare no conflict of interest.

References

1. Salvini, R. Modern technologies of geomatics applied to engineering geology. *Adv. Geo. Sci.* **2016**, *1*, 1–3. [CrossRef]
2. Alonso, E.E.; Gem, A.; Lloret, A. The landslide of Cortes de Pallas, Spain. *Géotechnique* **1993**, *43*, 507–521.
3. Friedli, E.; Presl, R.; Wieser, A. Influence of atmospheric refraction on terrestrial laser scanning at long range. In Proceedings of the 4th Joint International Symposium on Deformation Monitoring (JISDM), Athens, Greece, 15–17 May 2019.
4. Francioni, M.; Salvini, R.; Stead, D.; Coggan, J. Improvements in the integration of remote sensing and rock slope modelling. *Nat. Hazards* **2018**, *90*, 975–1004. [CrossRef]

5. James, M.R.; Quinton, J.N. Ultra-rapid topographic surveying for complex environments: The hand-held mobile laser scanner (HMLS). *Earth Surf. Process. Landf.* **2014**, *39*, 138–142. [CrossRef]
6. Tommaselli, A.M.G.; Moraes, M.V.A.; Silva, L.S.L.; Rubio, M.F.; Carvalho, G.J.; Tommaselli, J.T.G. Monitoring marginal erosion in hydroelectric reservoirs with terrestrial mobile laser scanner. *Int. Arch. Photogramm. Remote Sens. Spat. Inf. Sci. ISPRS Arch.* **2014**, *5*, 589–596. [CrossRef]
7. Jaboyedoff, M.; Oppikofer, T.; Abellán, A.; Derron, M.-H.; Loye, A.; Metzger, R.; Pedrazzini, A. Use of LIDAR in landslide investigations: A review. *Nat. Hazards* **2012**, *61*, 5–28. [CrossRef]
8. Lindenbergh, R.; Pietrzyk, P. Change detection and deformation analysis using static and mobile laser scanning. *Appl. Geomat.* **2015**, *7*, 65–74. [CrossRef]
9. D'Aranno, P.; Di Benedetto, A.; Fiani, M.; Marsella, M. Remote sensing technologies for linear infrastructure monitoring. In Proceedings of the GEORES 2019—2nd International Conference of Geomatics and Restoration, Milan, Italy, 8–10 May 2019; ISPRS: Hanover, Germany, 2019.
10. Francioni, M.; Salvini, R.; Stead, D.; Giovannini, R.; Riccucci, S.; Vanneschi, C.; Gulli, D. An integrated remote sensing-GIS approach for the analysis of an open pit in the Carrara marble district, Italy: Slope stability assessment through kinematic and numerical methods. *Comput. Geotech.* **2015**, *67*, 46–63. [CrossRef]
11. Michoud, C.; Carrea, D.; Costa, S.; Davidson, R.; Delacourt, C.; Derron, M.H.; Jaboyedoff, M.; Maquaire, O. Rockfall Detection and Landslide Monitoring Ability of Boat-based Mobile Laser Scanning along Dieppe Coastal Cliffs (Upper Normandy, France). In Proceedings of the Vertical Geology Conference 2014, Lausanne, Switzerland, 5–7 February 2014.
12. Ossowski, R.; Tysiąc, P. A new approach of coastal cliff monitoring using mobile laser scanning. *Pol. Marit. Res.* **2018**, *25*, 140–147. [CrossRef]
13. Jing, H.; Slatcher, N.; Meng, X.; Hunter, G. Monitoring capabilities of a mobile mapping system based on navigation qualities. In Proceedings of the XXIII ISPRS Congress, Prague, Czech Republic, 12–19 July 2016; ISPRS: Hanover, Germany, 2016.
14. Fuad, N.A.; Yusoff, A.R.; Zam, M.P.M.; Aspuri, A.; Salleh, M.F.; Ismail, Z.; Abbas, M.A.; Ariff, M.F.M.; Idris, K.M.; Majid, Z. Comparing the performance of point cloud registration methods for landslide monitoring using mobile laser scanning data. In Proceedings of the International Conference on Geomatics and Geospatial Technology (GGT 2018), Kuala Lumpur, Malaysia, 3–5 September 2018; ISPRS: Hanover, Germany, 2018.
15. Donker, J.; Van Maarseveen, M.; Ruessink, G. Spatio-temporal variations in foredune dynamics determined with mobile laser scanning. *J. Mar. Sci. Eng.* **2018**, *6*, 126. [CrossRef]
16. Nahon, A.; Molina, P.; Blázquez, M.; Simeon, J.; Capo, S.; Ferrero, C. Corridor Mapping of Sandy Coastal Foredunes with UAS Photogrammetry and Mobile Laser Scanning. *Remote Sens.* **2019**, *11*, 1352. [CrossRef]
17. Lim, S.; Thatcher, C.; Brock, J.C.; Kimbrow, D.R.; Danielson, J.J.; Reynolds, B. Accuracy assessment of a mobile terrestrial lidar survey at Padre Island National Seashore. *Int. J. Remote Sens.* **2013**, *34*, 6355–6366. [CrossRef]
18. Vaaja, M.T.; Hyyppä, J.; Kukko, A.; Kaartinen, H.; Hyyppä, H.; Alho, P. Mapping topography changes and elevation accuracies using a mobile laser scanner. *Remote Sens.* **2011**, *3*, 587–600. [CrossRef]
19. Williams, R.; Lamy, M.; Maniatis, G.; Stott, E. Three-dimensional reconstruction of fluvial surface sedimentology and topography using personal mobile laser scanning. *Earth Surf. Process. Landf.* **2020**, *45*, 251–261. [CrossRef]
20. Kumar, P.; Angelats, E. An automated road roughness detection from mobile laser scanning data. In Proceedings of the ISPRS Hannover Workshop: HRIGI 17–CMRT 17–ISA 17–EuroCOW 17, Hannover, Germany, 6–9 June 2017; ISPRS: Hanover, Germany, 2017.
21. Lienhart, W.; Kalenjuk, S.; Ehrhart, C. Efficient and Large Scale Monitoring of Retaining Walls along Highways using a Mobile Mapping System. In Proceedings of the 8th International Conference on Structural Health Monitoring of Intelligent Infrastructure, Brisbane, Australia, 5–8 December 2017; pp. 5–8.
22. De Moraes, M.V.A.; Tommaselli, A.M.G.; Santos, L.D.; Rubio, M.F.; Carvalho, G.J.; Tommaselli, J.T.G. Monitoring bank erosion in hydroelectric reservoirs with mobile laser scanning. *IEEE J. Sel. Top. Appl. Earth Obs. Remote Sens.* **2016**, *9*, 5524–5532. [CrossRef]
23. García-Asenjo, L.; Martínez, L.; Baselga, S.; Garrigues, P. Establishment of a multi-purpose 3D geodetic reference frame for deformation monitoring in Cortes de Pallás (Spain). In Proceedings of the 4th Joint International Symposium on Deformation Monitoring (JISDM), Athens, Greece, 15–17 May 2019.

24. Niemeier, W. Statistical tests for detecting movements in repeatedly measured geodetic networks. *Tectonophysics* **1981**, *71*, 335–351. [CrossRef]
25. Caspary, W.F. *Concepts of Network and Deformation Analysis*; School of Surveying, The University of New South Wales: Sydney, Australia, 1987; Monograph 11.
26. Bell, B. Workshop on the Use and Calibration of the Kern ME5000Mekometer. In *Proc. Stanford Linear Accelerator Center*; Stanford University: Stanford, CA, USA, 1992; pp. 1–80.
27. Rüeger, J.M. *Electronic Distance Measurement*; Springer Verlag: Berlin/Heidelberg, Germany, 1996.
28. Ciddor, P.E. Refractive index of air: New equations for the visible and near infrared. *Appl. Opt.* **1996**, *35*, 1566–1573. [CrossRef]
29. Ciddor, P.E.; Reginald, J.H. Refractive index of air. 2. Group index. *Appl. Opt.* **1999**, *38*, 1663–1667. [CrossRef]
30. Ciddor, P.E. Refractive index of air: 3. The roles of CO_2, H_2O, and refractivity virials. *Appl. Opt.* **2002**, *41*, 2292–2298. [CrossRef]
31. Puente, I.; González-Jorge, H.; Martínez-Sánchez, J.; Arias, P. Review of mobile mapping and surveying technologies. *Measurement* **2013**, *46*, 2127–2145. [CrossRef]
32. Paolanti, M.; Pierdicca, R.; Martini, M.; Di Stefano, F.; Morbidoni, C.; Mancini, A.; Malinverni, E.S.; Frontoni, E.; Zingaretti, P. Semantic 3D Object Maps for Everyday Robotic Retail Inspection. In *International Conference on Image Analysis and Processing*; Springer: Cham, Switzerland, 2019; pp. 263–274.
33. Zhang, W.; Qi, J.; Wan, P.; Wang, H.; Xie, D.; Wang, X.; Yan, G. An easy-to-use airborne LiDAR data filtering method based on cloth simulation. *Remote Sens.* **2016**, *8*, 501. [CrossRef]
34. Bronzino, G.P.C.; Grasso, N.; Matrone, F.; Osello, A.; Piras, M. Laser-visual-inertial odometry based solution for 3D heritage modeling: The Sanctuary of the Blessed Virgin of Trompone. In Proceedings of the 27th CIPA International Symposium "Documenting the Past for a Better Future", Avila, Spain, 1–5 September 2019; ISPRS: Hanover, Germany, 2019.

© 2020 by the authors. Licensee MDPI, Basel, Switzerland. This article is an open access article distributed under the terms and conditions of the Creative Commons Attribution (CC BY) license (http://creativecommons.org/licenses/by/4.0/).

Article

An Overall Deformation Monitoring Method of Structure Based on Tracking Deformation Contour

Xi Chu [1], Zhixiang Zhou [1,2,*], Guojun Deng [1], Xin Duan [1] and Xin Jiang [1]

1. State Key Laboratory of Mountain Bridge and Tunnel Engineering, Chongqing Jiaotong University, Chongqing 400074, China; chuxi1986@163.com (X.C.); dengguojun_cqjtu@163.com (G.D.); duanxin_cqjtu@163.com (X.D.); jiangxin_cqjtu@163.com (X.J.)
2. College of Civil and Transportation Engineering, Shenzhen University, Shenzhen 518060, China
* Correspondence: zhixiangzhou@cqjtu.edu.cn

Received: 4 September 2019; Accepted: 20 October 2019; Published: 25 October 2019

Featured Application: This study aims at the lack of sufficient data supporting structural damage identification, which is a general issue in traditional single-point measurement method. A novel method has been proposed for structural deformation monitoring based on digitalized photogrammetry, with improved efficiency and reduced cost. The method can be applied in the overall deformation monitoring of engineering structures, such as bridges. Furthermore, the method can provide a solid foundation for the estimation of structural health state.

Abstract: In structural deformation monitoring, traditional methods are mainly based on the deformation data measured at several individual points. As a result, only the discrete deformation, not the overall one, can be obtained, which hinders the researcher from a better and all-round understanding on the structural behavior. At the same time, the surrounding area around the measuring structure is usually complicated, which notably escalates the difficulty in accessing the deformation data. In dealing with the said issues, a digital image-based method is proposed for the overall structural deformation monitoring, utilizing the image perspective transformation and edge detection. Due to the limitation on camera sites, the lens is usually not orthogonal to the measuring structure. As a result, the obtained image cannot be used to extract the deformation data directly. Thus, the perspective transformation algorithm is used to obtain the orthogonal projection image of the test beam under the condition of inclined photography, which enables the direct extraction of deformation data from the original image. Meanwhile, edge detection operators are used to detect the edge of structure's orthogonal projection image, to further characterize the key feature of structural deformation. Using the operator, the complete deformation data of structural edge are obtained by locating and calibrating the edge pixels. Based on the above, a series of load tests has been carried out using a steel–concrete composite beam to validate the proposed method, with the implementation of traditional dial deformation gauges. It has been found that the extracted edge lines have an obvious sawtooth effect due to the illumination environment. The sawtooth effect makes the extracted edge lines slightly fluctuate around the actual contour of the structure. On this end, the fitting method is applied to minimize the fluctuation and obtain the linear approximation of the actual deflection curve. The deformation data obtained by the proposed method have been compared with the one measured by the dial meters, indicating that the measurement error of the proposed method is less than 5%. However, since the overall deformation data are continuously measured by the proposed method, it can better reflect the overall deformation of the structure, and moreover the structural health state, when compared with the traditional "point" measurements.

Keywords: structural engineering; overall deformation monitoring; perspective transformation; edge detection; close-range photogrammetry

1. Introduction

During the service life, engineering structures are subjected to various inherent deterioration processes of structure such as corrosion, fatigue, material creep, and so on. As a result, the deformation of the degraded structure will deviate from the original one. On this end, the structural deformation in eventually used as an important index in the structural health monitoring [1,2]. For instance, the external load and deformation of the structure system generally follows the below relation:

$$\{d\} = [K]^{-1}\{f\}$$

where $\{d\}$ stands for the deformation state; $[K]$ is the stiffness matrix of the structure; $\{f\}$ represent the effect induced by the external load. When any damage or deterioration occurs in the structure, the stiffness matrix $[K]$ will change correspondingly, which in turns lead to the inevitable change in the deformation state $\{d\}$. Therefore, the change in the deformation state can be utilized to evaluate the health state of the structure. The present study focuses on the direct extraction of the overall deformation, rather than approximating the deformation through the data measured at several discrete points. The major advantage of the overall deformation data is to eliminate the error in structural health evaluation caused by insufficient measurement.

Traditionally, the structural deformation can be measured by leveling, total station, GPS, vibration sensors, and other equipment. At present, these methods can accurately and rapidly measure the deformation information of structures. However, only the limited key points of the targeting structure can be measured using the above methods, which often lead to insufficient data and, moreover, the insensitivity to structural deterioration [3]. Obviously, the direct solution is to largely increase the number of sensors installed on the structure. However, it is both time and budget consuming, which is not applicable in engineering practices. Alternatively, the digital image full-field structural morphology measurement can be a very ideal solution, which can take the advantages of both the structural damage identification method and digital image processing technology. Therefore, it is very crucial to effectively utilize the structural image features to extract the full-field deformation information of the structure.

In recent years, digital image processing technology is eventually employed to measure the overall deformation of the structure. As a kind of remote sensing technique, photogrammetry does not need any contact with the objects, and this can be a great advantage in the deformation monitoring of structures. "Photogrammetry" is to set up a base station in a stable area on the front of the target, and then shoot the target, so as to get the shape and motion state of the target according to the image [4]. According to different imaging distances, photogrammetry can be divided into "space photogrammetry", "close-range photogrammetry", and "microscopic photogrammetry" [5]. In structural deformation monitoring, close-range photogrammetry has broad prospects for development [6], and "Close-range photogrammetry" means that the distance between the base station and the measured structure is within 300 m [7]. Feng et al. [8] presents a comprehensive review on the recent development of computer vision-based sensors for structural displacement response measurement and their applications for SHM. Importation issues critical to successful measurement are discussed in detail, including how to convert pixel displacements to physical displacements, how to achieve sub-pixel resolutions, and what to cause measurement errors and how to mitigate the errors. However, the article also clearly points out that in many respects, the vision-based sensor technology is still in its infancy. The majority studies have still been focused on measurements of small-scale laboratory structures or field measurements of large structures at a limited number of points for a short period of time. Rolands Kromanis et al. [9] introduces a low-cost robotic camera system (RCS) for accurate measurement collection of structural response. The low-cost RCS provides very accurate vertical displacements. The measurement error of the RCS is 1.4%. Serena Artese et al. [10] proposed a bridge monitoring system, which combines camera and laser indicator; the elastic line inclination is measured by analyzing the single frames of an HD video of the laser beam imprint projected on a flat target. The inclination of the elastic line at the support was obtained with a precision of 0.01 mrad. Ghorban et al. [11] measured the overall deformation of the masonry wall subjected to cyclic loads, using the 3D image correlation technology. The displacement, rotation, and interface slip between the reinforced concrete column and masonry were measured. Wang et al. [12] used the close-range photogrammetry technology to monitor the displacement of tunnel caverns. The measured results were compared with the values measured by mechanical convergence meter, and the difference between the two methods is no more than ±2 mm at the measuring distance of 8 m. This accuracy meets the requirement of general tunnel deformation monitoring. Reference [13] studied the application of sub-pixel displacement measurement method in soil strain monitoring. Based on the spatial correlation function iteration, the sub-pixel displacement of soils was measured. Zang et al. [14] applied the close-range photogrammetry technology in measuring bridge deflection and proved that a desirable accuracy can be achieved, i.e., ±1 mm. However, the accuracy is greatly affected by the positioning of the artificial marking points required by the method [15]. Although the above studies validated the feasibility of the application of close-range photogrammetry technology in structural deformation monitoring, the above methods are still unable to measure the structural overall deformation.

In order to explore the feasibility of photogrammetry in structural overall deformation monitoring, Ivan Detchev et al. [16] explored the use of consumer-grade cameras and projectors for the deformation monitoring of structural elements. A low-cost digital camera deformation monitoring system is proposed. Static load tests of concrete beams are carried out in the laboratory. The experiments proved that it was possible to detect sub-millimeter-level overall deformations given the used equipment and the geometry of the setup. However, this technology requires high texture characteristics of the structure surface and needs to project random pattern on the structure surface, which is difficult to achieve in the actual bridge structure deformation monitoring. On another hand, the close-range photogrammetry requires the measuring equipment to be located in the orthogonal projection position of the measured surface, which is usually difficult in engineering practice. Taking the bridge structure as an example, the surroundings near the bridge are usually complex, such as mountains, rivers, and trees, which makes it difficult for the camera to maintain the orthogonal projection position with respect to the measuring bridge.

In the deformation monitoring of structures, environmental factors must be considered. The complex geographical conditions of the structures means the photogrammetric camera is unable to work in the ideal measuring position. Therefore, it is necessary to study a new photogrammetric method for the overall deformation of structures under the condition that the camera is in the inclined position. In view of the actual needs of the structure deformation monitoring, this paper studies the overall deformation monitoring method under the condition of tilt photography. The steel truss concrete composite beam specimens were made in the laboratory. The static deformation images of the specimens were obtained by oblique photography. The overall deformation of the specimens was obtained by perspective transformation and edge detection technology. The error sources of this method were analyzed. This research is a comprehensive application of photogrammetry and digital image processing technology in the field of structure deformation monitoring. Its research foundation has been carefully verified and published in many publications [17,18]. The research results can alleviate the problem of insufficient deformation data in damage identification. In addition, compared with traditional photogrammetric methods, this study also highlights the advantages of flexible placement of camera positions in actual measurement work.

2. Orthogonal Projection and Global Deformation Acquisition Method of Structures

2.1. Perspective Transformation of Digital Image

On the basis of unchanged image content, the image pixel position is transformed, which is called image geometric transformation [19]. It mainly includes translation, rotation, zooming, reflection, and slicing. Usually, compound transformations, such as the perspective transformation, can be divided into a series of basic transformations. According to the perspective principle [20], when photographed under the condition of non-orthogonal projection, the image of the measured structure will deform. As a result, the true shape of the structure can be obtained only when the camera is in the orthogonal projection position of the measured surface. The process of mathematical transformation from oblique projection center to orthogonal projection center is called the perspective transformation. Figure 1 illustrates the basic model of perspective transformation.

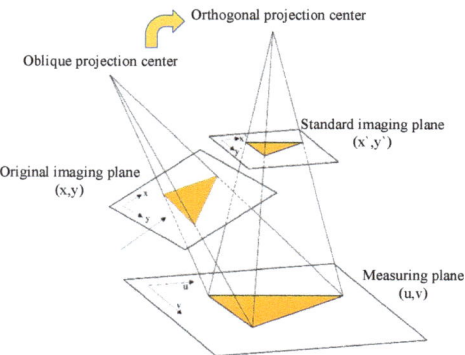

Figure 1. Perspective transformation.

A 3D Cartesian coordinate system can be established, in which the projection center of the camera is selected as the origin, called the camera coordinate. Meanwhile, the image plane is set as the x-y plane, and the focus of the plane is located at $[0, 0, f] (f > 0)$. A 2D Cartesian coordinate system can be established on the plane where the object is measured, called the measuring coordinate. The origin of the measuring coordinate system is $[x_0, y_0, z_0]^T$ in the camera coordinate. The unit vectors in the x-axis direction are $[u_1, u_2, u_3]^T$, the unit vectors in the y-axis direction are $[v_1, v_2, v_3]^T$, and the vector relation can be written as follows:

$$\begin{cases} u_1v_1 + u_2v_2 + u_3v_3 = 0 \\ u_1^2 + u_2^2 + u_3^2 = v_1^2 + v_2^2 + v_3^2 = 1 \end{cases}. \tag{1}$$

The points of coordinates $[u, v]^T$ in the measuring plane can be expressed in the camera coordinate system as the vector below,

$$u \begin{bmatrix} u_1 \\ u_2 \\ u_3 \end{bmatrix} + v \begin{bmatrix} v_1 \\ v_2 \\ v_3 \end{bmatrix} + \begin{bmatrix} x_0 \\ y_0 \\ z_0 \end{bmatrix}. \tag{2}$$

Assuming that the coordinate of the point in the original imaging plane is $[x, y, 0]^T$, $\exists k \in R$ the following expression can be derived,

$$u \begin{bmatrix} u_1 \\ u_2 \\ u_3 \end{bmatrix} + v \begin{bmatrix} v_1 \\ v_2 \\ v_3 \end{bmatrix} + \begin{bmatrix} x_0 \\ y_0 \\ z_0 \end{bmatrix} - \begin{bmatrix} 0 \\ 0 \\ f \end{bmatrix} = k \left(\begin{bmatrix} 0 \\ 0 \\ f \end{bmatrix} - \begin{bmatrix} x \\ y \\ 0 \end{bmatrix} \right). \tag{3}$$

Comparing the preceding formula, it yields

$$-k \begin{bmatrix} x \\ y \end{bmatrix} = u \begin{bmatrix} u_1 \\ u_2 \end{bmatrix} + v \begin{bmatrix} v_1 \\ v_2 \end{bmatrix} + \begin{bmatrix} x_0 \\ y_0 \end{bmatrix} = \begin{bmatrix} u_1 & v_1 & x_0 \\ u_2 & v_2 & y_0 \end{bmatrix} \begin{bmatrix} u \\ v \\ 1 \end{bmatrix} \tag{4}$$

$$kf = uu_3 + vv_3 + z_0 - f = \begin{bmatrix} u_3 & v_3 & z_0 - f \end{bmatrix} \begin{bmatrix} u \\ v \\ 1 \end{bmatrix}. \tag{5}$$

Equation (5) can be rewritten as the following,

$$-k = \begin{bmatrix} -\frac{u_3}{f} & -\frac{v_3}{f} & -\frac{z_0 - f}{f} \end{bmatrix} \begin{bmatrix} u \\ v \\ 1 \end{bmatrix}. \tag{6}$$

Combining Equations (6) and (4), it leads to

$$-k \begin{bmatrix} x \\ y \\ 1 \end{bmatrix} = \begin{bmatrix} u_1 & v_1 & x_0 \\ u_2 & v_2 & y_0 \\ -\frac{u_3}{f} & -\frac{v_3}{f} & -\frac{z_0 - f}{f} \end{bmatrix} \begin{bmatrix} u \\ v \\ 1 \end{bmatrix}. \tag{7}$$

For convenience, a parameter matrix M is introduced, as shown below,

$$M = \begin{bmatrix} u_1 & v_1 & x_0 \\ u_2 & v_2 & y_0 \\ -\frac{u_3}{f} & -\frac{v_3}{f} & -\frac{z_0 - f}{f} \end{bmatrix}. \tag{8}$$

If the focus $[0, 0, f]^T$ is not on the measuring plane, the matrix M is a nonsingular matrix. Under normal working conditions, the focus will not be on the measuring plane, so the matrix M can usually be treated as a nonsingular matrix. The focal length and spatial position of the camera will change when the camera moves to a new position to capture the target structure. It can also be considered that the camera imaging plane is fixed, the focal length and the actual spatial position of the structure are changed. Make the coordinates of the camera focus change to $[0, 0, f']^T$, and the original coordinates of the measuring plane change to $[x'_0, y'_0, z'_0]^T$. The unit vectors of the x, y axes in the measuring plane become $[u'_1, u'_2, u'_3]^T$, $[v'_1, v'_2, v'_3]^T$. Similarly, there is $\exists k \in R$, making the coordinate point $[u, v]^T$ on the measuring plane, and its corresponding imaging point $[x', y', 0]^T$ should satisfy:

$$-k' \begin{bmatrix} x' \\ y' \\ 1 \end{bmatrix} = \begin{bmatrix} u'_1 & v'_1 & x'_0 \\ u'_2 & v'_2 & y'_0 \\ -\frac{u'_3}{f'} & -\frac{v'_3}{f'} & -\frac{z'_0 - f'}{f'} \end{bmatrix} \begin{bmatrix} u \\ v \\ 1 \end{bmatrix}. \tag{9}$$

The parameter matrix M' is denoted as:

$$M' = \begin{bmatrix} u'_1 & v'_1 & x'_0 \\ u'_2 & v'_2 & y'_0 \\ -\frac{u'_3}{f'} & -\frac{v'_3}{f'} & -\frac{z'_0 - f'}{f'} \end{bmatrix}. \tag{10}$$

Comparison of Equations (7) and (9) leads to

$$-k' \begin{bmatrix} x' \\ y' \\ 1 \end{bmatrix} = M' \begin{bmatrix} u \\ v \\ 1 \end{bmatrix} = -kM'M^{-1} \begin{bmatrix} x \\ y \\ 1 \end{bmatrix} \tag{11}$$

assuming:

$$M' \cdot M^{-1} = \begin{bmatrix} m_{11} & m_{12} & m_{13} \\ m_{21} & m_{22} & m_{23} \\ m_{31} & m_{32} & m_{33} \end{bmatrix}. \tag{12}$$

Accordingly, the following expansions are introduced:

$$\begin{cases} k'x' = k(m_{11}x + m_{12}x + m_{13}) \\ k'x' = k(m_{21}x + m_{22}x + m_{23}) \\ k' = k(m_{31}x + m_{32}y + m_{33}) \end{cases}. \tag{13}$$

Therefore, there is:

$$\begin{cases} x' = \frac{m_{11}x + m_{12}x + m_{13}}{m_{31}x + m_{32}y + m_{33}} \\ y' = \frac{m_{21}x + m_{22}x + m_{23}}{m_{31}x + m_{32}y + m_{33}} \end{cases}. \tag{14}$$

The coordinates (x, y) are the imaging point of the original image, which is transformed into a new imaging point (x', y') after the perspective transformation. Based on the above analysis, the proposed process can convert the original oblique structural image into orthophoto-projection image, which provides technological foundation for monitoring the overall deformation of the structure.

2.2. Edge Detection

The edge of structural image is an important carrier of overall deformation information. Edge detection is a method to analyze the main features of images [21]; it can greatly reduce the amount of data, eliminate information not related to deformation monitoring, and retain the basic attributes of structure. The basic task of edge detection is to recognize the step change of the gray value of the structure edge in the image, which can be further used to obtain the feature edge of the structure. According to [22], the step edge is related to the peak value of the first-order derivative of the gray level of the image, and the degree of change of the gray value can be expressed by gradient. The gradient of image function is a vector with direction and size, as shown below:

$$G(x, y) = \begin{bmatrix} G_x \\ G_y \end{bmatrix} = \begin{bmatrix} \frac{\partial f}{\partial x} \\ \frac{\partial f}{\partial y} \end{bmatrix}. \tag{15}$$

It can be seen that the direction of the vector $G(x, y)$ is the change rate of the gray value of function $f(x, y)$.

The amplitude of the gradient can be expressed as:

$$|G(x, y)| = \sqrt{G_x^2 + G_y^2}. \tag{16}$$

In this paper, the absolute value is used to approximate the gradient amplitude:

$$|G(x, y)| \approx \max(|G_x|, |G_y|). \tag{17}$$

The direction of the gradient can be derived as:

$$\alpha(x, y) = \arctan(G_y/G_x) \tag{18}$$

where α is the angle between the direction vector and the x-axis.

From the above formulas, it is suggested that the degree of the change in gray levels can be detected by the discrete approximation function of gradient. At the edge of the structure, the gray value will change [23], resulting in the maximum value of the gradient function. On this end, the edge can be extracted through the above features.

From the above algorithm, we can see that the essence of the image edge is the point of discontinuous gray level, or where the gray level changes dramatically. The drastic change of gray level of edge means that near the edge point, the signal has high frequency components in the spatial domain. Therefore, the edge detection method is essentially to detect the high-frequency component of the signal, but it is difficult to distinguish the high-frequency component of the gray signal from the environmental noise of the actual structure photogrammetry, which makes it difficult to accurately extract the edge information of the structure. Taking one-dimensional signal of the structure image as an example, as shown in Figure 2, if point A is regarded as the edge point of the signal and there is a jump in the signal, then whether there is an edge at point B and point C needs to be treated with caution. In fact, point B and point C are probably the combination of the signal and some noise.

Figure 2. Illustration of the edge point and noise point.

Edges with continuous gradients, such as point A in Figure 2, are very rare in actual structure images. Most of the structural edge points will be accompanied by environmental noise, forming a large number of complex edge points such as points B and C. Therefore, it is necessary to study the false edges caused by noise in order to ensure the accuracy of structural deformation monitoring.

3. Static Test of the Beam

A static test has been carried out on a steel truss–concrete composite beam, to validate the proposed method, as shown in Figure 3.

The specimen is simply supported by two hinge bearings at the both ends, as shown in Figure 3a. Two hydraulic jacks have been used to apply the two-points bending load on the specimen. Three dial meters have been placed at the quarter-span and the midspan of the specimen, to measure the structural deflection.

The specimen has been loaded with a step-by-step prototype from 0 to 600 kN, with an increment of 100 kN. The loading protype is shown in Figure 4. It is worth stating that the measurement at each step has been made two minutes after the target load is reached, to allow the well-deformation of the specimen.

During the test, the digital image of the specimen is also collected using Canon EOS 5DS R low-cost digital camera; the camera and lens parameters are shown in Table 1. The spatial position of the camera in this experiment is set in the non-orthogonal projection position to simulate the normal condition in engineering practice, as aforementioned. During the whole process, the space position and azimuth of the camera should be maintained to ensure the consistency of projection centers of structure images in the whole process. In order to prevent the camera from being disturbed, shooting remote controller is used to control camera parameters and shutter shooting.

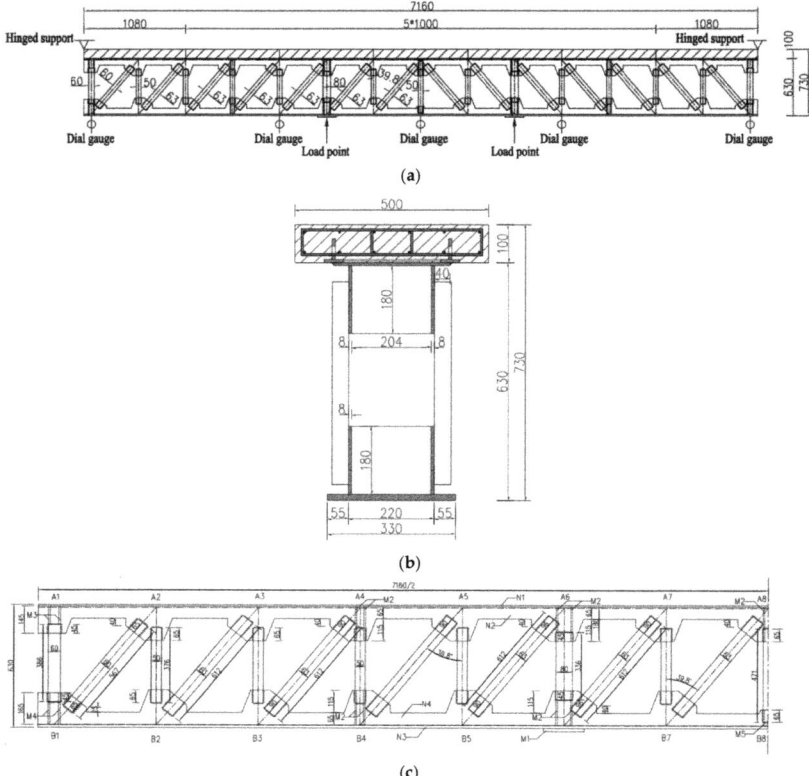

Figure 3. The tested steel–concrete composite beam (unit: mm); (**a**) elevational view; (**b**) sectional view; (**c**) detail size.

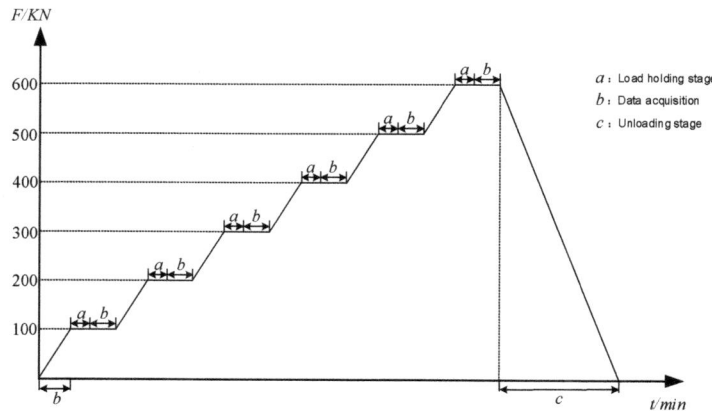

Figure 4. Test loading protype.

Appl. Sci. **2019**, *9*, 4532

Table 1. Parametric table of camera and lens.

Number of Pixels	Size of Sensor	Data Interface	Aspect Ratio	Photo-Sensors
50.6 million	36 × 24 mm	USB 3.0	3:2	CMOS
Image Amplitude	**Pixel Size**	**Lens Type**	**Focal Length**	**Lens Relative Aperture**
8688 × 5792	4.14 μm	EF 24–70 mm f/2.8 L	50 mm	F2.8–F22

As a common practice, the system error exists in the measurement due to the physic limitation of the applied hardware. Specifically, the accuracy of the photogrammetry-based method has a stronger dependence on the capacity of hardware when compared with the traditional methods. Therefore, the calibration of photogrammetry equipment is an essential part of the measurement. Generally, the largest part of error in photogrammetric hardware originates from lens distortion [24]. On this end, the checkerboard lattice calibration method [25,26] has been applied to calibrate the lens of photogrammetric camera. The calibration has been conducted with a total of 25 checkerboard lattice images, and the lens distortion parameters are obtained. Based on that, the photogrammetric images obtained in this paper have been corrected. The calibration process is shown in Figure 5.

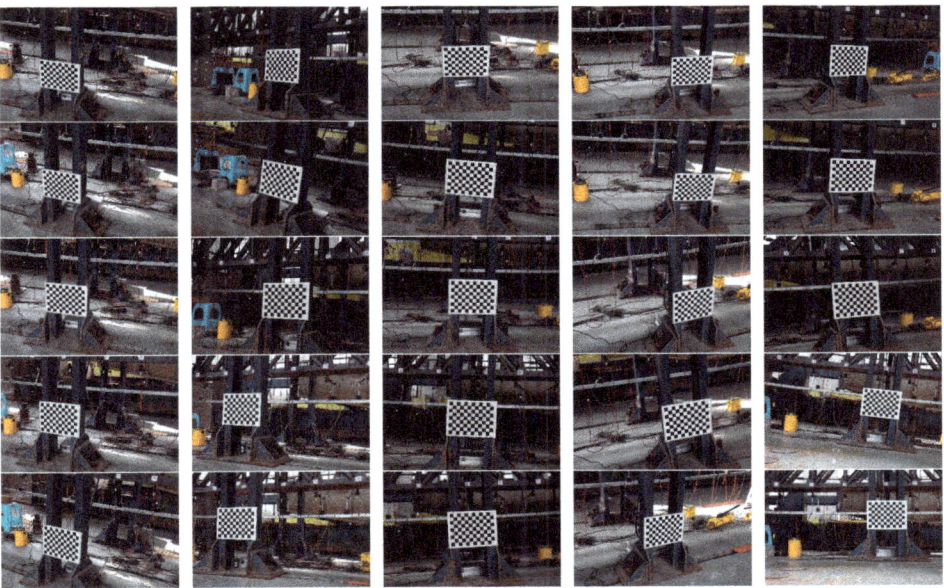

Figure 5. Camera calibration intersection photography.

The results of camera calibration parameters are shown in Table 2.

Table 2. Parametric table of camera and lens.

Outline Size	Phase Principal Coordinates X0	−0.0769 mm
	Phase Principal Coordinates Y0	0.0045 mm
	Camera Main Distance f	50.9339 mm
Radial Distortion Coefficient	K1	1.9644×10^{-8}
	K2	5.6287×10^{-6}
Eccentric Distortion Coefficient	P1	1.4683×10^{-5}
	P2	-3.8601×10^{-6}
Pixel Size		0.004096 mm
Image Size		5792 × 8668

In the calibration, the distortion correction formula [27] has been used to correct the structural image element by element. As a result, the ideal image without lens distortion has been obtained, which can be further used for the extraction of structural deformation in the next. The distortion correction effect of the image is shown in Figure 6.

Figure 6. Camera calibration distortion adjustment; (**a**) original image; (**b**) image after distortion correction.

In reality, the landform around the structure makes it difficult to obtain the orthogonal projection image, so that it can only be tilt photographed on both sides. According to the practical application requirements, the actual situation is simulated in the test, i.e., the camera takes pictures of the test beam at a fixed tilt angle. As shown in Figure 7a, the distance between the camera and the ground is about 3.5 m, and the distance between the camera and the test beam is about 3.0 m. Obviously, there is a large horizontal angle between the optical axis of the camera and the normal direction of the vertical plane of the test beam, and there is also an elevation angle in the vertical direction. This photogrammetric method simulates the possible inclination angle of the camera in the actual structure survey; unlike orthographic projection, this tilt photography will cause the structure image to be affected by the perspective relationship and present near-large-far-small imaging features, which will affect the extraction of structural deformation information. Figure 7 shows the image acquisition result of the specimen. It is worth noting in the image that the specimen near the right bearing is blocked by the reaction frame, which can reflect the usual monitoring conditions of actual structure. Thus, it is crucial to obtain the deformation data of the sheltered part of the structure, which is also a key part of the present study.

Appl. Sci. **2019**, *9*, 4532

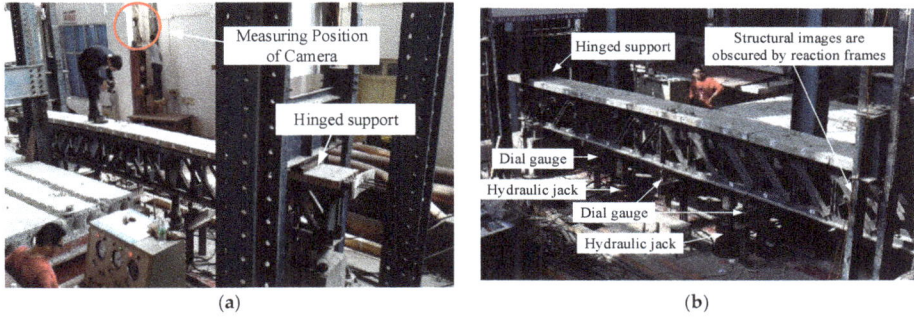

Figure 7. Field layout of static load test; (**a**) camera placement position; (**b**) collecting shape data of experimental beam.

4. Result and Discussion

4.1. Picture Perspective Transform of Test Beam

According to the principle of perspective transformation, the image of the test beam obtained under each load grades is processed, and the result of image processing under one of the load grades is shown in Figure 8. It can be seen that through perspective transformation, the image of the test beam changes from oblique projection to orthogonal projection, while all the details of the specimen have been well preserved. It is worth noting that the remain parts apart from the specimen are distorted by the perspective transformation. However, the distortion does not affect the acquisition of structural deformation data since the test aims at the overall deformation data of the specimen only.

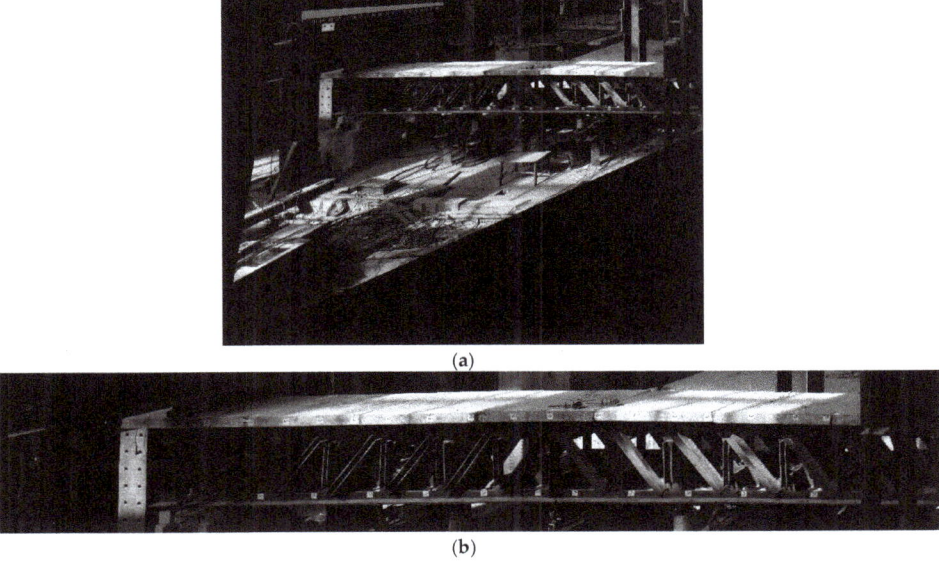

Figure 8. Perspective transformation of the tested beam; (**a**) after perspective transform; (**b**) integral drawing of specimen after projection transformation.

4.2. Edge Contour Extraction of Structures

Several types of operators can be employed for the edge detection, including the Sobel [28] operator, Prewitt operator [29], Roberts operator [30], and log operator [31], in which different methods are used to solve the gradient extremum. In this paper, all the five operators have been applied to detect the edge of the specimen using the image after perspective transformation, as shown in Figure 9. Compared with the Log operator, the edge detection results of the other four operators are not satisfactory due to the lack of edge information, which in turn has a negative impact on the accuracy. The advantage of the Log operator over the other methods is that the Gauss spatial filter is employed to smooth the original image, which minimizes the influence of noise on edge detection. The Log operator is a second-order edge detection operator [32], as shown in Equations (19) and (20):

$$\nabla^2 f = \frac{\partial^2 f}{\partial x^2} + \frac{\partial^2 f}{\partial y^2} \tag{19}$$

$$\begin{cases} \frac{\partial^2 f}{\partial x^2} = f(i, j+1) - 2f(i,j) + f(i, j-1) \\ \frac{\partial^2 f}{\partial y^2} = f(i+1, j) - 2f(i,j) + f(i-1, j) \end{cases} \tag{20}$$

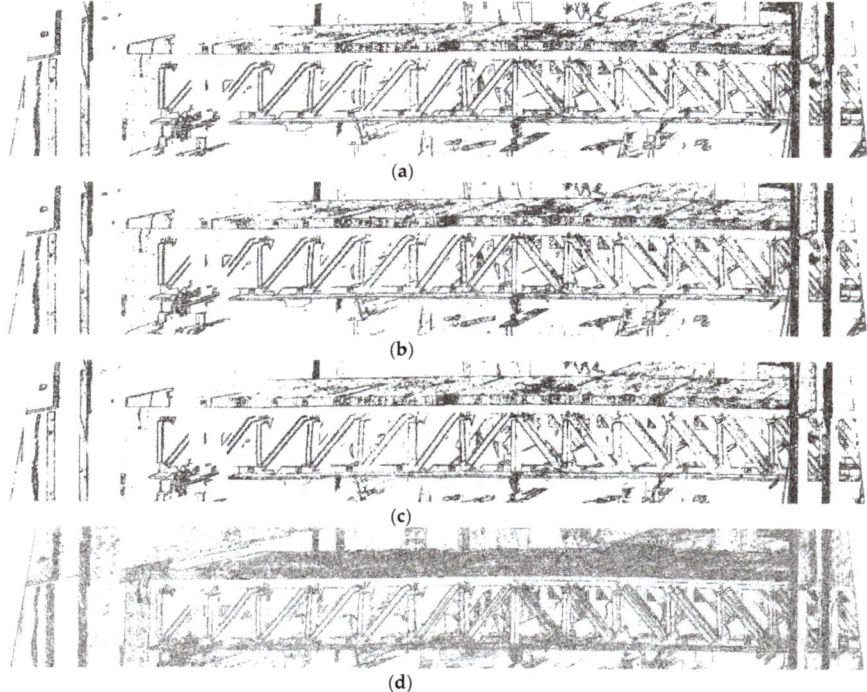

Figure 9. The applied five kinds of edge detection operators; (**a**) Sobel operator edge detection; (**b**) Prewitt operator edge detection; (**c**) Roberts operator edge detection; (**d**) log operator edge detection.

Based on the second-order differential of the image, the extreme points can be generated at the abrupt position of the gray value. According to these extreme points, the edge of the structure can be determined.

As shown in Figure 9, the distribution of the light intensity respecting the specimen is inconsistent, and the gray value of some edges does not change significantly, resulting in discontinuity in the detection of some edges. The discontinuous edges like Figure 9 are very normal in actual structure images. On the other hand, the distribution of edge pixels obtained by various edge algorithms is also different. The distribution density of edge pixels in this paper is shown in Figure 10.

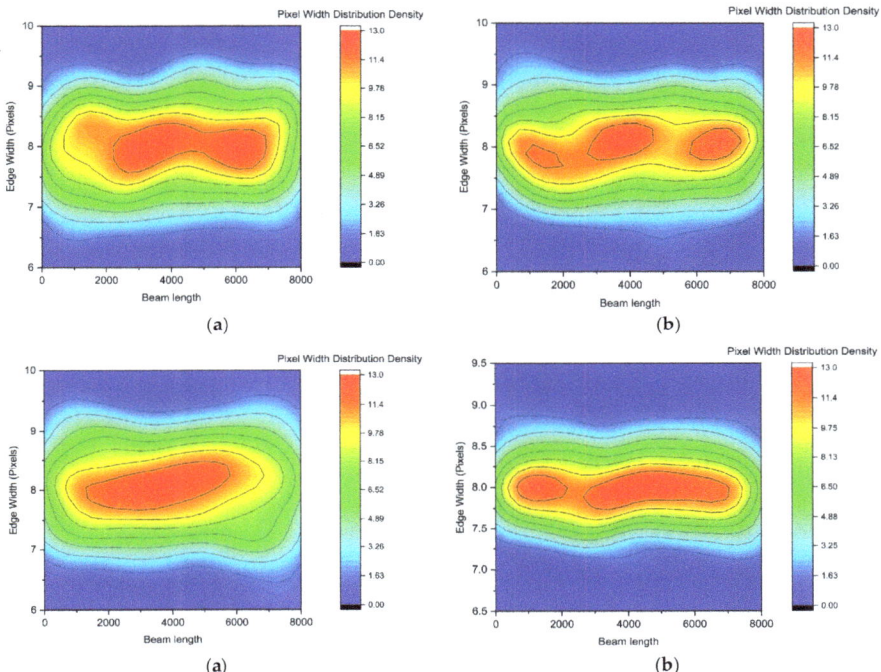

Figure 10. Distribution density of edge pixels by various edge detection operators; (**a**) edge width distribution degree of Sobel operator; (**b**) edge width distribution degree of Prewitt operator; (**c**) edge width distribution degree of Robert operator; (**d**) edge width distribution degree of log operator.

As shown in Figure 10, the edge features obtained by the above five operators are entirely different. For instance, the edge distributions in some operators are three pixels wide, while in some operators, e.g., the Log operator, the edges are just concentrated in one pixel. Because of the large scatter in the pixel distribution, it is difficult to determine the exact edge position of the structure. Naturally, an efficient edge detection operator should have the centralized pixel distribution and good pixel continuity. After comparison, it is found that the Log operator can extract the edges of the structure relatively intact and maintain a relatively centralized distribution of edge pixels. Thus, the log operator has been selected in extracting the edge of the structure.

The phenomenon of discontinuous edges and scattered edge distribution of the above-mentioned is similar to the detection issue in real structures, which is induced by the environment of measurement. In dealing with such kind of problem, the data processing and analysis process have been employed, as illustrated in the following. As an important part of the deformation, the edge of the structure is the key content of this paper. For the specimens in this paper, the upper and lower edges of the bridge deck and the lower edges of the specimens can be used as characteristic contours to analyze the overall deformation of the structure. From Figure 2, we can see that the noise caused by the environment will make the edge location confused. Points with continuous gradient change are very rare in the actual structure image. The presence of image noise can lead to the generation of pseudo edges. On the other hand, the real signal on the edge of the structure may also be smoothed out by the Gauss spatial filter, which will cause the edge of the structure to be discontinuous and the edge information missing as shown in Figure 11. Figure 11a shows that the lower edge contour of the bridge deck is relatively continuous. Therefore, the lower edge of the bridge deck is used as the characteristic contour of the test beam to extract the overall deformation of the structure; the position of edge contour extraction in this paper is shown in Figure 11b.

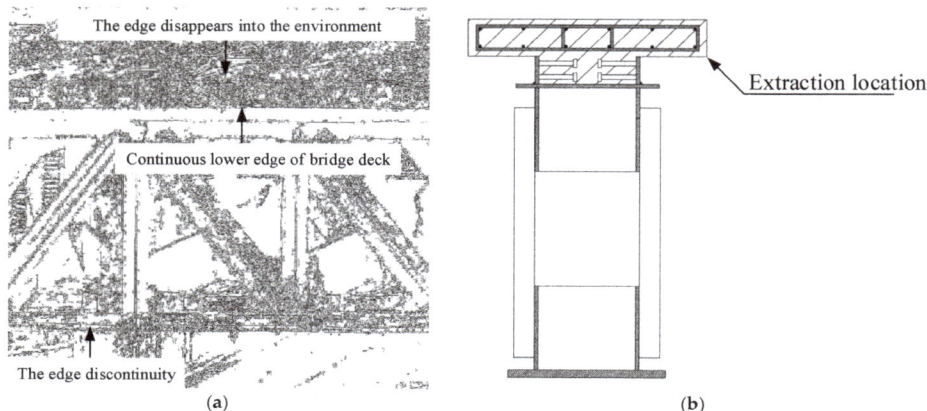

Figure 11. Feature contour extraction; (**a**) edge of the test beam affected by environment; (**b**) sketch map of edge extraction position.

Because the space position of the camera is fixed and the same perspective transformation method is used in each load grade, the edge contour before and after deformation can be directly extracted and compared without looking for fixed points in each load grade. The pixel coordinates of the lower edge of the bridge deck are extracted, and the original edge contour of the structure is obtained under each load grade, as shown in Figure 12.

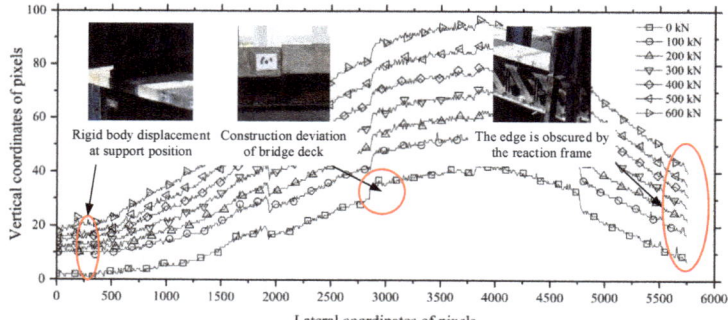

Figure 12. Edge line of lower edge of bridge deck.

The result shows that the deflection of the specimen increases with the load grade. Since the edge line of the specimen is obscured near the right bearing by the reaction frame, the edge information is partially missed. By contrast, the information of the left edge is well established. Considering the parallel change of the contour line near the bearing on the left, it can be inferred that the rigid displacement of the specimen exists under each load condition. When loaded from 0 to 100 kN, the rigid displacement reaches its largest. This large displacement is due to the fact that the bearings are not in close contact with the reaction frame before loading. After the load of 100 kN, the bearings will be tightly contacted with the reaction frame. However, due to the deformation of the reaction frame since it is not ideally stiff, a small amount of rigid body displacement still exists in all the load conditions. Besides, it can be found that the edge contour of the specimen is piecewise continuous under every single load condition, with the step change occurred at the four connections between the segments. During the fabrication, the specimen is first divided into five segments and each of the segments is manufactured independently. Then, the separated segments are assembled at the four points together, resulting in the inevitable assembly error. As a result, the shape of the specimen will change abruptly at those assembly points, which in turn lead to the step change as reflected in the edge contour.

4.3. Deformation Curve Obtained by Overlapping Difference of Contour Line

By calibrating each pixel in the image, the size of each pixel can be obtained. The size of each pixel is the theoretical limit of the accuracy in the proposed photogrammetric method. In the steel truss bridge tested, the vertical member is orthogonal to the pixels, as shown in Figure 13. Thus, the calibration of the vertical member is relatively simple. As shown in Table 3, by calibrating 13 visible vertical members, the measurement accuracy of this experiment is 1.12 mm. According to the calibration value, the contour of the pixel in Figure 12 can be transformed into the actual deformation value.

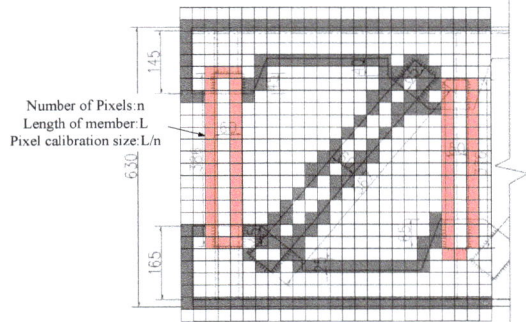

Figure 13. Pixel calibration schematic.

Table 3. Pixel size calibration table of vertical members.

Number of Vertical Members	Number of Pixels	Real Length of Members/mm	Calibration Value	Average Value
1	344	387	1.12	
2	342	379	1.10	
3	340	380	1.11	
4	345	377	1.09	
5	339	376	1.10	
6	292	332	1.12	
7	346	375	1.08	1.12 mm/px
8	421	473	1.12	
9	330	376	1.14	
10	295	331	1.22	
11	335	377	1.12	
12	351	376	1.07	
13	342	381	1.11	

The edge line shown in Figure 12 is notably discontinuous at the connecting points due to assembly error as analyzed before. However, the discontinuity is not caused by the structural deformation and will not change with the applied load. Therefore, the influence of the discontinuity can be eliminated by the overlapping difference method. Since the bearings contact the reaction frame closely after 100 kN, the edge line of the initial working condition can be overlap differenced by the edge line of the other load grades. Thus, the load-displacement curve of 100–600 kN can be derived, as shown in Figure 14.

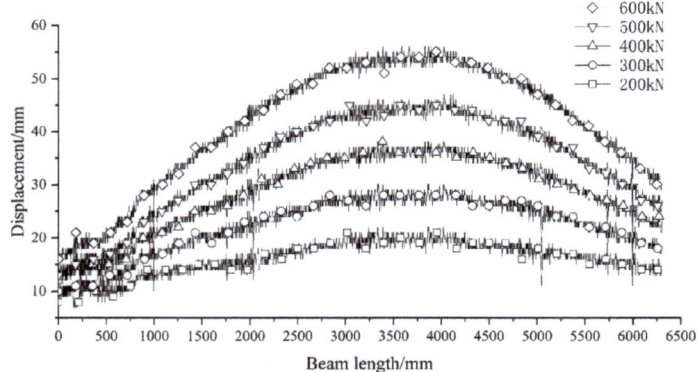

Figure 14. Load-displacement curves under various load grades.

It can be found that the load-displacement curve presents a zigzag feature. The reason for this problem is that the edge of the structure in the image is composed of one or more pixels, as shown in the red pixel of Figure 15. These pixels are arranged side by side to form a bandwidth. The final edge position (shown in deep red in Figure 15) is determined by the gradient change rate of the gray value of all pixels in the bandwidth. Under the influence of illumination conditions, the final edge and the actual edge will have errors, as shown in the dotted line of Figure 15. As a result, the zigzag phenomenon occurs in the load-displacement curve shown in Figure 14. Based on the above analysis, it can be indicated that the deformation extracted from the image will be distributed around the actual deformation of the structure. On this end, the load-deformation curve is polynomial fitted to approximate the actual deformation value of the structure, as shown in Figure 16. The measured comparison of three dial meters is shown in Figure 17.

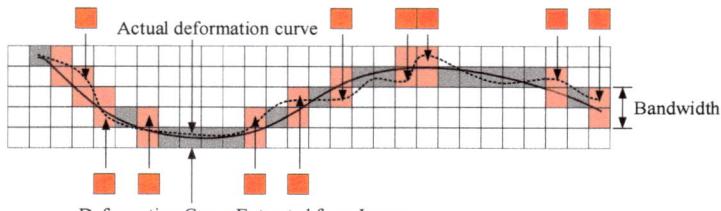

Figure 15. Difference between pixel edge and actual edge.

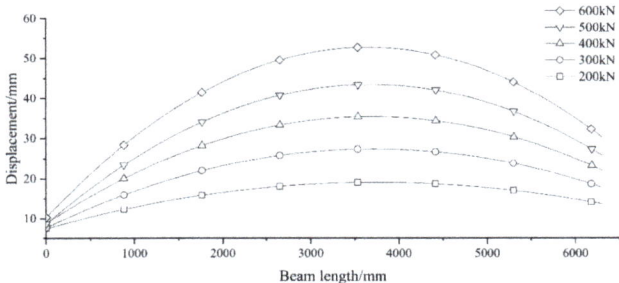

Figure 16. The load-displacement curve after fitting.

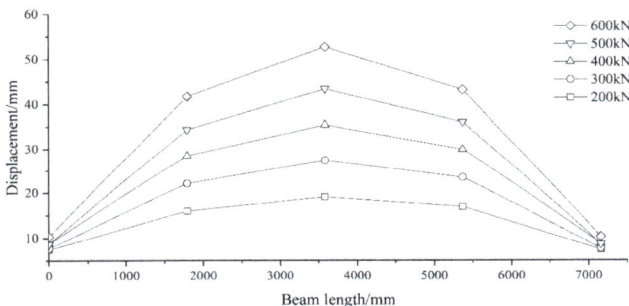

Figure 17. The load-displacement curve measured by dial meters.

4.4. Error Analysis of Photogrammetry

The overall deformation of the structure edge extracted by photogrammetry is compared with the dial gauge as shown in Figure 18. The numerical comparison results are shown in Table 4. From Figure 18 and Table 4, it can be seen that the structural deformation data obtained by photogrammetry are consistent with the data measured by the dial meters, and the maximum error is less than 5%. Compared with the traditional methods, the photogrammetry method has a wider observation range. Thus, the proposed method can obtain the displacement of any section without extensive efforts, which in turn can better reflect the overall displacement and deformation of the structure.

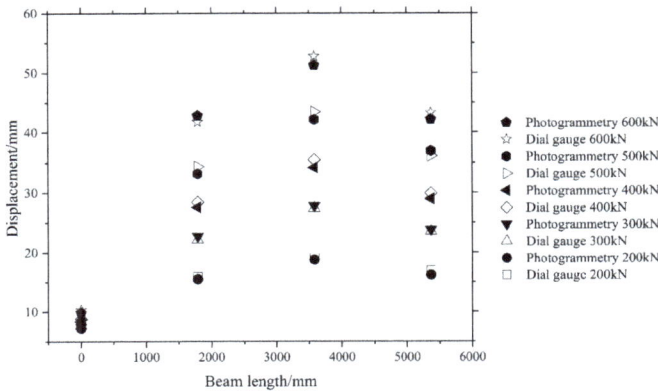

Figure 18. Error comparison chart.

Table 4. Comparison of displacement measurement error.

Loading Condition/kN	Dialgauge Position	Dialgauge Measured Value R1/mm	Photogrammetric Value R2/mm	|R2 − R1| = S/mm	Error S/R1/%	RMSE
200	Left support	7.50	7.25	0.25	3.33	
	L/4	15.95	15.50	0.45	2.82	
	2L/4	19.09	18.75	0.34	1.78	
	3L/4	16.90	16.19	0.71	4.20	
300	Left support	7.74	7.45	0.29	3.75	
	L/4	22.13	22.69	0.56	2.53	
	2L/4	27.37	27.80	0.43	1.57	
	3L/4	23.49	23.78	0.29	1.23	
400	Left support	8.82	8.47	0.35	3.97	0.82
	L/4	28.42	27.56	0.86	3.03	
	2L/4	35.45	34.15	1.30	3.67	
	3L/4	29.89	28.94	0.95	3.18	
500	Left support	8.72	8.36	0.36	4.13	
	L/4	34.34	33.14	1.20	3.49	
	2L/4	43.46	42.18	1.28	2.95	
	3L/4	36.08	36.91	0.83	2.30	
600	Left support	10.33	9.84	0.49	4.74	
	L/4	41.76	42.81	1.05	2.51	
	2L/4	52.75	51.26	1.49	2.82	
	3L/4	43.28	42.19	1.09	2.52	

5. Conclusions

Based on digital image processing technology, this paper presents a novel method for structural overall deformation monitoring. Using the proposed method, a series of deformation measurement experiments are carried out on a steel–concrete composite beam, and the main conclusions are as the following:

(1) Due to the limitation on camera sites, orthogonal projection images are usually difficult to be accessed for large engineering structures such as bridges. In dealing with the issue, the perspective transformation method is applied to acquire the orthogonal projection of structures from the originally inclined images. The experimental results show that the orthogonal projection image obtained by the proposed method can correctly reflect the overall deformation of the structure.

(2) In order to characterize the key feature of structural deformation, the edge detection operator is utilized to obtain the edge contour of the structure from the processed orthogonal images. Using the operator, the overall deflection curve of the structure can be obtained by locating and calibrating the edge pixels.

(3) The edge line of the structure acquired from the position of the pixel shows a notable zigzag effect. Further investigations have been carried out, and the result suggests that the illumination environment can be mainly attributed to the zigzag effect. Since the image edge of the structure has a certain bandwidth, the final position of edge pixels in the bandwidth range will be affected by the illumination environment, which eventually results in the fluctuation around the actual deflection curve. On this end, the fitting method is used to minimize the fluctuation and obtain the linear approximation of the actual deflection curve. After comparison with the data measured by the dial meters, it shows the error of the proposed method is less than 5%.

(4) Since the proposed method is based on digital images, the accuracy is dependent on the quality of available images even if some advanced image processing methods are utilized. For instance, a major limitation of the method is that the overall deformation cannot be directly obtained when

some parts of the measuring structures are obscured. Under such a situation, the postprocessing method, such as the fitting, can be applied to obtain the approximation data of the blocked parts.

Author Contributions: X.C. conceived this study, designed the computational algorithms, wrote the program code, and wrote the manuscript. Z.Z. proposed some valuable suggestions and guided the experiments. G.D., X.D. acquired the test images and performed the experiments. X.J. carried out the measurements and analyzed the experimental data.

Funding: This research was funded by the National Natural Science Foundation Projects (Grant No.: 51778094, 51708068).

Acknowledgments: Special thanks to J.L. Heng at the Southwest Jiaotong University.

Conflicts of Interest: The authors declare no conflict of interest.

References

1. Liu, Z.; Zhang, S.; Cai, S. Experimental Study on Deflection Monitoring Scheme of Steep Gradient and High Drop Bridge. *J. Highw. Transp. Res. Dev.* **2015**, *32*, 88–93.
2. Jiang, T.; Tang, L.; Zhou, Z. Study on Application of Close-Range Photogrammetric 3D Reconstruction in Structural Tests. *Res. Explor. Lab.* **2016**, *35*, 26–29.
3. Heng, J.; Zheng, K.; Kaewunruen, S.; Zhu, J.; Baniotopoulos, C. Probabilistic fatigue assessment of rib-to-deck joints using thickened edge U-ribs. *Steel Compos. Struct.* **2020**, *2*, 23–56.
4. Luhmann, T.; Robson, S.; Kyle, S.; Boehm, J. Close-Range Photogrammetry and 3D Imaging, 2nd Edition. *Photogramm. Eng. Remote Sens.* **2015**, *81*, 273–274.
5. Baltsavias, E.P. A comparison between photogrammetry and laser scanning. *ISPRS J. Photogramm. Remote Sens.* **1999**, *54*, 83–94. [CrossRef]
6. Smith, M.J. Close range photogrammetry and machine vision. *Emp. Surv. Rev.* **2001**, *34*, 276. [CrossRef]
7. Ke, T.; Zhang, Z.X.; Zhang, J.Q. Panning and multi-baseline digital close-range photogrammetry. *Proc. SPIE Int. Soc. Opt. Eng.* **2007**, *34*, 43–44.
8. Feng, D.; Feng, M.Q. Computer vision for SHM of civil infrastructure: From dynamic response measurement to damage detection—A review. *Eng. Struct.* **2018**, *156*, 105–117. [CrossRef]
9. Kromanis, R.; Forbes, C. A Low-Cost Robotic Camera System for Accurate Collection of Structural Response. *Inventions* **2019**, *4*, 47. [CrossRef]
10. Artese, S.; Achilli, V.; Zinno, R. Monitoring of bridges by a laser pointer: Dynamic measurement of support rotations and elastic line displacements: Methodology and first test. *Sensors* **2018**, *18*, 338. [CrossRef]
11. Ghorbani, R.; Matta, F.; Sutton, M.A. Full-field Deformation Measurement and Crack Mapping on Confined Masonry Walls Using Digital Image Correlation. *Exp. Mech.* **2015**, *55*, 227–243. [CrossRef]
12. Wang, G.; Ma, L.; Yang, T. Study and Application of Deformation Monitoring to Tunnel with Amateur Camera. *Chin. J. Rock Mech. Eng.* **2005**, *24*, 5885–5889.
13. Pan, B.; Xie, H.; Dai, F. An Investigation of Sub-pixel Displacements Registration Algorithms in Digital Image Correlation. *Chin. J. Theor. Appl. Mech.* **2007**, *39*, 245–252.
14. Zhang, G.; Yu, C. The Application of Digital Close-Range Photogrammetry in the Deformation Observation of Bridge. *GNSS World China* **2016**, *41*, 91–95.
15. Feng, D.; Feng, M.Q.; Ozer, E.; Fukuda, Y. A Vision-Based Sensor for Noncontact Structural Displacement Measurement. *Sensors* **2015**, *15*, 16557–16575. [CrossRef] [PubMed]
16. Detchev, I.; Habib, A.; El-Badry, M. Case study of beam deformation monitoring using conventional close range photogrammetry. In Proceedings of the ASPRS 2011 Annual Conference, ASPRS, Milwaukee, WI, USA, 1–5 May 2011.
17. Chu, X.; Zhou, Z.; Deng, G. Improved design of fuzzy edge detection algorithm for blurred images. *J. Jilin Univ. Sci. Ed.* **2019**, *57*, 875–881.
18. Chu, X.; Xiang, X.; Zhou, Z. Experimental study of Euler motion amplification algorithm in bridge vibration analysis. *J. Highw. Transp. Res. Dev.* **2019**, *36*, 41–47.
19. Chang, X.; Du, S.; Li, Y.; Fang, S. A Coarse-to-Fine Geometric Scale-Invariant Feature Transform for Large Size High Resolution Satellite Image Registration. *Sensors* **2018**, *18*, 1360. [CrossRef]

20. Jianming, W.; Yiming, M.; Tao, Y. Perspective Transformation Algorithm for Light Field Image. *Laser Optoelectron. Prog.* **2019**, *56*, 151003. [CrossRef]
21. Ren, H.; Zhao, S.; Gruska, J. Edge detection based on single-pixel imaging. *Opt. Express* **2018**, *26*, 5501. [CrossRef]
22. Bao, P.; Zhang, L.; Wu, X. Canny edge detection enhancement by scale multiplication. *IEEE Trans. Pattern Anal. Mach. Intell.* **2005**, *27*, 1485–1490. [CrossRef] [PubMed]
23. Yi-Bin, H.E.; Zeng, Y.J.; Chen, H.X. Research on improved edge extraction algorithm of rectangular piece. *Int. J. Mod. Phys. C* **2018**, *29*, 169–186.
24. Mamedbekov, S.N. Definition Lens Distortion Camera when Photogrammetric Image Processing. *Her. Dagestan State Tech. Univ. Tech. Sci.* **2016**, *35*, 8–13. [CrossRef]
25. Myung-Ho, J.; Hang-Bong, K. Stitching Images with Arbitrary Lens Distortions. *Int. J. Adv. Robot. Syst.* **2014**, *11*, 1–7.
26. Ronda, J.I.; Valdés, A. Geometrical Analysis of Polynomial Lens Distortion Models. *J. Math. Imaging Vis.* **2019**, *61*, 252–268. [CrossRef]
27. Bergamasco, F.; Cosmo, L.; Gasparetto, A. Parameter-Free Lens Distortion Calibration of Central Cameras. In Proceedings of the 2017 IEEE International Conference on Computer Vision (ICCV), Venice, Italy, 22–29 October 2017.
28. Li, J.; Di, W.; Yue, W. A novel method of Brillouin scattering spectrum identification based on Sobel operators in optical fiber sensing system. *Opt. Quantum Electron.* **2018**, *50*, 27. [CrossRef]
29. Bora, D.J. An Efficient Innovative Approach Towards Color Image Enhancement. *Int. J. Inf. Retr. Res.* **2018**, *8*, 20–37. [CrossRef]
30. Barbeiro, S.; Cuesta, E. Cross-Diffusion Systems for Image Processing: I. The Linear Case. *J. Math. Imaging Vis.* **2017**, *58*, 447–467.
31. Cobos, F.; Fernández-Cabrera, L.M.; Kühn, T. On an extreme class of real interpolation spaces. *J. Funct. Anal.* **2009**, *256*, 2321–2366. [CrossRef]
32. Liu, Y.; Cheng, M.M.; Hu, X. Richer Convolutional Features for Edge Detection. *IEEE Trans. Pattern Anal. Mach. Intell.* **2019**, *41*, 1939–1946. [CrossRef]

© 2019 by the authors. Licensee MDPI, Basel, Switzerland. This article is an open access article distributed under the terms and conditions of the Creative Commons Attribution (CC BY) license (http://creativecommons.org/licenses/by/4.0/).

Article

Determination of Axial Force in Tie Rods of Historical Buildings Using the Model-Updating Technique

Ivan Duvnjak, Suzana Ereiz, Domagoj Damjanović and Marko Bartolac *

Faculty of Civil Engineering, University of Zagreb, 10000 Zagreb, Croatia; ivan.duvnjak@grad.unizg.hr (I.D.); suzana.ereiz@grad.unizg.hr (S.E.); domagoj.damjanovic@grad.unizg.hr (D.D.)
* Correspondence: marko.bartolac@grad.unizg.hr

Received: 7 August 2020; Accepted: 28 August 2020; Published: 31 August 2020

Abstract: Tie rods are structural elements that transfer axial tensile loads and are typically used on walls, vaults, arches, and buttresses in historical buildings. To verify their load-bearing capacity and identify possible structural damage risks, the forces transferred by tie rods and the corresponding stresses must be determined. However, this is often a challenging task due to the lack of project documentation for historical buildings. Uncertainties like complex boundary conditions or unknown material and geometrical properties make it hard to assess the tie rods' load level. This paper presents a methodology for the determination of axial forces in tie rods that combines on-site experimental research and a numerical model-updating technique. Along with the common approach based on a determination of the natural frequency of tie rods, this paper presents an approach based on tie rods' mode shapes. Special emphasis is placed on the boundary conditions coefficient, which is a crucial parameter in the analytical solution for axial forces determination based on the conducted on-site experiments. The method is applied in a historical building case study.

Keywords: tie rod; structural health monitoring (SHM); natural frequencies; mode shapes; root-mean-square error (RMSE)

1. Introduction

Displacements that occur with historical buildings can be arrested using metal beams, or tie rods, which support the masonry walls, buttresses, arches, and vaults in the plane of bending out. Tie rods are subjected to axial tension and are an essential element in the control of horizontal forces (displacements) produced by static and dynamic loads related to seismic actions. In extreme cases, a tie rod can reach its maximum bearing capacity due to high stress or the pulling out of its anchor point. Both scenarios can lead to a loss of structural integrity. Therefore, the value of the internal tensile force in tie rods is a frequent subject of discussion. Figure 1 represents a typical tie rod in historically-important buildings such as cathedrals, churches, or castles.

Figure 1. Cathedral of St. James in Šibenik (Croatia). Tie rods are supporting the walls and arches.

Several uncertainties exist when determining the forces in tie rods, including complex boundary conditions [1–3] and geometrical and material properties [3,4]. Boundary conditions can vary, ranging from theoretically-fixed and pinned conditions to those that should be considered with spring elements (Figure 2). In practice, the length of the anchoring of the tie rod is associated with geometrical problems. Due to the limitations of inspection, material properties such as the Young's modulus are often unknown. Evaluation of tensile forces in tie rods can be achieved using static, dynamic, or mixed approaches.

Figure 2. Examples of complex (unclear) boundary conditions.

Two static approaches can be used for the determination of tensile forces in tie rods. The first involves loading of the tie rod in several stages and then measuring deflections and strains in representative locations [5–7]. The second approach is known as the residual stress method [8]. This involves attaching strain gauge rosettes to the surface, drilling a hole at the center of the gauge, and measuring the residual strain caused by the relaxation of the material surrounding the drilled hole. The several disadvantages to static approaches include the following:

- The tie rods are generally located at elevated positions;
- High-accuracy displacement sensors should be used due to small vertical displacements. These should be placed on a previously-determined referent location;
- The strain gauge installation can be complicated at elevated positions.

The aforementioned approaches require considerable time for application and they are slightly destructive. Based on the mentioned disadvantages, many authors have researched a dynamic approach based on numerical or analytical methods [9–11].

Vibration-based methods (dynamic approach), as nondestructive methods, are widely applied for the evaluation of axial forces in tensile elements of structures. These methods are mainly based on tie rods' natural frequencies, which are later used for the determination of forces. A distinction must be made between transverse oscillations of strings and beams. String theory is widely used for the determination of cable forces in structures such as bridges [12] due to the high ratio of length to cross-section dimensions. Unlike beam theory, string theory does not consider the flexural stiffness or boundary conditions [13,14].

Tensile forces in tie rods can also be evaluated using a mixed approach. This approach is a combination of previously-mentioned approaches—static- and vibration-based. In this case, tie rods are modeled as simply supported Euler beams with rotational springs with unknown stiffness on each edge [15]. Besides the stiffness of rotational springs, the second unknown parameter is force. These unknown parameters are obtained by solving a system of equations composed of static equations for deflection and dynamic equations for natural frequencies. This method requires data from two separate experiments. The first experiment involves measuring the deflection in representative locations on a beam and the second involves measuring the natural frequency. Although a mixed approach showed good results in laboratory conditions, measurement errors can cause a significant deviation in results. If the number of unknown parameters is larger, the probability of error is also greater. Although there are two such parameters in this case, they are enough to cause a significant deviation in the results [16–18].

In this study, a combination of experimental and numerical research was used for the determination of forces in tie rods based on the natural frequencies (f_n), flexural stiffness (EI), mass (m), and boundary conditions. The method was verified on a historical building case study. One characteristic tie rod in this building was analyzed in detail. Based on these analyses of the characteristics of a single tie rod and the experimentally-determined values of the natural frequencies on the other tie rods in the building, the force in all the tie rods was determined reliably.

The remainder of this paper is structured as follows: Section 2 presents an analytical solution for lateral vibration of a tie rod and provides an equation for the determination of the tensile force, Section 3 describes the proposed methodology in detail, and Section 4 presents the application of the proposed methodology on a real historical building.

2. Analytical Solution for Lateral Tie Rod Vibration

As discussed above, the natural frequencies of a tie rod depend on three basic assumptions—axial load, stiffness, and boundary conditions. The lateral vibrations of the tie rod are assumed as a superposition of two solutions—the lateral vibrations of the Euler-Bernoulli beam (considering stiffness) and the lateral vibration of string (considering the axial effect). The boundary conditions are considered in the proposed solutions of differential equations.

Figure 3a depicts small-amplitude free lateral vibration with a uniform cross-section of the beam with material density ρ. In the cross section, dx is the acting internal forces (P) with a positive orientation, including the weight of beam caused by vibrations (Figure 3b).

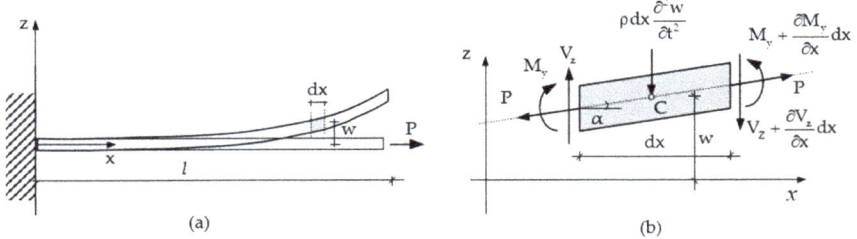

Figure 3. (a) Beam under lateral vibration and axial loading and (b) differential segment of beam representing the positive orientation of bending moments, shear forces, and axial and inertial forces of mass.

Shear forces (V_z) are acting in a vertical direction of an element and are responsible for balancing a weight of segment that varies in time along with element ($\rho dx \partial^2 w/\partial^2 t$). The sum of vertical forces is equal to the product of the mass of the element and acceleration (Equation (1)):

$$\sum z = 0 \rightarrow V_z - \rho dx \frac{\partial^2 w}{\partial t^2} - \left(V_z + \frac{\partial V_z}{\partial x} dx\right) = 0,$$
$$\frac{\partial V_z}{\partial x} = -\rho \frac{\partial^2 w}{\partial t^2}. \tag{1}$$

If the moments are taken about point C of the element dx (Equation (2a)), the term contains vertical forces (V_z and P_z) and axial bending moment (M_y). Substituting the bending moment with flexural stiffness (EI) and taking the second derivative of the deflection of beam ($M_y = -EI \partial^2 w/\partial x^2$) and a component of axial force ($P_z = Ptg\alpha = P \partial w/\partial x$, Equation (2b)) gives Equation (3)

$$\sum M_c = 0 \rightarrow M_y + V_z \tfrac{dx}{2} + P_z \tfrac{dx}{2} - \left(M_y + \tfrac{\partial M_y}{\partial x} dx\right) + \left(V_z + \tfrac{\partial V_z}{\partial x} dx\right)\tfrac{dx}{2} + P_z \tfrac{dx}{2} = 0,$$
$$2V_z \tfrac{dx}{2} + 2P_z \tfrac{dx}{2} - \tfrac{\partial M_y}{\partial x} dx = 0, \tag{2a}$$

$$V_z = \frac{\partial M_y}{\partial x} - P\frac{\partial w}{\partial x}, \tag{2b}$$

$$\frac{\partial V_z}{\partial x} = \frac{\partial^2 M_y}{\partial x^2} - P\frac{\partial^2 w}{\partial x^2}, \tag{2c}$$

$$\frac{\partial V_z}{\partial x} = -EI\frac{\partial^4 w}{\partial x^4} - P\frac{\partial^2 w}{\partial x^2}. \tag{3}$$

Finally, substituting Equation (3) into Equation (1) provides the basic equation for lateral vibration of the beam with inner constant axial force (Equation (4)):

$$EI\frac{\partial^4 w}{\partial x^4} + P\frac{\partial^2 w}{\partial x^2} - \rho\frac{\partial^2 w}{\partial t^2} = 0. \tag{4}$$

The solution of this equation can be expressed as a form of harmonic functions of w(x,t) (Equation (5)) or in terms of exponential functions. For the selected function, constants (A, B, C, and D) should be found considering various boundary conditions with applying known conditions to the deflection, slope, bending moment, and shear forces:

$$w = A(\cos \kappa x + \cosh \kappa x) + B(\cos \kappa x - \cosh \kappa x) \\ + C(\sin \kappa x + \sinh \kappa x) + D(\sin \kappa x - \sinh \kappa x) \tag{5}$$

Ultimately, a natural frequency for the nth mode shape in tie rods can be determined according to [19,20] as

$$f_n = \frac{\kappa^2}{2\pi l^2}\sqrt{\frac{EI}{m'}}\sqrt{1 + \frac{Pl^2}{EI\pi^2 n^2}}, \tag{6}$$

where n is the mode shape number, f_n is the nth natural frequency, l is a span of tie rod, m' is mass per unit length (m'= ρbh, b—width, and h—height of cross section) and κ is a boundary condition parameter, as presented in Table 1. By rearranging the previous equation, the boundary conditions (κ) can be determined:

$$\kappa = \sqrt{\frac{f_n 2\pi l^2}{\sqrt{\frac{EI}{m'} + \frac{Pl^2}{m'\pi^2 n^2}}}} \tag{7}$$

Table 1. Value of boundary condition parameter κ for the first two [19] and nth [20] natural frequencies having various boundary conditions.

Boundary Condition	Static System	Coefficient κ		
		1st Mode	2nd Mode	nth Mode
Hinge–hinge		3.142	6.283	$\cong n\pi$
Clamp–clamp		4.730	7.853	$\cong \frac{(2n+1)}{2}\pi$
Clamp–hinge		3.927	7.069	$\cong \frac{(4n+1)}{4}\pi$
Clamp–free		1.875	4.694	$\cong \frac{(2n-1)}{2}\pi$

Based on known natural frequencies (f_n), properties (EI, m'), and boundary conditions (κ), we can determine the axial force of a tie rod:

$$P = \frac{\pi^4 n^2}{\kappa^4} 4 f_n^2 m' l^2 - \pi^2 n^2 \frac{EI}{l^2}. \tag{8}$$

Although the previous equation is simple, it only considers ideal boundary conditions. Generally, in real life, the boundary conditions are quite complicated and very often nonsymmetrical problems. This is why we assessed the axial forces in tie rods using analytical solutions, experimental research, and numerical analysis.

3. Methodology for Boundary Conditions and Axial Load Identification

The proposed methodology is composed of three stages divided into experimental research on-site and numerical optimization, considering an analytical solution for the determination of axial force (Figure 4). The experimental research (stage 1) involved vibration-based measurement [21] of natural frequencies, f_n^{exp}, and mode shapes determination by using operational modal analysis (OMA) [22,23]. Based on the assumed boundary conditions, known materials, and geometric properties, an initial numerical model of a tie rod was developed (Figure 4, stage 2). Mode shapes from the numerical simulation and experiments were compared using normalized root-mean-square error (RMSE) [24], as presented in Equation (9):

$$\text{RMSE}_n = \frac{\sqrt{\frac{1}{k} \sum_{j=1}^{k} \left(\Phi_{n,j}^{num} - \Phi_{n,j}^{exp} \right)^2}}{\max(\Phi_n)}, \tag{9}$$

where n is the mode shape number, $\Phi_{n,j}^{num}$ is the numerically-obtained normalized mode shape vector, and $\Phi_{n,j}^{exp}$ is the experimentally-obtained operating deflection shape at the jth point on the tie rod. Based on the RMSE values, the numerical model was updated by adapting the boundary conditions (BC). When the RMSE reached its minimum value, it indicated that the experimental and numerical mode shapes were overlapping, which ultimately meant that boundary conditions were updated adequately (Figure 4, stage 2—updated model BC). Finally, based on known natural f_n^{exp}, f_n^{exp}, and a previously-updated numerical model, the axial force was tuned to a numerical model to match the natural frequency, f_n^{num}, with that from an experiment, f_n^{exp} (Figure 4, stage 2—updated model (BC + f_n^{num}). Using this procedure, considering experimental measurements, and updating the numerical model, the axial force (P) was determined (Figure 4, end of stage 2).

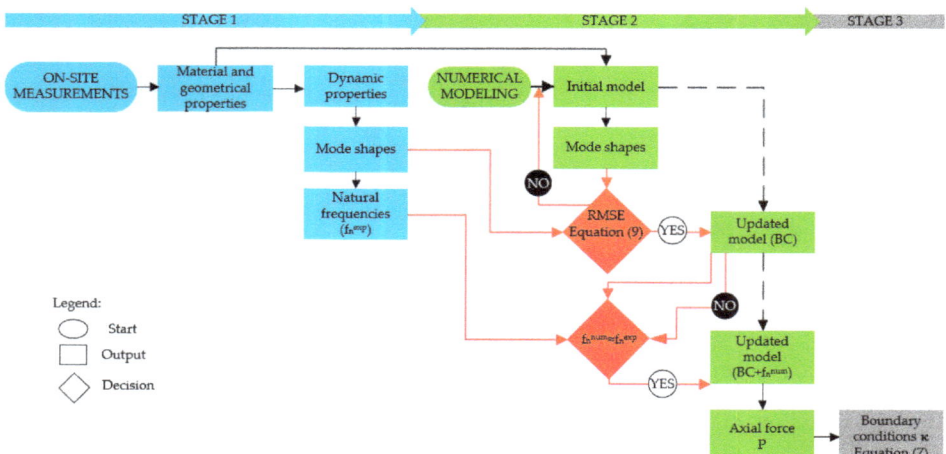

Figure 4. Methodology for boundary conditions determination.

Generally, the simulated boundary conditions in the numerical model are complicated and do not coincide with the basic support conditions given in Table 1. Hence, by applying known geometrical and material properties with the resulting axial force in an analytical solution (Equation (7)), we can determine the coefficient κ, which is associated with the boundary conditions (Figure 4, stage 3). Using this methodology on one characteristic tie rod, a boundary coefficient was determined and this value can then be applied to other tie rods with the same boundary conditions (Figure 5). Therefore, it is sufficient to perform on-site measurements of the natural frequencies in each tie rod to determine the axial force (Figure 5, end of stage 3). In case of varying boundary conditions for tie rods in the building, the numerical model updating should be applied to each tie rod (stages 1 and 2) without determination of coefficient κ.

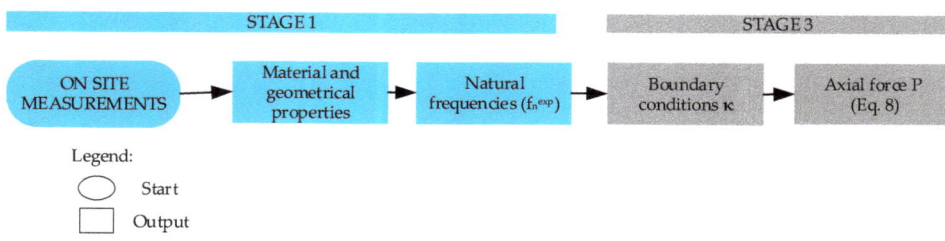

Figure 5. Methodology for axial load identification.

4. Case Study Using the Proposed Methodology

The previously-described methodology (Figures 4 and 5) was applied to a historical building case study. The Cathedral of St. James in Šibenik (Figure 6), Croatia, is a good example of a historical building with defined dynamic tie rod parameters (frequencies and mode shapes). For the sake of simplicity, one of the tested iron tie rods was taken as a reference for which we conducted a detailed analysis.

(a) (b)

Figure 6. Cathedral of St. James in Šibenik, Croatia. (**a**) West view of cathedral and (**b**) iron tie rods inside the cathedral.

4.1. Description of Structure

The Cathedral of St. James in Šibenik, Croatia, is a unique architectural work whose construction began in 1431 and lasted almost 100 years. The cathedral was built exclusively of stone and marble, without brick, wood, or concrete as binders. The cathedral is shaped like a Latin cross, divided by arcades into a three-nave structure. The medium-sized nave, which is tall and illuminated, holds 12 Gothic columns, whereas the lateral naves are darker and lower. Data regarding the geometry of the cathedral can be found in a previous study [25]. In 2001, the Cathedral of St. James was given UNESCO World Heritage status.

The system of tie rods in the Cathedral of St. James in Šibenik consists of iron and aluminum tie rods set in the three levels of the cathedral. At the abutment of the central barrel vault, nine iron tie rods are connected at level R4 (Figure 7). The columns and vaults of the gallery are connected by 24 aluminum tie rods shorter than the tie rods of the central barrel vault. Aluminum tie rods are placed in the transverse direction (axis 1–7, from A to B and from C to D) and in the longitudinal direction (axis B and C, from 1 to 7) at the R2 level (Figure 7). The iron tie rods were the focus of the research.

4.2. Experimental Identification of Dynamic Properties of Tie Rods

As a part of stage 1 of the proposed methodology, dynamic properties of tie rods were experimentally observed by using the operational modal analysis (OMA) method. Unlike the classical experimental modal analysis (EMA), the OMA method does not require known excitation. The reference [26] reports negligible differences between the dynamic parameters of simple structures obtained by these two methods. The obtained data (natural frequency, mode shapes, and damping coefficient) were used to improve the numerical model of the tie rods to estimate the tensile forces based on natural frequency. During the experiment, irregularities in the geometry of tie rods, their lengths, and the length of the anchoring to the wall were observed in the north and south galleries (Figure 2). Figure 8 shows the reference tie rod, 6B-C at level R4, with the geometric properties and arrangement of measuring points. As such, a total of nine measurement points were obtained for each tie rod, and the two endpoints were assumed to be fixed. For the purpose of determining the boundary conditions, the measuring points at the ends of the tie rods were more densely distributed.

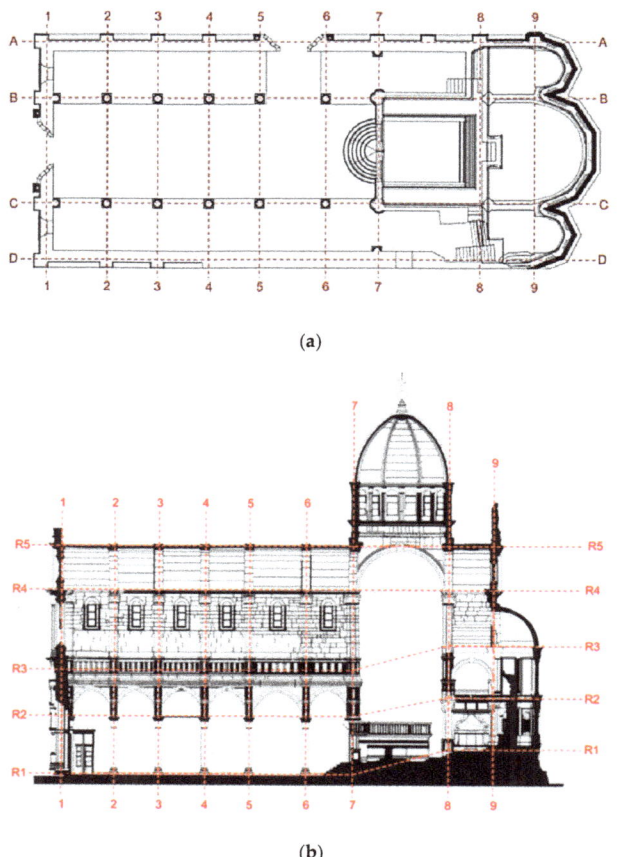

(a)

(b)

Figure 7. Cathedral of St. James in Šibenik. (**a**) Ground floor and (**b**) cross-section.

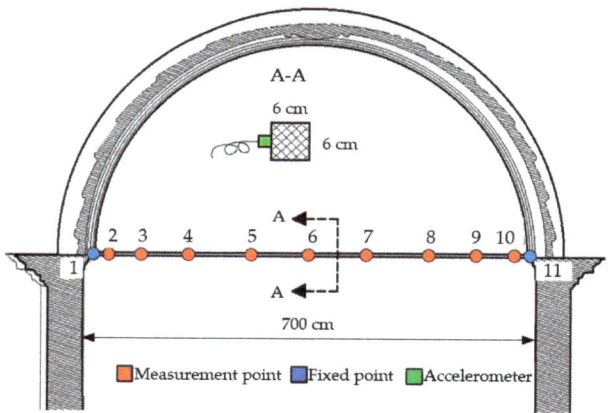

Figure 8. Reference tie rod (6B-C, R4) with measuring points and dimensions.

For simplicity, tie rod 6B-C was taken as the reference, and a more detailed analysis was performed on it, as follows: The on-site measurements were performed using five piezoelectric accelerometers (Brüell & Kjaer, Denmark, type 4508-B, nominal sensitivity 100 mV/g). The OMA was conducted by roving four accelerometers through two measuring stages, using one as a referent. The tie rods were excited by randomly using a rubber impact hammer. Frequency domain decomposition (FDD) and enhanced frequency domain decomposition (EFDD) were used for the estimation of modal parameters (Figure 9). The values of the experimentally-obtained frequencies for the concerned mode shapes were read from the characteristic record (Figure 10).

Figure 9. The first two experimentally-obtained mode shapes of the reference iron tie rod 6B-C.

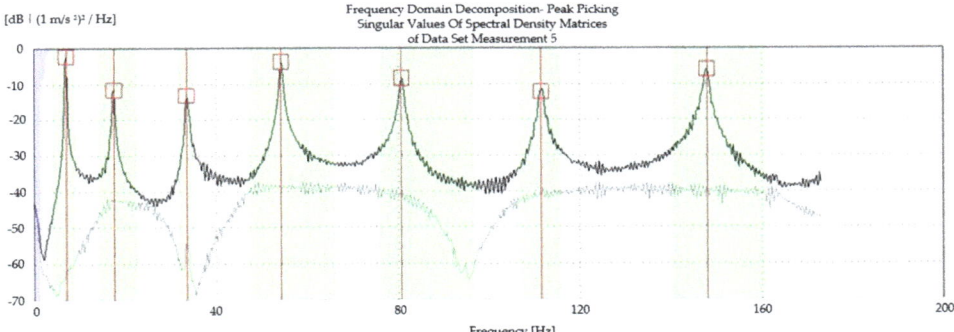

Figure 10. Characteristic record of frequency domain decomposition (FDD) for the determination of natural frequencies on reference iron tie rod 6B-C.

The natural frequencies of the remaining tie rods were measured using only one accelerometer placed at the node 4 (Figure 8). Geometric properties (L—length, b—width, and h—height) and the first two experimentally-obtained natural frequencies (f_n^{exp}, n = 1, 2) for the observed tie rods are presented in Table 2. Please note that the cross-section dimensions present an average value of five measurements performed in quarters of tie rods' length.

Table 2. Values of the first two frequencies of experimentally-observed tie rods in the Cathedral of St. James in Šibenik at level R4.

Tie Rod	L (m)	h (mm)	b (mm)	f_1^{exp} (Hz)	f_2^{exp} (Hz)
2B-C	6.84	55	55	7.25	17.94
3B-C	6.71	64	64	7.56	19.00
4B-C	6.81	60	60	7.31	18.69
5B-C	6.87	68	68	7.31	19.56
6B-C	6.90	61	61	6.94	17.50
7B-C	6.95	56	56	8.25	18.88
7-8B	6.97	56	56	8.13	19.38
7-8C	6.98	60	60	8.63	19.38

4.3. Numerical Simulation

The numerical model of the tie rod was developed as the Euler-Bernoulli beam element in SAP2000 software (Computers and Structures, Walnut Creek, CA, USA). The geometrical characteristics of tie rods—width and height—were assumed to be constant over their entire length and the rods were assumed to be homogenous throughout their volume. The material properties were based on literature review and taken as invariable; the elastic modulus E was considered as 185 GPa, the Poisson coefficient ν as 0.3 and material density ρ as 7850 kg/m^3. Once the modal parameters were experimentally obtained, a numerical model was updated while changing the boundary conditions from fixed to hinged. To construct a real-life model, boundary conditions were additionally tuned with spring coefficients (stage 2 of the proposed methodology).

4.3.1. Initial Model

The initial numerical model was constructed to determine the deviation of the real state of the structure from the ideal boundary conditions (clamp and hinge) and to evaluate which of those boundary conditions better described the real state. The finite element (FE) distribution of the numerical model corresponded to the arrangement of the measuring points shown in Figure 8. For each initial model, the first two natural frequencies and mode shapes were determined (Figure 11).

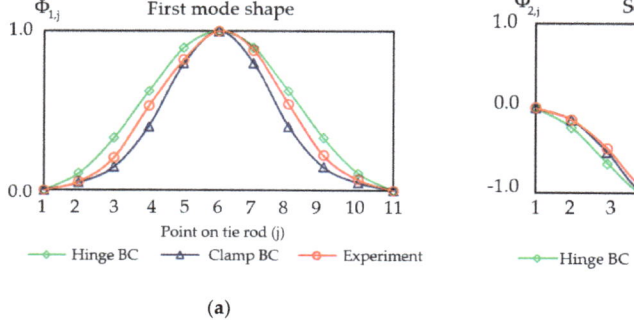
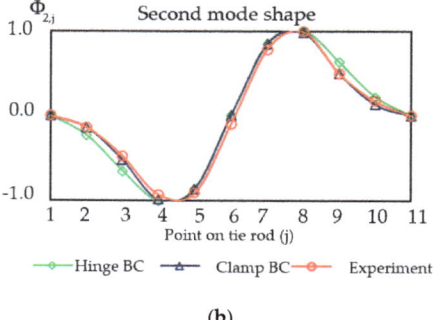

(a) (b)

Figure 11. Comparison of the experimental mode shapes of the reference tie rod 6B-C and the mode shapes associated initial numerical models for the (**a**) first and (**b**) second mode shapes.

To avoid subjective assessment of the boundary conditions, mode shapes from numerical simulation and experiments were compared using normalized RMSE (Equation (9)). For each initial numerical model and its corresponding mode shapes, the RMSE values were determined. Based on the RMSE values (Table 3), we observed that the initial model with hinged boundary conditions better correlated (values approaching 0) with the on-site measurements. Based on the results, hinged boundary conditions were identified as a good selection for the initial model and were further updated by adapting the spring stiffness on the boundary.

Table 3. Root-mean-square error (RMSE) values for initial numerical models for two mode shapes.

	RMSE Values	
Mode Number	Hinge–Hinge	Clamp–Clamp
1	7.06	9.15
2	8.50	10.66

4.3.2. Updated Model

When updating the numerical model of tie rods, rotational springs were added at each boundary, starting from the initial hinged model. Spring stiffness was determined by an iterative procedure. Their values were changed until the mode shapes of the numerical model entirely corresponded to the mode shapes obtained by on-site measurements. During the iteration, the change in the RMSE value was monitored depending on the change of the spring stiffness (k) (Figure 12). Through several iteration steps, due to the minimum value of RMSE, it was possible to select the appropriate spring stiffness for each mode shape. In this way, the finite element (FE) model was updated to overlap with the experimental model considering only the boundary conditions.

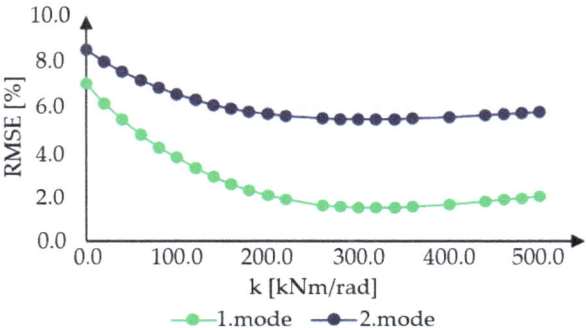

Figure 12. The values of the RMSE for the observed mode shapes and different spring stiffness values.

Table 4 shows the natural frequencies computed by SAP2000 (f_n^{num}) and the relative error between these frequencies and the frequency values determined by on-site measurement (f_n^{exp}). The error varied between 20% and 30% for the first two frequencies, which confirmed the assumptions about the axial force in the tie rods.

Table 4. Values of experimentally-measured natural frequencies and computed frequencies (f_n^{num}) for the updated numerical model.

Mode Number	f_n^{num} (Hz)	f_n^{exp} (Hz)	Relative Error (%)
1	4.83	6.94	30.20
2	13.95	17.50	20.29

Therefore, the next step was to adjust the natural frequencies of the numerical model to correspond to those obtained from on-site measurements. The natural frequencies were adjusted by changing the tension in the FE model of the tie rod. In addition, the procedure of tuning natural frequencies is iterative, and it was implemented for the first two mode shapes. The values of experimentally-measured natural frequencies (f_n^{exp}) were compared to the numerical model (f_n^{num}) and are presented in Figure 13. After tuning the tension for both mode shapes (P_1, P_2) in the FE model, the results of natural frequencies were in good agreement with those measured for both mode shapes (Figure 13). The tuning of dynamic properties of the FE model enabled us to determine the actual tension in the tie rod (Table 4).

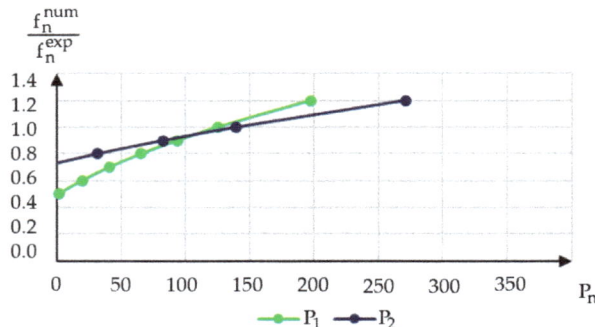

Figure 13. Change of force values depending on the ratio of numerical and experimental frequency values for two mode shapes (P_1, P_2).

Based on the proposed methodology, we continued to the final stage (stage 3) of determination of axial force and coefficient κ. With known axial force (from the updated numerical model) and geometrical and material properties (from on-site observations), coefficient κ could be determined by using Equation (7). As expected, the values of coefficient κ (Table 5) obtained for both mode shapes were within the theoretical boundaries of the hinge–hinge and clamp–clamp conditions previously presented in Table 1. Based on the force values obtained from two mode shapes indicated in Table 5, one can observe that there is a difference of around 12% between these values. It can be concluded that the force value varies depending on the number of the observed mode shapes as indicated in [27].

Table 5. The force values read for the first two mode shapes and the value of coefficient κ determined by the calculation of the updated model.

Mode Number	f_n^{num} (Hz)	P (kN)	σ (MPa)	κ
1	6.94	122.8	33.0	3.534
2	17.5	137.2	36.9	6.777

We assumed that the coefficient κ found by using this procedure could be applied to all remaining tie rods. The stated assumption was based on the on-site visual inspection and the observed geometrical and material properties.

Based on the defined coefficient κ, experimentally-determined natural frequencies, and analytical equations, the values of the axial forces and the stress levels (Table 6) were determined for the observed iron tie rods in the Cathedral of St. James in Šibenik, Croatia.

Table 6. The value of forces P_n (kN) and stress levels $σ_n$ (MPa) for the observed mode shapes in the tie rods observed in the Cathedral of St. James in Šibenik at level R4.

		2B-C		3B-C		4B-C		5B-C	
Mode Num.	κ	P_n (kN)	$σ_n$ (MPa)	P_n (kN)	$σ_n$ (MPa)	P_n (kN)	$σ_n$ (MPa)	P_n (kN)	$σ_n$ (MPa)
1	3.534	115.8	38.3	149.6	36.5	132.1	36.7	159.4	34.5
2	6.777	144.7	47.8	158.8	38.8	167.7	46.6	207.9	45.0
Mean values		130.3	43.1	154.2	37.7	149.9	41.6	183.6	39.7
		6B-C		7B-C		7-8B		7-8C	
Mode Num.	κ	P_n (kN)	$σ_n$ (MPa)	P_n (kN)	$σ_n$ (MPa)	P_n (kN)	$σ_n$ (MPa)	P_n (kN)	$σ_n$ (MPa)
1	3.534	122.8	33.0	170.8	54.5	166.3	53.0	215.2	59.8
2	6.777	137.2	36.9	188.7	60.2	208.1	66.4	219.6	61.0
Mean values		130.0	34.9	179.8	57.3	187.2	59.7	217.4	60.4

The shown stress levels were determined as the arithmetic values of the stress determined for the first two mode shapes (Table 6). Based on Figure 14, we concluded that the stress levels in the observed tie rods at level R4 in the Cathedral of St. James are far below the yield strength. Similar values were presented in study [28].

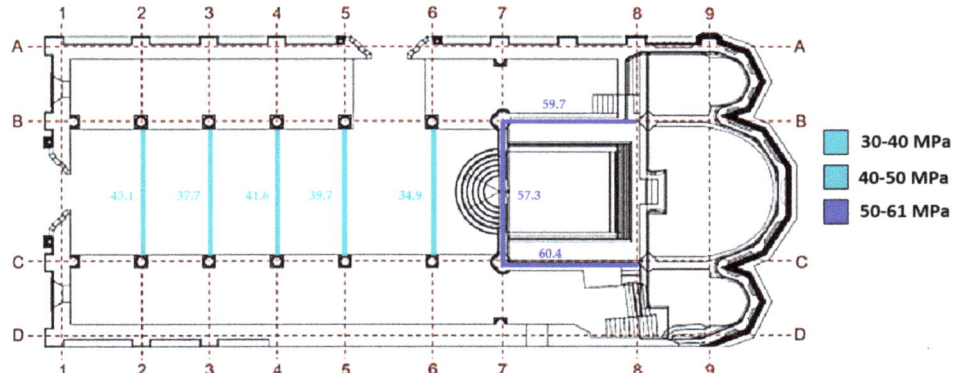

Figure 14. Stress levels measured on the tie rods in the Cathedral of St. James in Šibenik at level R4.

5. Conclusions

Tie rods are used in historical buildings to prevent horizontal displacement and resulting structural damage to elements like walls, buttresses, arches, and vaults. In their usual application, tie rods transfer axial tensile loads. Due to the importance of preventing damage to historical buildings, determining these loads and the corresponding stress levels is crucial, and can be achieved using several approaches based on static, dynamic, and mixed methods. This paper presented a methodology for the determination of axial forces in tie rods of historical buildings based on the model-updating technique. This methodology is composed of three stages. The first stage involves an on-site experimental study. In addition to the determination of the geometrical and material properties of the selected tie rod, this stage results in the determination of its natural frequencies and mode shapes. In the second stage, an initial numerical model is developed based on the experimentally-determined geometrical and material properties and the assumed boundary conditions. The mode shapes from the numerical model and the experiment are then compared by using the root-mean-square error. A minimum value of error indicates that the numerical and experimental findings overlap and the boundary conditions are adequately updated. Axial forces are then tuned in the numerical model to match the experimentally-determined natural frequencies from the first stage. In the third stage, the geometrical and material properties are combined with the tuned axial force into an analytical solution, resulting in the determination of the boundary conditions coefficient κ. This coefficient can then be applied to other tie rods with similar geometrical and material characteristics in their respective historical buildings without repeating the second stage of the methodology. The outlined methodology was verified using a historical building case study. Notably, the boundary coefficient κ indicated in this paper is valid only for the analyzed building. Therefore, the values obtained for this coefficient are not universal and the tie rods of each building should be investigated individually based on the proposed methodology. The importance of the proposed methodology lies in its nondestructive nature, a very important feature in case of historical buildings. Also, the proposed method is relatively simple and quick to implement on site. We considered two mode shapes in the determination of tie rod axial load. To investigate the possibility of axial load determination from higher-order mode shapes, we recommend applying a denser disposition of acceleration measurement positions.

Author Contributions: Conceptualization, I.D., D.D., and M.B.; methodology, I.D and D.D.; formal analysis, S.E. and I.D.; investigation, I.D. and D.D.; writing—original draft preparation, S.E.; writing—review and editing, S.E. and M.B.; visualization, I.D.; supervision, I.D. and D.D.; project administration, I.D.; funding acquisition, I.D., D.D., and M.B. All authors have read and agreed to the published version of the manuscript.

Funding: This research was funded by the European Union through the European Regional Development Fund's Competitiveness and Cohesion Operational Program, grant number KK.01.1.1.04.0041, project "Autonomous System for Assessment and Prediction of infrastructure integrity (ASAP)." The APC was funded by the authors' affiliation.

Conflicts of Interest: The authors declare no conflict of interest. The funders had no role in the design of the study; in the collection, analyses, or interpretation of data; in the writing of the manuscript, or in the decision to publish the results.

References

1. Li, S.; Reynders, E.; Maes, K.; De Roeck, G. Vibration-based estimation of axial force for a beam member with uncertain boundary conditions. *J. Sound Vib.* **2013**, *332*, 795–806. [CrossRef]
2. Li, D.S.; Yuan, Y.Q.; Li, K.P.; Li, H.N. Experimental axial force identification based on modified Timoshenko beam theory. *Struct. Monit. Maint.* **2017**, *4*, 153–173.
3. Li, S.; Josa, I.; Cavero, E. Post Earthquake Evaluation of Axial Forces and Boundary Conditions for High-Tension Bars. In Proceedings of the 16th World Conference of Earthquake, Santiago, Chile, 3–9 January 2017.
4. Calderini, C.; Piccardo, P.; Vecchiattini, R. Experimental Characterization of Ancient Metal Tie-Rods in Historic Masonry Buildings. *Int. J. Arch. Herit.* **2019**, *13*, 416–428. [CrossRef]
5. Briccoli, S.B.; Ugo, T. Experimental Methods for Estimating in Situ Tensile Force in Tie-Rods. *J. Eng. Mech.* **2001**, *127*, 1275–1283.
6. Tullini, N.; Rebecchi, G.; Laudiero, F. Bending tests to estimate the axial force in tie-rods. *Mech. Res. Commun.* **2012**, *44*, 57–64. [CrossRef]
7. Tullini, N. Bending tests to estimate the axial force in slender beams with unknown boundary conditions. *Mech. Res. Commun.* **2013**, *53*, 15–23. [CrossRef]
8. Duvnjak, I.; Damjanović, D.; Krolo, J. Structural health monitoring of cultural heritage structures: Applications on Peristyle of Diocletian's palace in Split. In Proceedings of the 8th European Workshop on Structural Health Monitoring, Bilbao, Spain, 5–8 July 2016.
9. Collini, L.; Garziera, R.; Riabova, K. Vibration Analysis for Monitoring of Ancient Tie-Rods. *Shock. Vib.* **2017**, *2017*, 7591749. [CrossRef]
10. Lai, J.-W.; Mahin, S. Strongback System: A Way to Reduce Damage Concentration in Steel-Braced Frames. *J. Struct. Eng.* **2015**, *141*, 04014223. [CrossRef]
11. Lagomarsino, S.; Calderini, C. The dynamical identification of the tensile force in ancient tie-rods. *Eng. Struct.* **2005**, *27*, 846–856. [CrossRef]
12. Casas, J. A Combined Method for Measuring Cable Forces: The Cable-Stayed Alamillo Bridge, Spain. *Struct. Eng. Int.* **1994**, *4*, 235–240. [CrossRef]
13. Geier, R.; De Roeck, G.; Flesch, R. Accurate cable force determination using ambient vibration measurements. *Struct. Infrastruct. Eng.* **2006**, *2*, 43–52. [CrossRef]
14. Irawan, R.; Priyosulistyo, H.; Suhendro, B. Evaluation of Forces on a Steel Truss Structure Using Modified Resonance Frequency. *Procedia Eng.* **2014**, *95*, 196–203. [CrossRef]
15. Gentilini, C.; Marzani, A.; Mazzotti, M. Nondestructive characterization of tie-rods by means of dynamic testing, added masses and genetic algorithms. *J. Sound Vib.* **2013**, *332*, 76–101. [CrossRef]
16. Blasi, C.; Sorace, S. Determining the Axial Force in Metallic Rods. *Struct. Eng. Int.* **1994**, *4*, 241–246. [CrossRef]
17. Sorace, S. Parameter Models for Estimating In-Situ Tensile Force in Tie-Rods. *J. Eng. Mech.* **1996**, *122*, 818–825. [CrossRef]
18. Vasić, M. A Multidisciplinary Approach for the Structural Assessment of Historical Construction with Tie-Rods. Ph.D. Thesis, Politecnico di Milano, Milano, Italy, 2015.
19. Stokey, W.F. Vibration of systems having distributed mass and elasticity. In *Harris' Shock and Vibration Handbook*, 5th ed.; Harris, C.M., Piersol, A.G., Eds.; McGraw-Hill1: New York, NY, USA, 2002; Volume 15, pp. 238–287.

20. Nugroho, G.; Priyosulistyo, H.; Suhendro, B. Evaluation of Tension Force Using Vibration Technique Related to String and Beam Theory to Ratio of Moment of Inertia to Span. *Procedia Eng.* **2014**, *95*, 225–231. [CrossRef]
21. Rak, M.; Krolo, J.; Herceg, L.; Čalogović, V.; Šimunić, Ž. Monitoring for special civil engineering facilities. *Građevinar* **2010**, *62*, 897–904.
22. Zhang, L.; Brincker, R.P. Andersen, An overview of major developments and issues in modal identification. In Proceedings of the IMAC XXII: A Conference on Structural Dynamics, Dearborn, MI, USA, 26–29 January 2004; pp. 1–8.
23. Cakir, F. Determination of dynamic parameters of double- layered brick arches. *Građevinar* **2015**, *67*, 123–130.
24. Bakhshizade, A.; Ashory, M.R. Root mean square error criterion using operational deflection shape curvature for structural damage detection. *Math. Models Eng.* **2015**, *1*, 96–101.
25. Gelo, D.; Meštrović, M. Discrete dome model for St. Jacob cathedral in Šibenik. *Građevinar* **2016**, *68*, 687–696.
26. Orlowitz, E.; Brandt, A. Comparison of experimental and operational modal analysis on a laboratory test plate. *Measurement* **2017**, *102*, 121–130. [CrossRef]
27. Amabili, M.; Carra, S.; Collini, L.; Garziera, R.; Panno, A. Estimation of tensile force in tie-rods using a frequency-based identification method. *J. Sound Vib.* **2010**, *329*, 2057–2067. [CrossRef]
28. Gentile, C.; Poggi, C.; Ruccolo, A.; Vasic, M. Dynamic assessment of the axial force in the tie-rods of the Milan Cathedral. *Procedia Eng.* **2017**, *199*, 3362–3367. [CrossRef]

© 2020 by the authors. Licensee MDPI, Basel, Switzerland. This article is an open access article distributed under the terms and conditions of the Creative Commons Attribution (CC BY) license (http://creativecommons.org/licenses/by/4.0/).

Article

TLS and GB-RAR Measurements of Vibration Frequencies and Oscillation Amplitudes of Tall Structures: An Application to Wind Towers

Serena Artese [1,2,*] and Giovanni Nico [3,4]

1. Department of Civil Engineering, University of Calabria, 87036 Rende (CS), Italy
2. Spring Research s.r.l., University of Calabria, 87036 Rende (CS), Italy
3. Istituto per le Applicazioni del Calcolo, Consiglio Nazionale delle Ricerche, 70126 Bari, Italy; g.nico@ba.iac.cnr.it
4. Institute of Earth Sciences, Saint Petersburg State University, 199034 Saint Petersburg, Russia
* Correspondence: serena.artese@unical.it

Received: 5 March 2020; Accepted: 23 March 2020; Published: 25 March 2020

Abstract: This article presents a methodology for the monitoring of tall structures based on the joint use of a terrestrial laser scanner (TLS), configured in line scanner mode, and a ground-based real aperture radar (GB-RAR) interferometer. The methodology provides both natural frequencies and oscillation amplitudes of tall structures. Acquisitions of the surface of the tall structure are performed by the TLS with a high sampling rate: each line scan provides an instantaneous longitudinal section. By interpolating the points of each line, oscillation profiles are estimated with a much better precision than each single point. The amplitude and frequency of the main oscillation mode of the whole structure are derived from the TLS profiles. GB-RAR measurements are used to measure the vibration frequencies of higher oscillation modes which are not caught by the TLS due its lower precision in the measurement of displacements. In contrast, the high spatial resolution of TLS measurements provides an accurate description of oscillation amplitude along the tower, which cannot be caught by the GB-RAR, due to its poorer spatial resolution. TLS and GB-RAR acquisitions are simultaneous. The comparison with the analytical solution for oscillation modes demonstrates that the proposed methodology can provide useful information for structural health monitoring (SHM). The methodology does not require the use of targets on the structure and it can be applied during its normal use, even in presence of dynamic loads (wind, traffic vibrations, etc.). A test was carried out on a wind tower where the synergistic use of TLS and GB-RAR made it possible to fully describe the spectral properties of the tower and at the same time measure the amplitude of the first oscillation mode along the tower with a high spatial resolution.

Keywords: terrestrial laser scanner (TLS); ground-based real aperture radar (GB-RAR); line scanner; vibration frequency; spectral analysis; displacement; structural health monitoring (SHM)

1. Introduction

In the last decade, a growing sensitivity has developed towards problems related to the integrity of structures, both in the construction sector and in that of road and railway infrastructures, as well as in that of industrial plants and products.

To check the functionality and health of these structures, natural vibration frequencies and oscillation amplitudes are among the most important parameters. This is even truer for slender or tall structures, such as very tall buildings, piles and spars of bridges and towers. For this kind of structure, it is mandatory to know the behavior when dynamic loads are applied (e.g., strong winds, vibrations due to traffic, earthquake-induced shaking). The ever more widespread presence of tall structures in urbanized and even densely inhabited areas makes their control more useful, which also helps avoid risks to the safety of people. For this reason, the development of simple methodologies for the quick inspection of slender or tall structures is of interest for SHM.

Wind towers are high cantilever structures, which are subject at one end to strong horizontal loads and cyclical stresses due to the rotors; these stresses are reflected in particular on the foundations and on the foundation soil. Thus, for these structures, the control of the frequencies is of fundamental importance.

Several techniques have been adopted for SHM [1]. Among these, techniques based on the measurement of the structural vibration are increasingly used; a review of vibration-based methods can be found in [2,3]. Among the various techniques proposed for measuring the frequencies of structures, those based on the acquisition of accelerometers and velocimeters have been widely used for decades [4,5]. The use of micro-electro-mechanical systems (MEMS)-based sensors and wireless connected sensors is growing, in order to set up even very dense control networks [6,7]. Global navigation satellite system (GNSS) receivers offer the possibility of extracting vibration frequencies [8]; as for the aforementioned techniques, the limit of GNSS is the point-like property of its measurements; moreover, it is necessary to position the GNSS receiver on the point to be monitored. Total stations (TS) are also used, thanks to their increased sampling rate [9] and the possibility of performing long-range monitoring with high precision using appropriate atmospheric correction techniques [10]. Another non-contact technique, continuously growing thanks to the increasing resolution of charge-coupled devices (CCD) and complementary metal–oxide–semiconductor (CMOS) sensors, and to the high frame rate of the most recent cameras, is based on photogrammetry and image processing [11,12].

Ground-based real aperture radar (GB-RAR) offers the possibility to monitor the dynamic characteristics of the structures while ensuring, at the same time, precise measurements of vibration frequencies and displacements. This long-range, non-contact methodology is even more widely adopted for monitoring large structures, for post-disaster interventions and for SHM [13–18].

TLS is a recent technique, still in the evolution phase, used in a wide range of applications and in particular for structural monitoring. Its strength lies in its ability to acquire a very large number of points in a short time, thus making it possible to survey even very complex objects, and to obtain detailed 3D models, thanks to point cloud processing software that is increasingly evolved. TLS, usually mounted on a tripod in fixed positions, is mainly used to monitor the deformations of the structures by measuring the differences between the surfaces scanned at different times [19]. In the last decade, several TLS applications can be counted for the monitoring of dynamic phenomena and the measurement of structure vibrations. Kim and Kim [20] used a Riegl VZ-400® to perform dynamic displacement measurement. The smoothing of the acquired data was performed by using a kriging approach. Two experimental tests were conducted in laboratory on a small object and under ideal conditions. Neitzel et al. [21] showed a comparison between a TLS and a sensor system based on low cost MEMS accelerometers for measuring the vibrations of a bridge. They used a Zoller + Froehlich Imager 5003® in 1D mode; thus, measurements were performed on a single point. A GB-RAR was used for reference purposes. In this case, the accelerometers showed a better accuracy than TLS.

Schill and Eichhorn investigated the movements and frequencies of two wind towers. For each test, they used two Zoller+Fröhlich® 9012 profilers, with sight lines on the rotor plane and perpendicular to it [22].

In this work, a technique is proposed for the measurement of natural vibration frequencies and oscillation amplitudes of structures, based on the TLS acquisition of scan lines and GB-RAR interferometry. The technique used to process the TLS acquisitions, in order to obtain a better precision than the instrumental one, is explained. A test is carried out on a wind tower and data is acquired both during normal activity and during the deactivation phase of the wind turbine. The frequencies and the oscillation amplitudes obtained from the processing of the collected data are compared with the theoretical ones, resulting from an analytical solution. A comparison is made with the results obtained by GB-RAR, which acquires data simultaneously with the TLS. The results obtained, and the comparison with the theoretical solution and with GB-RAR, are shown also through figures and tables and the synergistic use of the two instruments is discussed.

The structure of the article is as follows: Section 2 introduces the TLS technique in line scanner mode and the technique used for data processing, in order to optimize the results, along with a brief introduction to the GB-RAR technique. Section 3 shows the test results. Section 4 provides discussion of the results. Finally, in Section 5, some conclusions are drawn.

2. Materials and Methods

This section introduces the basic principles of TLS and GB-RAR techniques and describes the proposed methodology to measure the vibration frequencies and the oscillation amplitudes of a tall structure, with particular reference to wind towers.

2.1. TLS Basic Principles

TLS is a non-destructive testing (NDT) technique based on the emission of a laser beam of known direction, used to quickly measure the position of points on the surface of a surveyed object, in a reference system with the origin in the center of the instrument. Two principles are adopted to measure distances: time-of-flight (ToF) and phase shift. In our case, a TLS ToF is used. Figure 1 illustrates the operating principle of the ToF. A short pulse is emitted by the TLS through a photodiode. The distance is obtained through the ToF of the emitted impulse, which is the time interval Δt taken by the impulse to reach a target and go back. Then, the distance d from TLS to target is given by Equation (1):

$$d = \frac{c\Delta t}{2} \tag{1}$$

where c is the speed of light.

As for the direction of the laser beam, this is obtained by the scanning system, usually consisting of a multi-facet mirror that rotates around a horizontal axis for vertical scanning, while the entire instrument rotates around a vertical axis through a motor, like the alidade of a motorized TS. In the case of a line scanner configuration, rotation around the vertical axis is disabled. The maximum range can reach 6000 m, while the attainable precision for single points varies from 1 to 20 mm. Better values are obtained for shorter distances [23].

For our application, we made use of a TLS Riegl VZ-1000®, RIEGL Laser Measurement Systems GmbH, Horn, Austria, whose characteristics are summarized in Table 1.

Table 1. Characteristics of TLS Riegl VZ-1000®.

Accuracy of Single Point	± 8 mm
Precision of single point	± 5 mm
Range	from 1 m to 1400 m.
Sampling frequency	until 122.000 points/sec.
Field of view	100° (Vertical +60° −40°)/360° (Horizontal).
Scan Speed	3 lines/sec to 120 lines/sec
Angular Stepwidth	0.0024° to 0.288° between consecutive shoots
Angle Measurement Resolution	0.0005° (1.8 arcsec)
Timestamp	Yes

The Riegl VZ 1000® TLS is a versatile instrument. Along with the possibility to operate in line scan mode, it makes it possible to perform long-range surveys and, therefore, to obtain 3D models of objects and of the surrounding environment.

When line scan mode is selected, the laser beam describes a line repeatedly, being addressed only by the multi-faceted mirror (see Figure 1). Using the maximum scanning speed of 120 lines per second, it would be theoretically possible to measure vibration frequencies up to 60 Hz. Taking into account the precision of measurement, it is necessary to consider a strong noise in the time series of samples, which would make it very difficult to measure oscillations with a peak-to-peak difference less than 1 cm. This limit can be overcome if information on the geometry of the surveyed object is used.

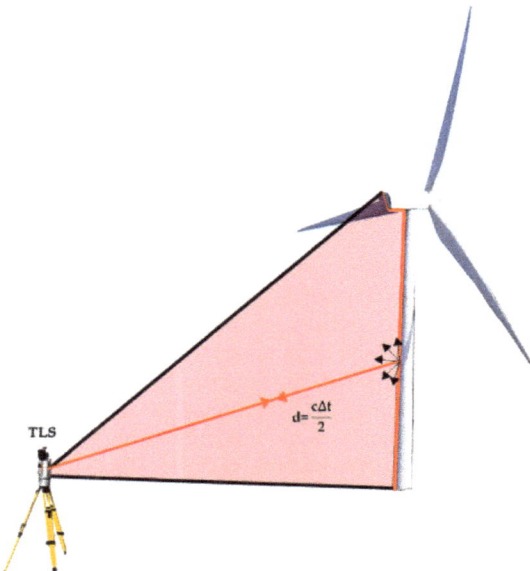

Figure 1. Sketch of the TLS functioning in line scan mode. The distance d is computed in Equation (1).

2.2. GB-RAR Basic Principles

A ground-based real aperture radar (RAR) system is a stepped-frequency continuous-wave (SF-CW) radar that emits a continuous wave with different progressive frequencies within a given frequency band. The frequency bandwidth B provides the range resolution ΔR of the radar according to the relationship, Equation (2):

$$\Delta R = \frac{c}{2B} \qquad (2)$$

where c is the speed of electromagnetic waves in vacuum. The corresponding echoes, backscattered by the scene, give rise to the raw data. Table 2 summarizes the main technical specifications of the IBIS-F/L® GB-RAR, IDS S.p.a, Pisa, Italy, used in this experiment.

The inverse Fourier transform of the raw signal acquired by the radar, normalized and converted to a logarithmic scale, provides the normalized radar cross-section (NRCS) profile, which shows the amplitude of the radar signal backscattered by targets located within the scene, discriminating them with the range resolution (2).

The interferometric phase is computed as follows:

$$\Delta \varphi_{1,2} = \mathrm{atan}\{S_2 \cdot conj(S_1)\} \qquad (3)$$

where S1 and S2 are two coherent complex-values radar data acquired at times t_1 and t_2, respectively, during the data acquisition. The line-of-sight (LoS) displacement $D_{1,2}$ of a point P, occurring in the time interval $[t_1, t_2]$, is related to the interferometric phase $\Delta \varphi_{1,2}$, computed in Equation (3), by the relationship:

$$D_{1,2} = \frac{\lambda}{4\pi} \Delta \varphi_{1,2} \qquad (4)$$

where λ is the radar wavelength. The precision of displacement measurements depends on the precision of phase measurements and it is a fraction of millimeter.

In the case of RAR interferometric applications, the radar acquires data with a sampling time in the order of a fraction of second, usually of a few milliseconds. This means that the radar can accurately track in time the deformation profile of the target. The vibration frequency spectrum is obtained by a spectral analysis of the displacement profile, computed in Equation (4). The vibration frequency spectra can be visualized as 2D maps. For each target, discriminated in range with a range resolution ΔR, the frequency spectra and displacement profiles are displayed vs. frequency and time, respectively.

Table 2. Technical Specifications of IBIS-F/L GB-RAR.

Range Resolution	0.75 m
Central frequency	17.2 GHz
Frequency bandwidth B	200 MHz

2.3. Methodology

This section describes the methodology proposed in this paper for the accurate characterization of dynamical behavior of tall structures. The methodology goes through the following steps:

(1) Co-location of TLS and GB-RAR in the same local reference system, and high-resolution geometric survey of the tall structure by TLS, for the co-registration of TLS and GB-RAR measurements;

(2) Model-based processing of TLS data acquired in line-scan mode, to increase the precision of TLS displacement measurements and consequent estimation of displacements and frequencies of oscillation modes having an amplitude lower than the precision achieved by TLS measurements;

(3) Interferometric processing of GB-RAR data, with 2D visualization of the frequency spectra of the structure oscillations.

The comparison and integration of TLS-based vibration frequencies with those provided by GB-RAR makes it possible to identify instrument artifacts and to extend measurements to higher oscillation modes, with an amplitude smaller than the enhanced TLS precision attained at point (2).

Below, we will examine the previous points to describe in detail each step of the proposed methodology applied to a wind tower.

2.3.1. Co-Location of TLS and GB-RAR Instruments and Co-Registration of Their Measurements

The co-location of TLS and GB-RAR is carried out by a TS topographic survey providing the coordinates of the TLS and GB-RAR, as well as of the base of the wind tower in the same local reference system. Figure 2 shows the layout of the TS survey (a) and a sketch of the local reference system with the positions of the TS, TLS, GB-RAR and wind tower (b). The slant ranges $R_{TS,TLS}$ and $R_{TS,GB-RAR}$ of TLS and GB-RAR with respect to the TS are indicated, along with the slant distances $R_{TS,WT-B}$ and $R_{TS,WT-T}$ of the base and top of the wind tower. The TS survey makes it possible to compute the coordinates of each point P of the tower, in the same local reference system, and hence their slant ranges $R_{P,TLS}$ and $R_{P,GB-RAR}$ with respect to the TLS and GB-RAR. The topographic survey is not necessary if the entire structure can be surveyed by the TLS, which in this case allows us to obtain, along with the mutual positioning of the structure and GB-RAR, a 3D model of the structure and of the surrounding environment.

Figure 3 displays a high-resolution 3D digital model of the tower provided by the TLS with the slant ranges $R_{P,GB-RAR}$ and $R_{P,TLS}$. As a result, the spatial co-registration of TLS and GB-RAR measurements is obtained in a seamless way, taking into consideration the different spatial resolutions of TLS and GB-RAR measurements. Figure 4 shows the TLS measurements points mapping the shape of the wind tower and the points corresponding to the center of each of 0.75 m-resolution cell of the GB-RAR. For each range cell of GB-RAR data, the corresponding TLS measurements are easily identified and compared or merged with GB-RAR data. The co-registration of the data in time is provided by a synchronization of TLS and GB-RAR acquisitions.

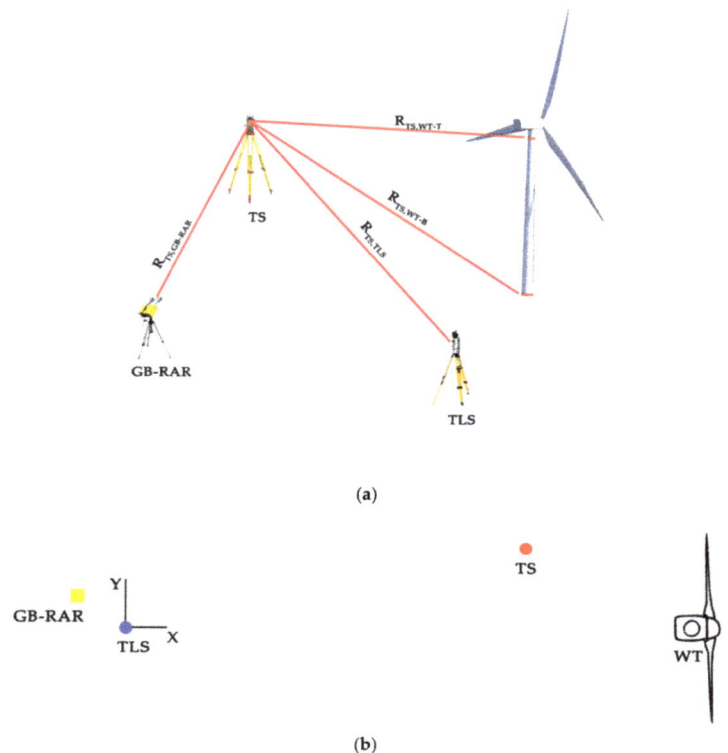

Figure 2. The TS survey layout with slant distances of TLS, GB-RAR, top and bottom of wind tower (**a**); The local reference system (TLS centered) used to co-locate TLS and GB-RAR measurements (**b**).

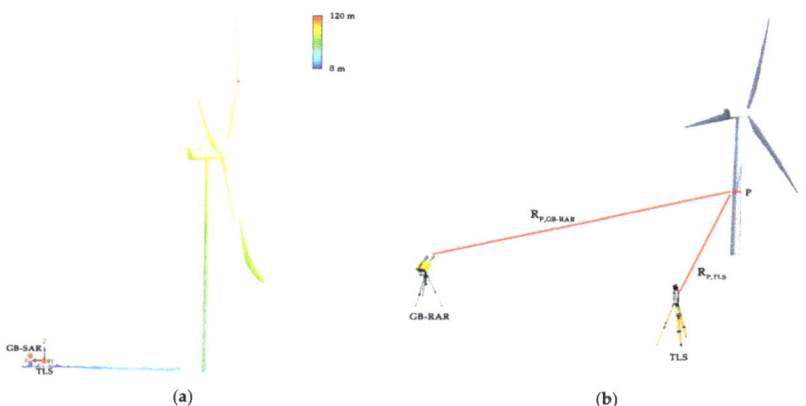

Figure 3. 3D digital surface model of the wind tower (**a**); The distances RP, GB-RAR and RP, TLS of point P on the tower with respect to TLS and GB-RAR (**b**).

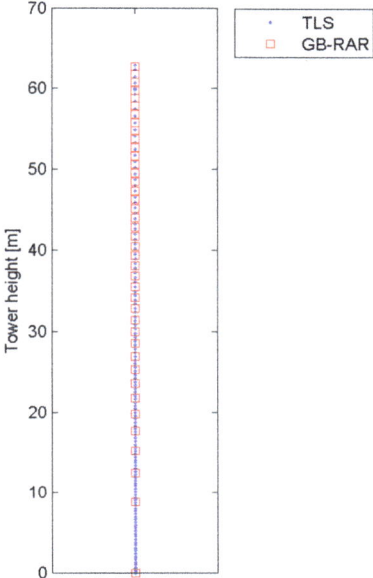

Figure 4. TLS (**blue dots**) and GB-RAR (**red square**) measurement points along the wind tower. The range distance and resolution of each GB-RAR measurement point is given by Equation (2).

2.3.2. Model-Based Processing of TLS Data and Estimation of Oscillation Amplitude and Frequency

Model-based processing of TLS data is widely used to determine with high precision the position of points materialized through known geometries. A typical example is given by the targets (spherical, cylindrical or plane) used for the registration of the scans carried out for the surveying of a 3D object [24,25].

If a stable target having a known geometry is surveyed, the parameters of the fitting surface corresponding to this geometry can be estimated from the cloud of n points acquired during the scan, through a least-squares procedure. The RMS of surface fit residuals exhibits a linear trend as a function of sampling resolution, i.e., of the square root of the number n of sampled points [26]. The deviations between the fitting surfaces, obtained from different scans of the same target, are proportional to $n^{-1/2}$ and, therefore, drastically lower than the single point precision of the instrument.

The model-based processing, in our case, is based on the use of interpolation to reduce noise and increase the precision of displacement measurements [20]. This approach is particularly suited for linear structures such as wind towers. In [19] the authors showed that, using an interpolating curve, the precision achieved in measuring the deformation of a beam can be 20 times higher than that of a single point. These conclusions had been confirmed in [27,28]. The knowledge of the geometry of the wind tower shown in Figure 3a is needed. A wind tower is a truncated cone with a circular section and a decreasing diameter upwards. Therefore, a generatrix is a straight line. The elastic line is certainly a continuous curve with a continuous derivative; in static conditions, for a load applied to the free end, it can be approximated by a third-order polynomial [29]. It is worth noting that the lines scanned by the VZ-1000 in line scanner mode do not belong to the same vertical plane, due to an imperfect orthogonality between the emitted laser beam and the rotation axis of the multi-facet mirror. This can also be inferred by observing the different horizontal angle associated with each scanned point. The laser beam would describe a conic section on a vertical surface instead of a straight line, as in the classic case of the theodolite line-of-sight, not orthogonal to the secondary axis. Given the size of the misalignment, there is a maximum deviation of about 0.4 m at the top of the tower compared to the average generatrix. In any case, this effect does not affect the results obtained, since they are based on the differences between the scanned lines, which have the same deviation. To take into account these effects without making the calculation too heavy, the trace of the laser beam on the tower is approximated by a fourth-order polynomial in a (D,H)-plane, where D is the ground distance from the TLS and H is the height from the tower base. The following procedure is adopted for each line scan: (a) computation of coordinates of points of the TLS scanned line; (b) estimation of interpolating polynomial coefficients; (c) shift of the polynomial curve in order to make it pass through the base of the tower; (d) computation of the distance D for each selected height H along the tower.

Figure 5 shows the polynomial fitting of the points of a single line scan along the tower. The procedure is repeated for all the TLS scanned lines. This provides the distance D(H,t), of a point P with a height H at the acquisition time t and hence the time series of the distance D of point P with respect to the TLS. Time series are used to obtain the amplitude of the oscillations. As a final step, the wind tower's vibration frequencies are obtained by a spectral analysis.

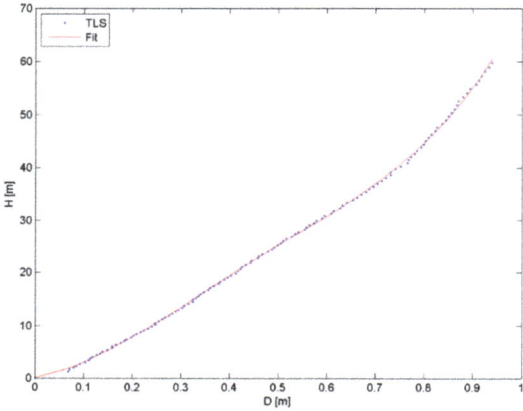

Figure 5. Polynomial fitting of points along the wind tower of a single line scan.

2.3.3. Interferometric Processing of GB-RAR Data and 2D Visualization of Oscillation Frequency Spectra

The stack of complex-valued GB-RAR acquisitions is processed by means of the interferometric radar technique by computing the phase differences of each pixel with respect to the first GB-RAR acquisition. The phase is first unwrapped and then mapped to propagation delay. Hence, the contribution of delay due to propagation in the atmosphere is modelled and removed, resulting in the measurement of the range $R_{P,GB-RAR}$ between the GB-RAR location and a target on the wind tower. A spectral analysis of the stack of range vectors $\{R_{P,GB-RAR}(t)\}_{P \in TOWER}$, results in the 2D spectrum of the vibration frequencies. This procedure has been described in [18] and applied to the monitoring of bell-towers and monuments in [30]. An important step of the adaption of the estimation of the GB-RAR estimation of vibration frequencies to the study of the characterization of the dynamical properties of a wind tower is the co-registration of TLS and GB-RAR spectra as sketched in Figure 6. After co-registering TLS and GB-RAR measurements, it is possible to compare vibration spectra obtained by TLS and GB-RAR by a spectral analysis of the time series of distances. It is worth noting that due to the different spatial resolutions of TLS and GB-RAR data, more TLS spectra correspond to the a given frequency spectrum provided by GB-RAR interferometry. In fact, a GB-RAR range resolution of 0.75 m corresponds to a coarser spatial resolution along the wind tower, depending on the radar looking angle and the antenna irradiation pattern (Figure 4).

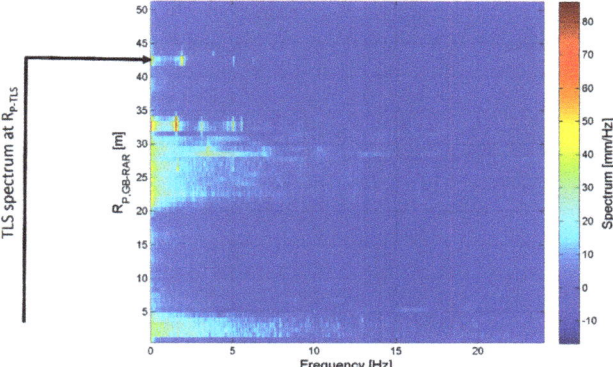

Figure 6. Sketch of proposed methodology to estimate vibration frequencies by the joint use of TLS and GB-RAR techniques.

3. Test and Results

This section summarizes the results obtained in the experimental test. It includes two subsections: Section 3.1 describes the case study and the measurement set up. Section 3.2 shows the results obtained by the proposed methodology.

3.1. Case Study

For the case study, a wind farm was chosen located in Tarsia, Southern Italy. Figure 7 displays the geographical location (a), an aerial Google© image (b) and a photo of the wind farm (c).

Figure 7. Google© images of the study area: (**a**) geographic position; (**b**) wind farm area (the investigated tower is circled in red); (**c**) panoramic view.

The wind farm is made up of 22 2.0 MW wind turbines, mounted on 65 m tall towers. The turbines are Gamesa® G90, characterized by a three-blade rotor and a diameter of $\phi = 90$ m. The swept area is $A = 6362$ m^2 and the cut-in wind speed is $v = 3.0$ m/s. The tower has a truncated cone shape, the diameter is $\phi_b = 4.034$ m at the base and $\phi_t = 2.314$ m at the top; the height is $H = 65$ m.

The monitored wind tower is located at the most eastern side, where on the acquisition day the wind speed was greater than the threshold of v = 3 m/sec necessary to activate the wind turbines. The wind direction was almost constant during data acquisition, with a prevalent direction from the north-east, with an azimuth angle of 69°. The rotations of the nacelle during the data acquisition reached a maximum value of 5°, so they had no appreciable effect on the results.

A topographic survey was carried out in order to obtain a local reference system for all acquisitions. The survey made it possible to have a single reference also for the heights; zero was fixed at the base of the tower, not visible from the TLS and the GB-RAR up to the height of 1.80 m. For the alignment of the acquisitions, a TS Leica TP 1201+®, Leica Geosystem AG, Heerbrugg, Switzerland (angular accuracy 1″, distance accuracy ± 1 mm ± 1 mm/km) was used. The acquisitions from TLS and GB RAR were carried out simultaneously, from 12.30 to 14.10 Central European Time (CET) on 17 July 2019.

The TLS, in line scanner configuration, was placed with the main axis vertical, so that the zenith angles of the laser beam varied between 30° and 130 °. The maximum scan speed of 120 lines/sec. was selected. The time taken per scan is 3.326×10^{-3} sec, while the time between two consecutive lines is 8.324×10^{-3} sec, with a time gap of 4.998×10^{-3} sec due to the geometry of the multi-facet mirror. The time taken to scan the only tower is 1.126×10^{-3} sec. Each line scanned describes a section comprising the wind tower and the ground between the TLS and the tower. Each TLS scan line provides a set of 140 point-like distance measurements, covering the whole tower. Given the scan speed of 120 lines per second, several thousand lines were acquired for each acquisition session. For the processing of the lines, measurement points near the connection of the nacelle were excluded to avoid outliers due to the pitch motion. The best-interpolating polynomial was obtained using the procedure described in Section 2.3.2. At this point, for a given height, it was possible to compute the value of the horizontal distance from the TLS for each line, approximately every 8 msec. The positions of the points on the same line, given the line scan time of about 1 msec, can be considered contemporary.

GB-RAR data were acquired using an elevation angle of 10° and antennas with a main irradiation lobe having an aperture of 39° in elevation and 11° in azimuth. As for time synchronization, the TLS is equipped with a GPS receiver, which makes it possible to convert timestamps to Central European Time (CET). For GB-RAR, the recorded acquisition time was converted using the CET provided by the operating system of the computer used as a data logger.

Acquisitions were made under various operating conditions of the wind turbine: (a) fully operational with the turbine active; (b) during the deactivation of the turbine; (c) during the stop without the action of the wind; (d) during the restart; (a) once again fully operational after the restart. Table 3 shows the start and end times for the seven acquisitions performed, the corresponding timestamps, and the operating conditions.

Table 3. TLS sampling periods.

Operating Condition	Initial Timestamp	Final Timestamp	Initial Time (CET)	Final Time (CET)
a	408.67	430.35	13.37.50	13.38.12
a	664.55	678.75	13.41.57	13.42.11
a	736.23	754.69	13.43.08	13.43.27
b	766.81	846.19	13.43.39	13.44.59
c	888.28	917.17	13.45.41	13.46.10
d	937.18	1054.41	13.46.29	13.48.27
a	643.17	708.68	14.00.00	14.01.06

3.2. Results

Acquisitions corresponding to operating condition c (rotor blades motionless, no action of the wind) were used to obtain the mean value of the horizontal distance from TLS. For a point P on the wind tower located at H = 65 m height from the tower base, the mean ground distance of D = 47.293 m was obtained. The procedure described in the Section 2.3.2 was applied. The comparison between raw data and the values obtained by the procedure can be observed in Figure 8.

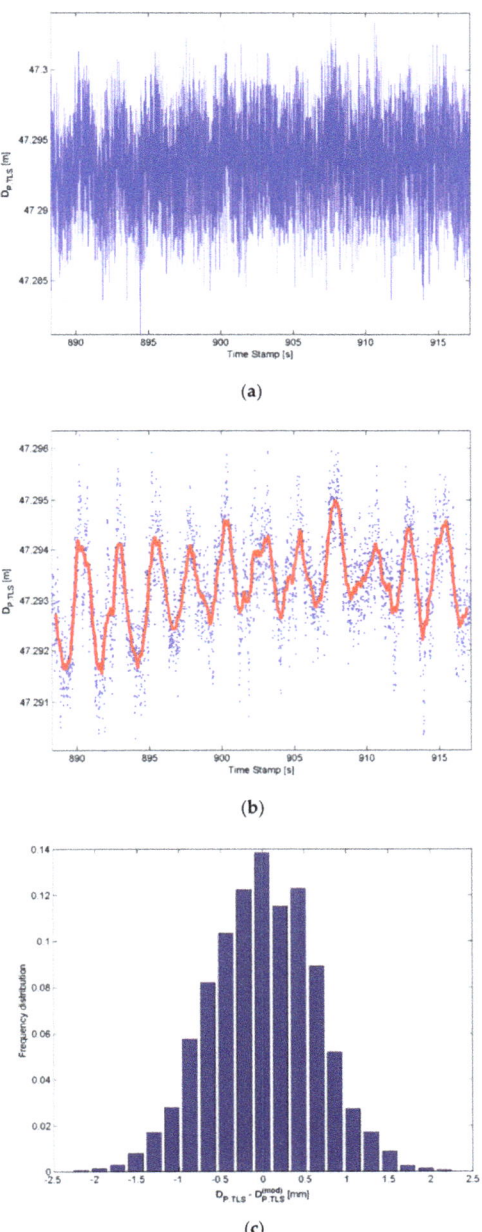

Figure 8. (a) Distance DP,TLS of point P on the wind tower located at H = 65 m during acquisition in operating condition c (see Table 3); (b) result after applying the proposed methodology; (c) frequency distribution of differences between modelled DP,TLS distances (blue dots) and moving average (red line). The standard difference between measured and model DP,TLS distances is 0.6 mm.

The time series of TLS distances of point P, located at the top of the wind tower, is plotted in panel 8(a). Panel 8(b) displays the results obtained by applying the procedure in Section 2.3.2 to the detrended measurements of Panel 8(a). Results provided by the proposed methodology and the mean curve obtained by applying a moving average operator are plotted in Figure 8b, while the frequency distribution of their differences is plotted in Figure 8c. The standard deviation of this frequency distribution provides the precision of TLS estimates of distances equal to 0.6 mm.

Figure 9 shows the plot of TLS ground distances vs. timestamps of four points P on the wind tower located at different heights above the base, obtained by the proposed procedure. Data were collected from 13.41.57 till 13.42.11 CET. These results were obtained in condition a (see Table 3) with a fully operational wind turbine. The peak-to-peak amplitude of oscillations decreased from 15 mm to 2 mm when moving from H = 60 m to H = 15 m on the wind tower.

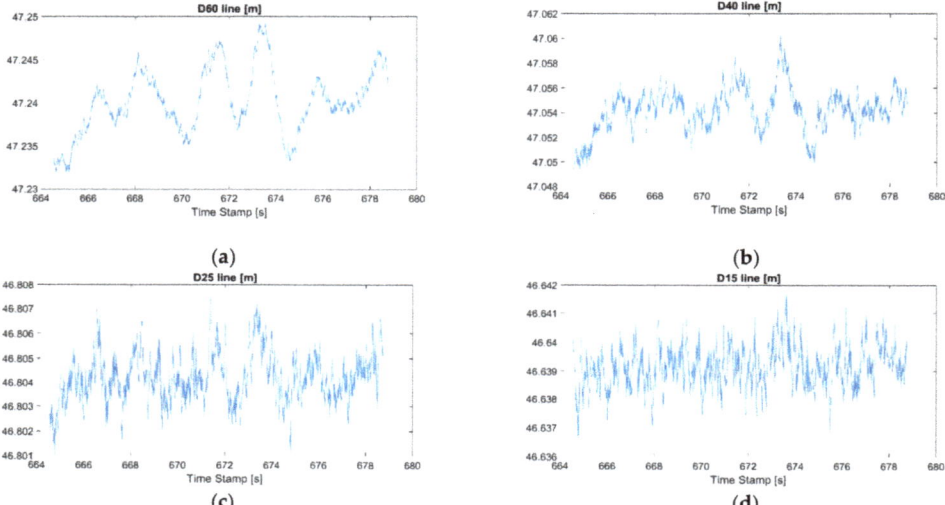

Figure 9. Ground distances from TLS of points at a height of 60 m (d_{60} line), 40 m (d_{40} line), 25 m (d_{25} line) and 15 m (d_{15} line) on the tower base. Abscissae are the timestamps in sec. The wind turbine is fully operational.

Figure 10 shows the plot of TLS ground distances vs. timestamps of a point P on the wind tower located at H = 60 m above the base. Data were collected from 13.43.39 till 13.44.59 CET. These results were obtained in condition b (see Table 3) with the wind turbine in deactivation mode. Damped oscillations had an amplitude decreasing from 90 mm to a few mm. The average ground distance from the TLS increased and tended to stabilize around a value of 47.293 m.

The corresponding spectrum of vibration frequencies are shown in Figure 11a,b, along with the spectrum obtained for data collected during the stop of the wind turbine in Figure 11c,d. In both cases, two peaks are observed at 0.4 Hz and 40 Hz.

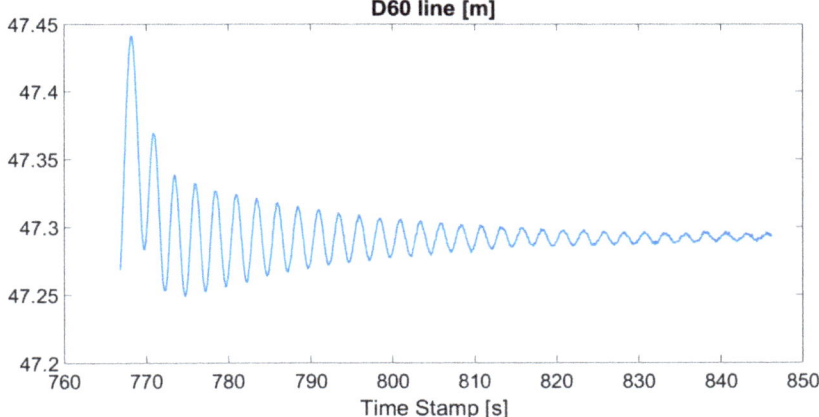

Figure 10. Ground distance D_{60} line from TLS of a point 60 m high on the tower base during the deactivation of the turbine.

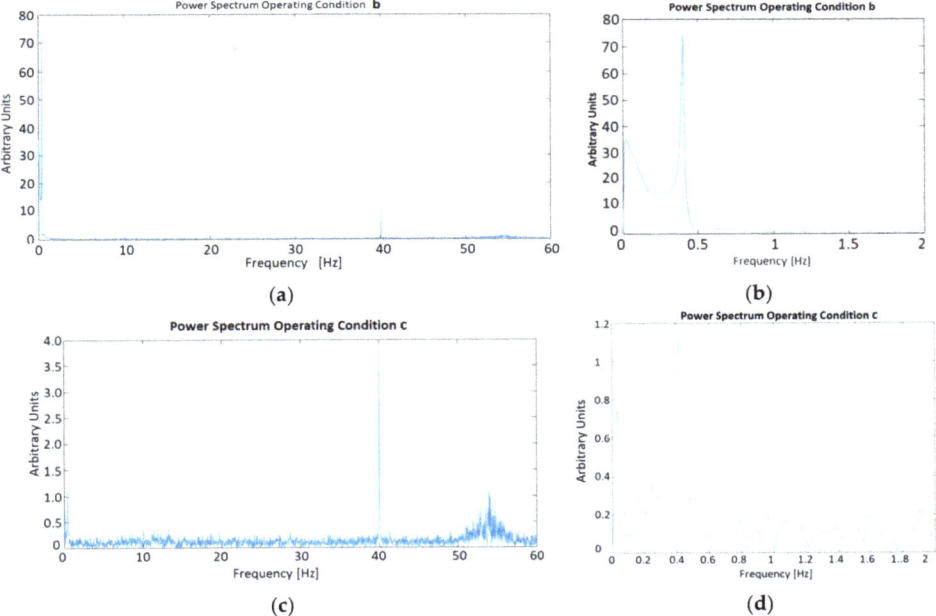

Figure 11. Spectra of vibration frequencies of point P on the wind tower located at H = 60 m. Spectrum for the data collected during the deactivation of the wind turbine (**a**). Enlargement of the low frequencies (**b**). Spectrum for the data collected during the stop of the wind turbine (**c**). Enlargement of the low frequencies (**d**).

Figure 12 shows the spectrum of vibration frequencies for a point P on the wind tower located at H = 25 m above the base, obtained for data collected during the deactivation of the wind turbine. Additionally, in this case, two peaks can be observed at 0.4 Hz and 40 Hz.

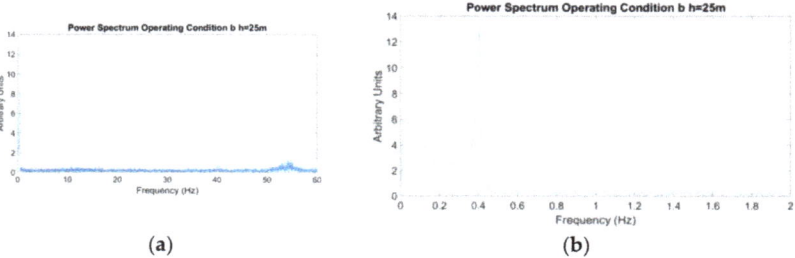

Figure 12. Spectrum of vibration frequencies of point P on the wind tower located at H = 25 m. Data have been collected during the deactivation of the turbine (**a**). Enlargement of the low frequencies (**b**).

The spectrum of vibration frequencies obtained by processing GB-RAR data acquired during the stop of the wind turbine is reported in Figure 13. The spectrum refers to a point located at a height H = 25 m above the tower base, corresponding to a slant distance R = 54 m from the radar. Two peaks are observed at 0.4 Hz and 3.1 Hz. It worth noting that besides the common peak at frequency f = 0.4 Hz, the two spectra had different characteristics. In particular, it should be noted that the peak at f = 0.4 Hz measured by GB-RAR in operating condition c was measured by TLS at H = 60 m but not at H = 25 m. Probably, this was due to the different amplitude of oscillations at the top and bottom of the tower.

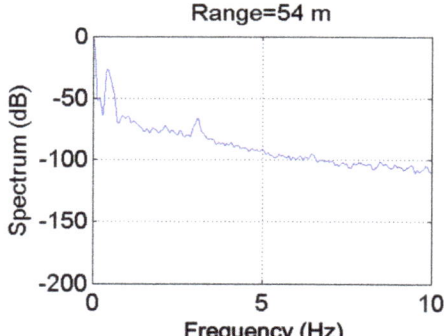

Figure 13. Spectrum of vibration frequencies of point P on the wind tower located at H = 25 m. Data have been collected during the stop of the turbine.

4. Numerical Analysis and Discussion

This section discusses the results presented in Section 3.

4.1. Evaluation of Displacements Due to the Wind

During the test, anemometric data were acquired, which showed a constant wind direction. The wind speed data make it possible to obtain the thrust on the tower. Given the geometric and mechanical characteristics of the tower, it was possible to calculate the displacements of the points at various heights. The wind pressure can be obtained by the Bernoulli Equation:

$$p = \frac{1}{2}\rho v^2 \qquad (5)$$

where:

p = is the wind pressure [N/m^2]

ρ = air density [kg/m³]
v = air speed [m/sec]

The air pressure, applied to the swept area of the rotor, must be multiplied to a coefficient less than 0.6 due to the Betz limit [31]. In our case, given the low wind speed, which implies a non-optimal rotor efficiency, we can use the value 0.4. It is also necessary to take into account the interference factor b (i.e., the ratio of the downstream speed v_2 to the upstream speed v_1, which in our case can be considered equal to 0.577 [32]). Given that the speed measured by the anemometers and used for our test (installed on the nacelle and behind the rotor) is the downstream speed, the thrust F acting on the top of the tower, considered as a cantilever, can therefore be evaluated as [32]:

$$F \cong 0.4\left(\frac{1}{2}\rho 3 v_2^2 \pi r^2\right) = 0.60\left(\rho v_2^2 \pi r^2\right) \quad (6)$$

where r is the radius of the rotor. Geometric and material data on the tower can be found in [33]. The displacement δ at the top of the tower, in correspondence of the connection with the nacelle, is obtained by:

$$\delta = F\frac{L^3}{3EJ} \quad (7)$$

where:

L = cantilever length
E = Young's modulus
J = moment of inertia

For our calculations, we assumed ρ = 1.2 kg/m³, L = 65 m, E = 2.1 × 10¹¹ N/m², J = 0.323 m⁴. Actually, the lowering changes with the wind speed, and other effects affect the final results, so only a rough estimate can be obtained with Equation (7). Table 4 shows the displacements of a point 65 m high on the base of the tower computed for different wind speeds. By adding to the displacements, the mean value of the ground distances measured in the operating condition c, i.e., during the stop of the rotor blades, one can obtain the ground distances from TLS for the tower subject to the wind load.

Table 4. Theoretical displacements.

Wind Speed [m/s]	Thrust [N]	E [N/m²]	J [m⁴]	L [m]	Displacement d [m]
2	18,321	2.1 × 10¹¹	0.323	65	0.0247
2.5	28,627	2.1 × 10¹¹	0.323	65	0.0387
3	41,223	2.1 × 10¹¹	0.323	65	0.0557
3.5	56,109	2.1 × 10¹¹	0.323	65	0.0758
4	73,285	2.1 × 10¹¹	0.323	65	0.0990
4.5	92,751	2.1 × 10¹¹	0.323	65	0.1253
5	114,508	2.1 × 10¹¹	0.323	65	0.1547

4.2. Numerical Analysis of Natural Frequencies

This section provides a simple numerical analysis for the computation of natural frequencies of a cantilever subject to free oscillations. An approximated value of the circular natural frequencies, in case of a fixed cross section, is given by [34]:

$$\omega_n = \alpha_n^2 \sqrt{\frac{EJ}{mL^4}} \quad (8)$$

where:

ω_n = n-th circular natural frequency
α_n = 1.875, 4.694, 7.885

E = Young's modulus
J = moment of inertia
m = mass per length unit
L = cantilever length

The natural frequency f_n can be obtained from the circular frequency using the equation:

$$f_n = \frac{\omega_n}{2\pi} \qquad (9)$$

To bring back the problem to the case of a massless cantilever with a discrete effective mass applied to the free end, we use the effective mass for the n-th frequency $m_{eff}^{(n)}$, given by:

$$m_{eff}^{(n)} = \frac{3EJ}{L^3 \omega_n^2} \qquad (10)$$

This way we can add a m_{end} mass actually positioned at the free end of the cantilever and consider a total mass M_n at the free end:

$$M_n = m_{eff}^{(n)} + m_{end} \qquad (11)$$

The n-th natural frequency will be:

$$\omega_n = \alpha_n^2 \sqrt{\frac{3EJ}{M_n L^4}} \qquad (12)$$

In our case, given the characteristics of the tower, its mean section and the wind turbine data [33] we can assume $E = 2.1 \times 10^{11}$ N/m^2, J = 0.322 m^4, m = 2800 kg/m, L = 65 m. By using Equations (9) and (10), we obtain a discrete effective mass of 40,740 kg for the first frequency and of 1120 kg for the second one. We must add the weight of rotor and nacelle, equal to about 8×10^4 kg. By using Equations (11) and (12), we get, for the first three natural frequencies, the values f_1 = 0.394 Hz, f_2 = 3.10 Hz, f_3 = 8.596 Hz, and the relevant periods T_1 = 2.54 sec, T_2 = 0.32 sec, and T_3 = 0.11 sec.

4.3. Measured vs. Theoretical Displacements

In the following, we consider the oscillation amplitudes of a point located at H = 65 m above the base of the tower. The choice of a point near the top of the tower allows us to consider displacements of greater amplitude. This is also useful, given the wind speed, which was not high during the test. Figure 14 shows the wind downstream speed during the test. The wind direction was constant during data acquisition.

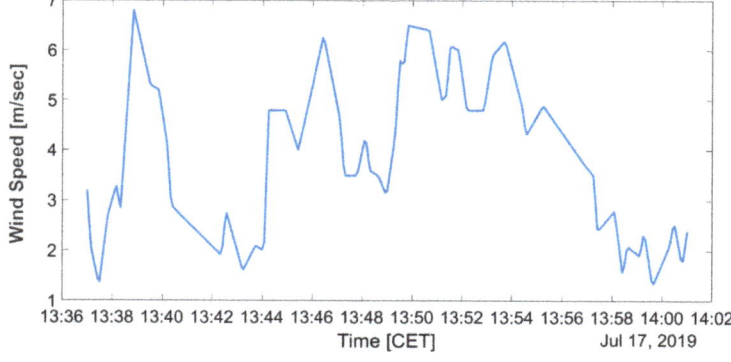

Figure 14. Wind speed in m/sec (downstream) measured by the wind turbine anemometers during the test.

In Table 5 we can see a comparison between the values of the theoretical ground distances and the values measured by TLS in different acquisitions. The theoretical distances have been obtained by subtracting the computed displacements to the mean ground distance D = 47.293 m, obtained in case of motionless rotor blades (see Figure 8a and condition c in Table 3).

Table 5. Theoretical and measured ground distances from TLS of a point 65 m high.

Start Time [CET]	End Time [CET]	Wind Speed [m/sec]		Measured D [m]		Theoretical D [m]	
		min	max	min	max	min	max
13.37.50	13.38.12	2.72	3.25	47.247	47.225	47.247	47.227
14.00.00	14.01.06	2.20	2.40	47.262	47.255	47.260	47.254

Figure 15 shows the displacements of the considered point during the activity of turbine. In the lower part of the figure, the wind speed is represented. Acquisition times are synchronized (see Table 3); the wind direction was constant.

It can be observed that the increase in wind speed leads to a greater deformation and, therefore, to an approach of the point to the TLS located on the opposite side of the rotor. The deformation values δ obtained are in accordance with those obtained with Equation (7).

A continuous oscillation of the point is observed which has a period of around two seconds. This value is not constant and is affected by the noise of the measurements, as well as by the variability of the wind thrust. The amplitude of the oscillations varies from 10 to 20 mm.

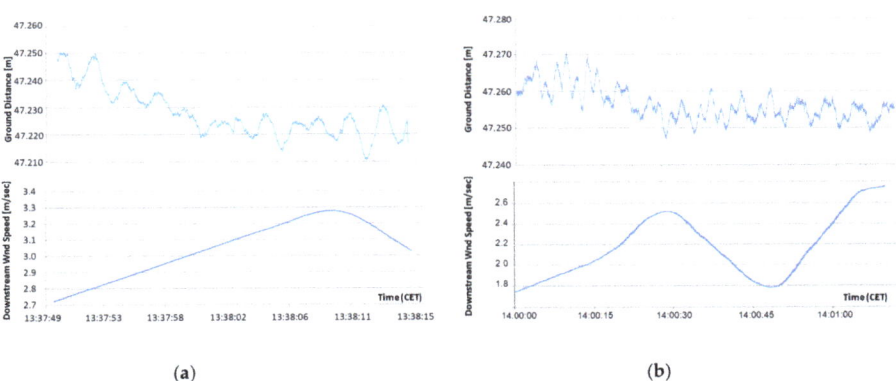

Figure 15. Ground distance from TLS of a point P on the tower located at H = 65 m during the activity of the turbine, from 13:37:50 to 13:38:12 (**a**) and from 14:00:00 to 14:01:06 (**b**). Below the graphics of the distances, the wind speed in m/sec. X axis is the time, both for upper and lower figures.

4.4. Measured vs. Theoretical Frequencies

In this section, we discuss the properties of the measured vibration spectra and compare them to the theoretical analysis of the wind tower reported in Section 4.2. The 0.4 Hz frequency measured by both TLS and GB-RAR is in agreement with the value obtained using the analytical formulas for the first natural frequency. The frequency f = 3.1 Hz of the second oscillation mode is measured by GB-RAR but not by TLS, probably due to lower precision of TLS when compared to GB-RAR.

The frequency peak at f = 40 Hz found in all TLS measurements is an artifact due to the rotating mirror. In fact, the laser beam used for measurements in the TLS VZ1000 is addressed and collected by a rotating three-sided mirror, i.e., by the lateral faces of a triangular-based prism. When a scan speed of 120 lines per second is selected, the three-sided mirror rotates with an angular speed of 40 revolutions per second. Even a slight vibration of the rotation axis can cause oscillations in the results at a frequency of 40 Hz (Figure 16a). To confirm this hypothesis, the results of the measurements performed with a slower scanning speed of 60 lines per second were analyzed.

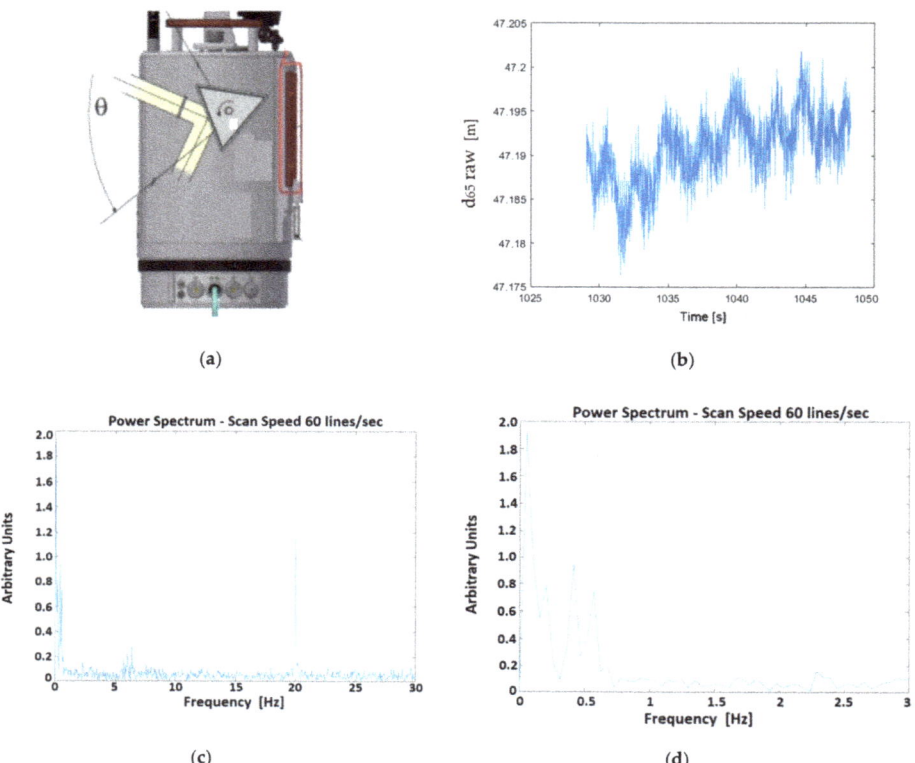

Figure 16. The scheme of rotating mirror of VZ1000 TLS (**a**). Ground distance from TLS of a point 65 m high on the tower base during the acquisition with scan speed 60 Hz (**b**). Power Spectrum obtained by FFT for the ground distance from TLS (**c**). Enlargement of Figure 16c (**d**).

Figure 16b shows the displacements of a point 65 m high on the tower base during the acquisitions. Figure 16c shows the power spectrum for the ground distance from TLS of a point 65 m high on the tower base. Two peaks are present: the first at a low frequency, the second at 20 Hz. In Figure 16d the first peak is enlarged, corresponding to a frequency of 0.4 Hz. The results confirm the above hypothesis, so we can attribute the peak at the highest frequency to the vibration of the rotating mirror of the TLS.

5. Conclusions

A methodology has been proposed for measuring the vibration frequencies and the amplitude of the oscillations of tall structures by joint use of TLS and GB-RAR. The methodology exploits the possibility of using a TLS in line scan mode, thus extracting a large number of scan lines per second. The use of lines interpolating the individual scanned points makes it possible to obtain results with a precision better than that of the instrument. For a given point on the structure, the stack of TLS lines is used to derive the temporal profile of displacements and from this the spectrum of vibration frequencies. It has been observed the TLS estimate of vibration frequencies is limited by the precision of displacement measurements. For oscillations whose amplitude is much smaller than the TLS precision, the proposed methodology based on TLS data cannot estimate the vibration frequency. This problem is overcome by the joint use of co-located GB-RAR. The higher precision of GB-RAR measurements of tower displacements provides vibration frequencies corresponding to smaller oscillation amplitudes.

Results provided by the proposed methodology based on the joint use of co-located TLS and GB-RAR systems have been compared to the output of a numerical analysis of tower displacements, modelling the tower as subject to a wind load derived from the anemometric data.

The comparison of results obtained by the numerical analysis and measurements leads to the following conclusions: (a) the proposed processing technique based on the geometrical modelling of the structure can enhance the precision of distance measurements below the 5 mm declared for the TLS instrument; (b) the joint use of co-located TLS and GB-RAR can provide richer information on the dynamical behavior of the structure, with the sub-centimeter spatial resolution of TLS, and on the whole spectral characterization of vibrations, including those having an amplitude much smaller than the TLS precision, thanks to the sub-millimeter precision of GB-RAR measurements.

The proposed methodology can be a useful and cheap support for non-contact inspection and monitoring of tall structures.

Author Contributions: Conceptualization, S.A.; methodology, S.A.; software, S.A. and G.N.; investigation, S.A. and G.N.; writing—original draft preparation, S.A. and G.N.; writing—review and editing, S.A. and G.N. All authors have read and agreed to the published version of this manuscript.

Funding: This research received no external funding.

Acknowledgments: Authors acknowledge EON Italia Spa for support during data acquisition. The authors would like to thank the entire DIAN S.R.L. team for the acquisition and processing of GB-RAR data.

Conflicts of Interest: The authors declare no conflict of interest.

References

1. Zinno, R.; Artese, S.; Clausi, G.; Magarò, F.; Meduri, S.; Miceli, A.; Venneri, A. Structural Health Monitoring (SHM). In *The Internet of Things for Smart Urban Ecosystems*; Cicirelli, F., Guerrieri, A., Mastroianni, C., Spezzano, G., Vinci, A., Eds.; Springer International Publishing: Cham, Switzerland, 2019; pp. 225–249. [CrossRef]
2. Fan, W.; Qiao, P. Vibration-based damage identification methods: A review and comparative study. *Struct. Health Monit.* **2011**, *10*, 83–111. [CrossRef]
3. Carden, E.P.; Fanning, P. Vibration based condition monitoring: A review. *Struct. Health Monit.* **2004**, *3*, 355–377. [CrossRef]
4. Kim, D.H.; Feng, M.Q. Real-time structural health monitoring using a novel fiber-optic accelerometer system. *IEEE Sens. J.* **2007**, *7*, 536–543. [CrossRef]
5. Bui, Q.B.; Hans, S. Dynamic behaviour of an asymmetric building: Experimental and numerical studies. *Case Stud. Nondestruct. Test. Eval.* **2014**, *2*, 38–48. [CrossRef]
6. Sabato, A.; Niezrecki, C.; Fortino, G. Wireless MEMS-Based Accelerometer Sensor Boards for Structural Vibration Monitoring: A Review. *IEEE Sens. J.* **2016**, *17*, 226–235. [CrossRef]
7. Artese, G.; Perrelli, M.; Artese, S.; Meduri, S.; Brogno, N. POIS, a Low Cost Tilt and Position Sensor: Design and First Tests. *Sensors* **2015**, *15*, 10806–10824. [CrossRef]
8. Yu, J.X.; Meng, X.; Shao, X.; Yan, B.; Yang, L. Identification of dynamic displacements and modal frequencies of a medium-span suspension bridge using multimode GNSS processing. *Eng. Struct.* **2014**, *81*, 432–443. [CrossRef]
9. Lienhart, W.; Ehrhart, M.; Grick, M. High Frequent Total Station Measurements for the Monitoring of Bridge Vibrations. In Proceedings of the 3rd Joint International Symposium on Deformation Monitoring (JISDM), Vienna, Austria, 30 March–1 April 2016; Available online: https://www.fig.net/resources/proceedings/2016/ (accessed on 2 September 2019).
10. Artese, S.; Perrelli, M. Monitoring a Landslide with High Accuracy by Total Station: A DTM-Based Model to Correct for the Atmospheric Effects. *Geosciences* **2018**, *8*, 46. [CrossRef]
11. Xu, Y.; Brownjohn, J.M.W. Review of machine-vision based methodologies for displacement measurement in civil structures. *J. Civ. Struct. Health Monit.* **2018**, *8*, 91. [CrossRef]
12. Artese, S.; Achilli, V.; Zinno, R. Monitoring of Bridges by a Laser Pointer: Dynamic Measurement of Support Rotations and Elastic Line Displacements: Methodology and First Test. *Sensors* **2018**, *18*, 338. [CrossRef]

13. Pieraccini, M. Monitoring of Civil Infrastructures by Interferometric Radar: A Review. *Sci. World J. Vol.* **2013**, 786961. [CrossRef] [PubMed]
14. Anghel, A.; Tudose, M.; Cacoveanu, R.; Datcu, M.; Nico, G.; Masci, O.; Dongyang, A.; Tian, W.; Hu, C.; Ding, Z.; et al. Compact Ground-Based Interferometric Synthetic Aperture Radar: Short-range structural monitoring. *IEEE Signal Process. Mag.* **2019**, *36*, 42–52. [CrossRef]
15. Nico, G.; Borrelli, L.; Di Pasquale, A.; Antronico, L.; Gullà, G. Monitoring of an ancient landslide phenomenon by GBSAR technique in the Maierato town (Calabria, Italy). *Eng. Geol. Soc. Territ.* **2015**, *2*, 129–133. [CrossRef]
16. Nico, G.; Cifarelli, G.; Miccoli, G.; Soccodato, F.; Feng, W.; Sato, M.; Miliziano, S.; Marini, M. Measurement of pier deformation patterns by ground-based SAR interferometry: Application to a bollard pull trial. *IEEE J. Ocean. Eng.* **2018**, *43*, 822–829. [CrossRef]
17. Di Pasquale, A.; Nico, G.; Pitullo, A.; Prezioso, G. Monitoring Strategies of Earth Dams by Ground-Based Radar Interferometry: How to Extract Useful Information for Seismic Risk Assessment. *Sensors* **2018**, *18*, 244. [CrossRef]
18. Atzeni, C.; Bicci, A.; Dei, D.; Fratini, M.; Pieraccini, M. Remote Survey of the Leaning Tower of Pisa by Interferometric Sensing. *IEEE Geosci. Remote Sens.* **2010**, *7*, 185–189. [CrossRef]
19. Gordon, S.J.; Lichti, D.D. Modeling Terrestrial Laser Scanner Data for Precise Structural Deformation Measurement. *J. Surv. Eng.* **2007**, *133*, 72–80. [CrossRef]
20. Kim, K.; Kim, J. Dynamic displacement measurement of a vibratory object using a terrestrial laser scanner. *Meas. Sci. Technol.* **2015**, *26*, 045002. [CrossRef]
21. Neitzel, F.; Niemeier, W.; Weisbrich, S.; Lehmann, M. Investigation of Low-Cost Accelerometer, Terrestrial Laser Scanner and Ground-Based Radar Interferometer for Vibration Monitoring of Bridges. In Proceedings of the Sixth European Workshop on Structural Health Monitoring, Dresden, Germany, 3–6 July 2012.
22. Schill, F.; Eichhorn, A. Investigations of low- and high-frequency movements of wind power plants using a profile laser scanner. In Proceedings of the Joint International Symposium on Deformation Monitoring JISDM 2016, Vienna, Austria, 30 March–1 April 2016.
23. Mukupa, W.; Roberts, G.W.; Hancock, C.M.; Al-Manasir, K. A review of the use of terrestrial laser scanning application for change detection and deformation monitoring of structures. *Surv. Rev.* **2017**, *49*, 99–116. [CrossRef]
24. Urbančič, T.; Roškar, Z.; Fras, M.K.; Grigillo, D. New Target for Accurate Terrestrial Laser Scanning and Unmanned Aerial Vehicle Point Cloud Registration. *Sensors* **2019**, *19*, 3179. [CrossRef]
25. Bolkas, D.; Martinez, A. Effect of target color and scanning geometry on terrestrial LiDAR point-cloud noise and plane fitting. *J. Appl. Geod.* **2017**. [CrossRef]
26. Lichti, D.D.; Glennie, C.L.; Jahraus, A.; Hartzell, P. New Approach for Low-Cost TLS Target Measurement. *J. Surv. Eng.* **2019**, *145*, 04019008. [CrossRef]
27. Artese, S. The Survey of the San Francesco Bridge by Santiago Calatrava in Cosenza, Italy. *Int. Arch. Photogramm. Remote Sens. Spat. Inf. Sci.* **2019**, 33–37. [CrossRef]
28. Artese, S.; Zinno, R. TLS for dynamic measurement of the elastic line of bridges. *Appl. Sci.* **2020**, *10*, 1182. [CrossRef]
29. Chen, W.F.; Lui, E.M. *Handbook of Structural Engineering*, 2nd ed.; CRC Press: Boca Raton, FL, USA, 2005; ISBN 9780849315695.
30. Nico, G.; Masci, O.; Panidi, E. Non-destructive monitoring strategies of historical constructions and tangible cultural heritage based on ground-based SAR interferometry. In Proceedings of the InterCarto/InterGIS, International Conference, Petrozavodsk, Russia, July 19–1 August 2018; Volume 24, pp. 528–535. [CrossRef]
31. Betz, A. The Maximum of the theoretically possible exploitation of wind by means of a wind motor. *Wind Eng.* **2013**, *37*, 441–446. [CrossRef]
32. Ragheb, M.; Ragheb, A.M. Wind Turbines Theory—The Betz Equation and Optimal Rotor Tip Speed Ratio. In *Fundamental and Advanced Topics in Wind Power*; Carriveau, R., Ed.; IntechOpen Limited: London, UK, 2011; ISBN 978-953-307-508-2. [CrossRef]
33. Gamesa G90. Available online: https://en.wind-turbine-models.com/turbines/763-gamesa-g90 (accessed on 20 January 2020).
34. Thomson, W.T. *Theory of Vibration with Applications*; Kindersley Publishing, Inc.: London, UK, 2007.

© 2020 by the authors. Licensee MDPI, Basel, Switzerland. This article is an open access article distributed under the terms and conditions of the Creative Commons Attribution (CC BY) license (http://creativecommons.org/licenses/by/4.0/).

Article

TLS for Dynamic Measurement of the Elastic Line of Bridges

Serena Artese [1,2,*] and Raffaele Zinno [3]

1. Department of Civil Engineering, University of Calabria, Via Pietro Bucci cubo 45B, 87036 Rende (CS), Italy
2. Spring Research s.r.l., Spin-off of University of Calabria, Via Pietro Bucci cubo 45B, 87036 Rende (CS), Italy
3. Department of Informatics, Modeling, Electronic and System Engineering, University of Calabria, Via P. Bucci cubo 42C, 87036 Rende, Italy; raffaele.zinno@unical.it
* Correspondence: serena.artese@unical.it; Tel.: +39-09-84-496-763

Received: 4 October 2019; Accepted: 6 February 2020; Published: 10 February 2020

Abstract: The evaluation of the structural health of a bridge and the monitoring of its bearing capacity are performed by measuring different parameters. The most important ones are the displacements due to fixed or mobile loads, whose monitoring can be performed using several methods, both conventional and innovative. Terrestrial Laser Scanner (TLS) is effectively used to obtain the displacements of the decks for static loads, while for dynamic measurements, several punctual sensors are in general used. The proposed system uses a TLS, set as a line scanner and positioned under the bridge deck. The TLS acquires a vertical section of the intrados, or a line along a section to be monitored. The instantaneous deviations between the lines detected in dynamic conditions and the reference one acquired with the unloaded bridge, allow to extract the displacements and, consequently, the elastic curve. The synchronization of TLS acquisitions and load location, obtained from a Global Navigation Satellite System GNSS receiver or from a video, is an important feature of the method. Three tests were carried out on as many bridges. The first was performed during the maneuvers of a heavy truck traveling on a bridge characterized by a simply supported metal structure deck. The second concerned a prestressed concrete bridge with cantilever beams. The third concerned the pylon of a cantilever spar cable-stayed bridge during a load test. The results show high precision and confirm the usefulness of this method both for performing dynamic tests and for monitoring bridges.

Keywords: laser scanner; line scanner; structure monitoring; deformation; dynamic measurements

1. Introduction

Among the technical parameters for assessing the structural health, the displacements are probably the most relevant. The structural suitability is verified by loading a structure with known loads and comparing the actual displacements with those expected in the design phase.

As for bridges, some trucks of known weight, arranged in one or more rows, are positioned on the deck in predetermined positions. Levels or total stations are used to measure beam deflections; additional dynamic tests are performed, depending on the importance of the structure [1].

For Structural Health Monitoring (SHM), along with the deformations, different physical and mechanical parameters are measured. A broad overview of the methodologies used can be found in [2]. Below is a short list of noteworthy techniques: (a) The structural vibration data are increasingly used; a review of vibration-based methods can be found in [3], where natural frequency-based, mode shape-based and curvature/strain mode shape-base methods are described; (b) Instrumented vehicles are more and more used for indirect bridge monitoring. This technique allows measurements without the need to place sensors on the deck or on the piers; references can be found in [4]; (c) Acoustic Emission (AE) is an effective technique, used since several decades [5]; this method, as the X-ray methods, requires to roughly know the position of damages, along with the accessibility for performing tests; (d)

Magnetic sensing is used in particular in structures that contain ferrous elements such as reinforced concrete [6]; (e) Ground Penetrating Radar (GPR) is increasingly adopted both to identify hidden lesions and to find discontinuities in structural materials [7].

For detecting and monitoring structural displacements, both conventional and innovative methods are used; below are some of the most used, with their pros and cons. (1) Digital levels are still among the most used tools, thanks to their high accuracy. Their main limit is the need to measure one target at a time, so they cannot be used to dynamically monitor multiple points; (2) Total robotic stations are characterized by high precision and measurement automation. Thanks to the capability to transfer data even via the Internet and be managed remotely, they are used to monitor slopes and large structures [8,9]. The sampling frequency up to 7 Hz allows its use for dynamic measurements [10,11], but it is not possible to track more points at the same time; (3) GNSS satellite surveying is often used for tall buildings and large span bridges [12–14]. It is possible to obtain an accuracy of a few millimeters, while the acquisition frequency reaches 20 Hz. The main limitation lies in the need to place an antenna on each point one must monitor; (4) The increase in the resolution of digital cameras facilitated the development of computer vision based systems. Among the techniques used, digital image correlation (DIC) has proved to be effective for measurements of bridge deflection [15–20]; (5) Dial gauges, widely employed for measurements of floor slabs deflections, are used for bridges with limited height and in absence of water; (6) The use of inclination parameters obtained from microelectromechanical systems (MEMS) has been proposed to derive the deflection [21]. The high S/N ratio, typical of these devices, strongly influences the results of dynamic tests; (7) Currently, the measurement of deflections by using laser beams is widespread. Some systems for the measurement of deflections and displacements with the use of laser beams are described in [22,23]; (8) An effective methodology for measuring displacements of large structures is offered by Ground-Based Synthetic Aperture Radar (GB-SAR). Used in RAR mode (Real Aperture Radar) it allows measurements of vibration frequency and displacements associated with dynamic loads. In SAR mode (Synthetic Aperture Radar) the instrument is mounted on a slide and allows to obtain the mapping of the area investigated at different times and to get the deflection of each point. In both cases the precision of the displacement measurements is about 0.1 mm. This technique is increasingly used, but is still limited by the high cost of the instrumentation [24,25].

Nowadays TLS is a widespread and reliable technique for monitoring bridges in static conditions [26]. Comparing the scans acquired in different epochs, we can get, e.g., the maps of deviations between the surface points of a bridge subject to various conditions (loads, temperature, etc.). As for dynamic monitoring, it is possible to take advantage of the high sampling rate of TLS; specifically, we can measure the deflections of a bridge superstructure dynamically. In this regard a point-surface based method has been proposed in [27].

We can see that, even if the coordinates of the single points obtained by a TLS are of low precision (±2 mm to ±20 mm), a 2D/3D model of the entire point cloud can be effective for detecting the shape of a structure and its changes. Therefore, if the goal of a TLS survey is to model a line or a surface starting from a number of points, we can observe that the best interpolating line/surface has generally a rather better precision than any single point; for this reason, the reconstructed lines/surfaces show a better precision than the one declared by the manufacturers. In other words, a modeled surface will represent an object more precisely than the unmodeled observations.

Starting from this remark, Gordon and Lichti [26] have developed a methodology devoted to the measurement of the deformation of structures. At the basis of the method are theoretical elements of mechanics of beam. For its implementation, least-squares are used. This method is based on obtaining analytical models that represent the physical bending of a structure and, in particular, of a beam.

The models, derived from the governing differential equation for the elastic curve, are represented, in the case of simple static schemes, by low order polynomials. The coefficients, treated as unknown parameters, are estimated by solving a least squares procedure. Such a method is effectively applicable when two conditions are satisfied: (a) the theoretical model and the real structure are really similar and

(b) the structure had not experienced events such as yielding of foundation or phenomena such as relaxation and creep.

In this paper, in light of all the above considerations, a methodology has been developed that allows to obtain the displacements and, consequently, the elastic curve of a structure using a TLS. The instrument must be configured as a line scanner and able to provide line scans, along with the timestamp for the detected points [28]. The displacements at time t are given by the difference between the best fitting line of the points acquired at the same time for a bridge with a loaded deck, and the fitting line obtained for the unloaded bridge. These displacements provide the instant elastic curve of the structure subject to the mobile load present at the time t.

The procedure can also be used for vertical structures subjected to loads with a horizontal component (tall buildings, pylons, etc.). The main characteristics of the method are: (1) Ease of installation; (2) The accuracy required to monitor the deflection of the bridges; and (3) High acquisition rate (up to 120 lines per second) useful for monitoring phenomena characterized by rapid changes.

The methodology for dynamic monitoring of displacements and the elastic curves of the structures presented below, is characterized by high precision, is easily implementable and can be effectively used within the SHM. It is worth mentioning that the steps of Structural Health Monitoring are: (1) Determination of damage existence; (2) Determination of damage's geometric location; (3) Quantification of damage severity and (4) Prediction of remaining life of the structure. The methodology presented concerns the first and third steps. To perform the others, it is generally necessary to integrate various information, coming from different types of sensors, with an accurate model of the structure to be analyzed. We should stress that TLS is a non-destructive technique performed remotely. Such techniques usually require a priori knowledge of the damaged region. As for the first and the third phases, the determination of damage existence and its quantification can be obtained by comparing the deflections measured with those expected in a design phase or obtained by previous measurements.

One of the main features of this methodology lies in the synchronization of measurement of deflections and positioning of mobile loads. Three experimental tests performed on as many bridges demonstrated its effectiveness and precision.

The topics covered in this paper are: (1) the methodology description; (2) the hardware components (Terrestrial Laser Scanner, Total Station, Digital Camera, GNSS receiver, Computer); (3) the software implemented, (4) the in-field tests; and (5) the discussion of results.

2. Materials and Methods

2.1. The Methodology

The method described exploits the capability of some TLS models to function as a line scanner and to provide the timestamp of each detected point, synchronized with the GPS time.

It is necessary to identify a significant line to be surveyed, which allows to obtain the difference between the elastic curve of the unloaded structure (dead load) and that of the structure subject to mobile or static loads. To this purpose, it is generally chosen to scan a line at the bottom of the deck, on the longitudinal axis or parallel to it.

For this reason, the TLS must be correctly aligned. Since the instrument does not project a visible beam, which would allow to check the alignment, it is possible to use a total station and/or exploit the features present on the surface of the structure or specifically positioned targets.

As a preliminary step, a line scan of the unloaded structure is performed, in correspondence with the selected section, which allows to have the reference shape. This shape is given by the best fitting line of the points measured (usually 10 to 100 points per meter at a 100 m distance from TLS). To obtain instant displacements, line scans are performed during the normal activity of the structure, with a sampling rate of up to 120 lines per second. For each line scan, the interpolation line is obtained and the scan time t, provided by the instrument timestamp, is memorized. The displacements at time t, are given by the deviation between the interpolation line at time t and the reference line. These displacements provide the elastic curve due to the load present at time t.

TLS is equipped with a GNSS receiver, which allows you to transform the timestamp into GPS time. The instantaneous position of a vehicle can be obtained from a GNSS receiver with which the vehicle is equipped. A video of the structure during the activity can also be used to determine the position of moving loads. Nowadays, several cameras are equipped with accessories that can provide Coordinated Universal Time (UTC) and, therefore, allow us to synchronize TLS and camera acquisitions.

2.1.1. Time Synchronization

The TLS must be able to provide a timestamp, in order to have the acquisition time of each point and assign a time to each surveyed line. The use of the GNSS receiver provided with the TLS, can allow us to express the timestamp as GPS time; in this way the acquisitions obtained with different instruments (videos, point sensors, etc.) can be synchronized and, above all, by positioning a receiver on a mobile load, its positions can be correlated to the acquired deformations.

The TLS are able to scan many lines per second (the instrument used, for example, can scan up to 120 lines per second), so it is possible to assign to a line the time derived from the timestamp of a point of the central zone with the precision of about 0.01 s.

As for the TLS/Video synchronization, we need to compose the approximation of the time assigned to the lines with the frame rate which, typically, is 30 fps. Therefore, we can consider conservatively a mean synchronization error equal to $(0.0167^2 + 0.01^2)^{1/2} = 0.0194$ sec., and a 95% confidence interval of 0.039 sec. With regard to the TLS/GNSS synchronization, given that the time given by the GNSS receiver can be easily obtained with a precision of 0.001 sec, we can assume a mean error equal to $(0.01^2 + 0.001^2)^{1/2} = 0.010$ sec., and a 95% confidence interval of 0.020 sec. The synchronization error involves a positioning error proportional to the speed of the mobile load. For the TLS/GNSS combination, by taking into account a speed of 20 m/sec, an error of ±0.20 m is obtained, with a maximum of ±0.40 m. For the TLS/video pair, an error of ±0.39 m is obtained, with a maximum of ±0.78 m. This error, that can be dramatically reduced by using higher frame rates (60 to 120 fps), should be composed with the error due to the approximation in the positioning of the mobile load in the image, depending on the Ground Sample Distance (GSD). This last error is negligible (a few cm) even for frames taken from a distance of 100 m, given the high resolution of the recent cameras.

Ultimately we can say that the error due to imperfect synchronization of acquisitions produces an error in the positioning of the load on average equal to a few decimetres. This implies a variation of the displacement from 1 to 3 percent for medium-span bridges (from 20 to 50 m) and lower for larger spans.

2.1.2. Use of Theoretical Models to Extract the Elastic Curves

From a theoretical point of view, if we have the detailed geometric and physical data, we could obtain analytical models that represent the physical bending of a structure and, in particular, of a beam. For a beam, the governing differential equation for the elastic curve, is given by:

$$\frac{\partial^2 z}{\partial x^2} = \frac{M(x)}{EI} \tag{1}$$

where:

z = deflection
x = abscissa (along the longitudinal axis)
M = bending moment;
E = Modulus of Elasticity;
I = Area moment of Inertia cross-section.

The displacement values are a function of the structural scheme and of the position in which the loads are applied [29]. In many cases, the displacement model can be assimilated to polynomials. With reference to Figure 1, e.g., taking into account a simply supported beam with uniform section, a structural scheme currently used in many bridges, a punctual load produces a displacement given by:

$$\delta_x = \frac{Fax(l^2 - a^2 - x^2)}{6lEI} \quad (2)$$

for $0 < x < b$, and by:

$$\delta_x = \frac{Fb(l-x)\left[l^2 - b^2 - (l-x)^2\right]}{6lEI} \quad (3)$$

for $b < x < l$
where:

F = Force acting on the beam;
δx = displacement at a distance x from the support 1;
a = distance from the load to the support 2;
b = distance from the load to the support 1;
l = distance between supports.

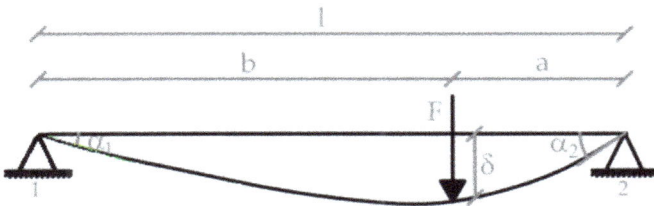

Figure 1. The elastic line of a simply supported uniform cross-section beam subject to a point load.

The generalized form of the cubic polynomial is given in Equations (4) and (5) [30].

$$\delta_x = a_{30}x^3 + a_{10}x + a_{00} + a_{01}y \quad (4)$$

for $0 < x < b$, and by:

$$\delta_x = b_{30}x^3 + b_{20}x^2 + b_{10}x + b_{00} + a_{01}y \quad (5)$$

for $b < x < l$, where:

y = horizontal distance from the longitudinal axis.

The best fitting line of the 2D point cloud provided by TLS in line scanner mode will be obtained by finding the coefficients a_{i0} and b_{i0} of the previous formulas with a least-squares procedure. The coefficient a_{01} has been introduced in [30] to take into account rotation about x axis.

2.1.3. Structures with Complex Shape

The use of a polynomial interpolation line is not possible for those structures characterized by complex structural patterns, or by non-uniformity in the geometric and physical characteristics of the structural elements, or that have suffered yielding or phenomena such as relaxation and creep.

In this cases structural calculations cannot be based on simple analytical expressions, but are performed through Finite Element Method (FEM); for this aim a 3D survey performed by TLS can be very useful when the project and the *as built* of the structures to be monitored are not available. The discrete values of the displacements obtained through FEM analysis will be interpolated using splines for longitudinal or transverse sections. Splines will be used also to obtain the interpolating lines of the 2D point cloud provided by the TLS in line scanner mode. For new structures, which do not have local irregularities due, for example, to material detachments, it may be considered to eliminate the points with a distance from the spline greater than the precision indicated by the instrument manufacturer (2 sigma). For dated structures, this procedure could worsen the results.

2.1.4. Elastic Curves

The instant elastic curves are given by the difference between the spline relative to the point cloud acquired on the structure subject to load, and the spline relative to the unloaded structure. As for the spline relative to the loaded structure, it will be necessary to respect the structural constraints (no displacements of the supports of a simply supported beam, no displacement and rotation at the fixed support of a cantilever beam, etc.). Comparing the displacements obtained by the TLS surveying and those obtained by means of the FEM analysis, it is possible to verify whether the behavior of a structure under various loading conditions is that expected in the design phase.

It is worth noting that the spline for the unloaded structure will be obtained from the entire point cloud sampled by the TLS in line scanner mode, composed of several line scans. The same observation can be made in the case of static loads.

In the case of dynamic loads, the goal of the surveying is to obtain instantaneous displacements, so we should derive a spline for each single sampled line. Therefore, we must first extract all the lines scanned by the TLS, with their timestamps. The software provided with TLS allows automatic extraction of the lines, once the instrument has been used in line scanner mode. This software generally returns the lines extracted in graphic form, but not as a sequence of 2D coordinates of the acquired points. In this way it is not possible to carry out most of the elaborations useful for understanding the behavior of the monitored structure.

It is therefore necessary to extract the individual lines from the overall file supplied by the instrument; for this purpose a code has been created in Matlab® which returns a file in text format, for each extracted line, containing, for each acquired point, the 3D coordinates. Once the files with the points of the single scanned lines are obtained, the interpolating lines to be used for the elaborations are found. Due to the large number of points detected, a high computing time is required.

Using the GNSS receiver supplied with the TLS, it is possible to associate to each extracted line also the position of the mobile load, obtained from synchronized videos or from a GNSS receiver mounted on the load itself. In this way a displacement line is obtained for each position of the mobile load.

The deformation lines can also be used to identify the natural frequencies of a monitored structure, if these are much lower than the sampling rate of the lines The elastic lines obtained by a Riegl VZ 1000 TLS, with a sampling rate of up to 120 lines per second, can be effectively used to derive the natural frequencies of high structures. This topic is outside the aims of this article, and it is object of a different investigation.

2.2. The Hardware Components

The hardware components are: (a) a Terrestrial Laser Scanner; (b) a Total Station for the alignment of the TLS; (c) a Digital Camera to get the video of the moving load; (d) a GNSS receiver. A medium end computer is used for data processing.

2.2.1. The Terrestrial Laser Scanner

A laser scanner RIEGL VZ 1000 was used. Table 1 shows its main characteristics.

Table 1. TLS Riegl VZ 1000 Main Technical Specifications.

Technical Feature	Values/Availability
Max Measurement Range	1400 m
Effective Measurement Rate	29,000 to 122,000 meas/s
Precision	5 mm
Accuracy	8 mm
Vertical Scan Angle Range	$+60°/-40°$
Scan Speed	3 lines/sec to 120 lines/s
GPS Receiver	Integrated L1 antenna
Compass	Integrated
Internal Sync Timer	Integrated real time synchronized time stamping of scan data
Multiple Target Capability	Yes
Laser Plummet	Integrated
Beam Divergence	0.3 mrad

2.2.2. The Total Station

A Total Station Leica 1201+ has been used for TLS alignment and control measurements; its main characteristics are reported in Table 2. We note that a low-cost total station can also be used, provided it is equipped with a visible Laser Pointer.

Table 2. Total Station Leica 1201+ Main Technical Specifications.

Technical Feature	Values/Availability
Angular accuracy	1″
Pinpoint EDM accuracy	1 mm + 1 ppm
Measurement range without prism	1000 m
Measurement range with a prism	3500 m
Automatic Target Recognition (ATR) accuracy	1″
Visible Laser Pointer	YES

2.2.3. The Digital Camera

For the second test, a video of the mobile loads was acquired using a NIKON D610 camera with a 55 mm previously calibrated NIKKOR lens. The main features are shown in Table 3.

Table 3. Characteristics of the NIKON D610 digital camera.

Feature	Value
Type	Single-lens reflex digital camera
Effective pixels	24.3 million
Image sensor	Nikon FX format 35.9 × 24.0 mm—DX format 24 × 16 mm
File format	NEF (RAW), JPEG, NEF (RAW) + JPEG
Lens	NIKKOR 18–55 mm f/3.5–5.6 G VR
Shutter	Electronically-controlled vertical-travel focal-plane shutter
ISO sensitivity	ISO 100 to 6400 in steps of 1/3 or 1/2 EV
HD frame and frame rate	1920 × 1080 pixels; 30 p (progressive), 25 p, 24 p
GP-1 unit providing Coordinated Universal Time (UTC).	YES

2.2.4. The GNSS Receiver

The GNSS receiver used for the mobile load of the third test is a Leica GS15, mounted on a strong magnetic base that allows the fixing on the roof of a vehicle. It can provide raw and rinex data and allows a recording rate up to 20 Hz. For our aims, the receiver was configured to track GPS and

GLONASS; the Continuously Operating Reference Station (CORS) of the University of Calabria was used in order to perform the post processing.

2.3. The Software Implemented

For the processing of data obtained from TLS in line scan mode, a code has been developed in Matlab®. Starting from the point cloud, the best fitting cubic spline is obtained in the case of unloaded structure. This spline represents the reference line for obtaining the displacement curves.

In the case of static load, the same calculation is performed using the point cloud acquired by TLS during the permanence of the load on the structure. The spline extracted is corrected by imposing compliance with structural constraints. The difference between the obtained spline and the reference line gives the displacements due to the load.

As for mobile loads, the following operations are performed: (1) the single line scans are obtained by detecting the negative increase of the TLS zenith angle; (2) for each line scan, the average of the first and the last point's timestamps is assumed as line time; (3) the best fitting cubic spline is calculated and corrected by imposing compliance with structural constraints; (4) the position of the mobile load is obtained using the previous and following GPS times with respect to the timestamp. The difference between the spline obtained for each line and the reference line gives the displacements due to the mobile load in the position obtained by GNSS receiver.

If a FEM analysis is available, a comparison between the displacements obtained by TLS measurements and those expected in the design phase can be performed.

The flowchart of the implemented software is shown in Figure 2.

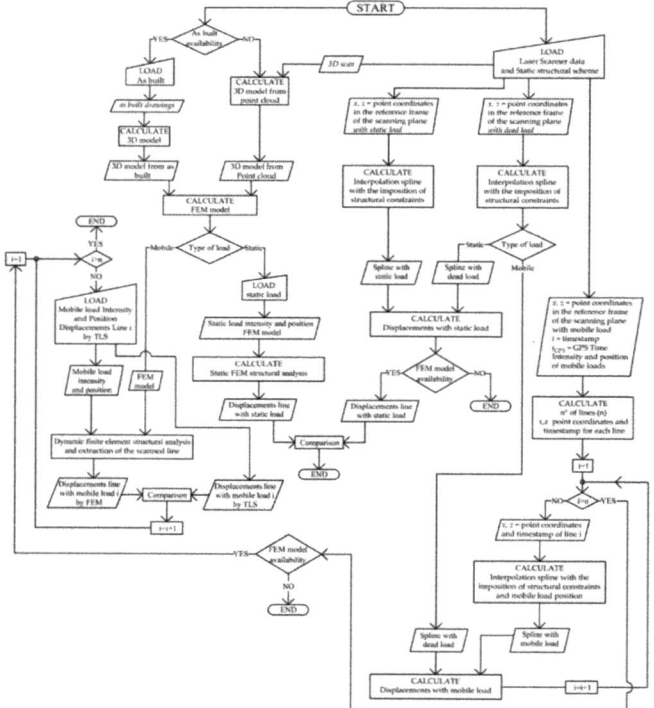

Figure 2. The Flowchart of the implemented software for structural analysis.

3. Results

3.1. The First Test: The Bridge at University of Calabria

The first test was performed on a double deck bridge at the University of Calabria, Italy. The bridge materializes the South-North central axis, that characterizes the campus layout. All the buildings are lined up on the sides of the bridge. The lower deck is intended for pedestrians. The upper one is used occasionally for vehicular traffic. The main structure is a simply supported truss beam, characterized by three longitudinal tubular elements; the lower deck is hung from the upper ones (Figure 3).

Figure 3. The bridge at the University of Calabria: (**a**) the pedestrian deck; (**b**) the upper deck.

In Figure 4 the cross sections of the simply supported truss beam at the bearings are shown. The red line represents the path followed by the laser scanner during the acquisitions in line-scan mode.

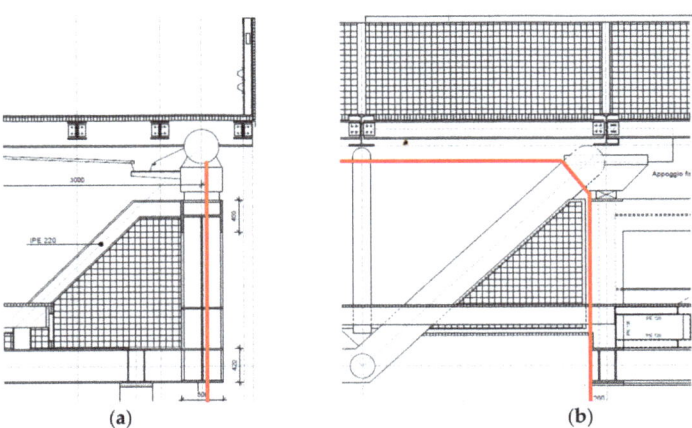

Figure 4. The bridge's Cross Sections at the bearings: (**a**) transversal; (**b**) longitudinal.

Figure 5 shows the test layout. The test was performed during the parking of a truck elevator, which weight was about 260 kN (Figure 6). The TLS position was chosen below the deck. In this way it was possible to acquire the points of the upper element of the deck structure, realized with a 3D truss. To minimize the oscillations of the rotation axis, the minimum sampling rate was chosen, equal to 70,000 points per second. The distance between the bearings of the bridge is 40.30 m. Measuring the position of the wheel axles, the center of gravity of the truck was obtained, located 11.07 m from the south support.

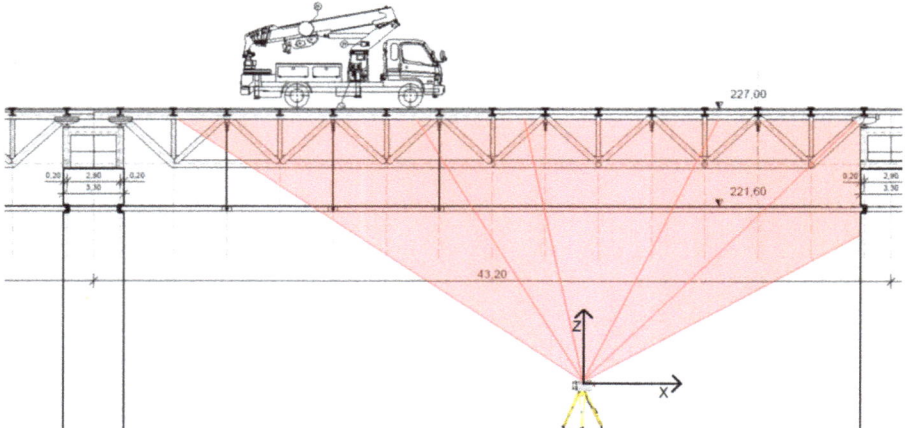

Figure 5. The Layout of the first Test.

Figure 6. The truck elevator, used as static load.

RiSCAN PRO®, the software provided with the TLS, was used for the first processing of data. The coordinates of the scanned points were obtained, with reference to the TLS centre and expressed in meters. The software described in Section 2.3 was then used; the following figures show the results.

Figure 7a shows, on an arbitrary scale, a view of the entire span and of the lines obtained with the bridge loaded and without loads. The origin of the reference system is the center of TLS. One can observe that the deck has a vertical displacement, with respect to the horizontal direction, of about 134 mm in the middle of the span, with only the dead load. The circles indicate the enlargements shown in Figure 8. Figure 7b shows the deflections due to the truck load, obtained from raw data and the trend-line. In this case, a cubic polynomial, of the type described in Equations (4) and (5), provides an effective fitting line.

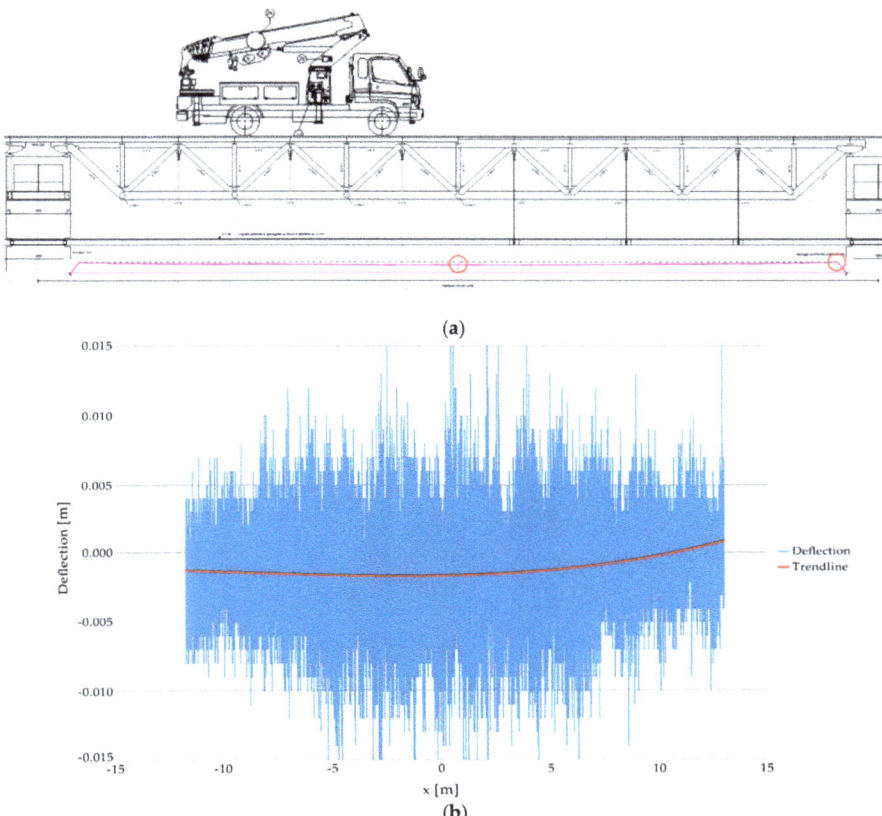

Figure 7. (a) The elevation of the bridge and the lines obtained from TLS. The circles indicate the zooms shown in Figure 8; (b) The deflections obtained from the row data (blue) and the trendline (red).

Figure 8 shows two enlargements of raw data and interpolation lines. It is possible to observe that the lines obtained with the loading and unloading bridge coincide on the bearing (Figure 8a), while in the central part of the beam there is a difference in height of about 2 mm (Figure 8b). Also in this case the coordinates are referred to the intrinsic reference system of the instrument.

Using a FEM program developed at the University of Calabria [31], it was obtained an accurate evaluation of vertical displacements. The computed deflection was 1.95 mm, 2.5% less than the value attained by using the proposed methodology. Furthermore, by means of a total station Leica 1201+, a precise measurement was performed. A prism was placed in the middle of the span and was repeatedly monitored from a distance of about 22 m. The automatic target recognition function (ATR) has been activated, which has an accuracy of 1 s; therefore the vertical displacements are obtained with an accuracy of about 0.1 mm. The variation of the bridge deflection thus obtained was 1.9 mm, 5% less than that obtained using the proposed methodology.

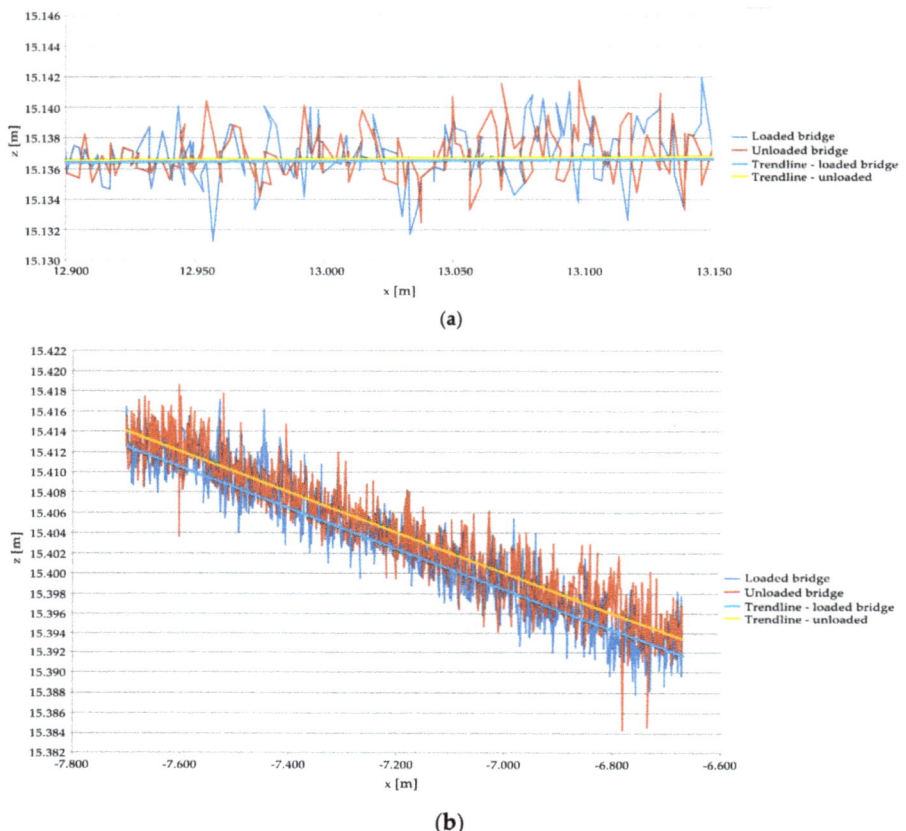

Figure 8. Points and trendlines obtained with loading and unloading bridge: (**a**) on the bearing; (**b**) on the midspan.

3.2. The Second Test: The Cannavino Bridge at Celico

The Cannavino bridge (Figure 9a) has been realized of prestressed concrete. The structural scheme makes use of opposing cantilevers, whose ends recently have been subjected to large deflections (Figure 9); for this reason the technicians of ANAS, the Italian National Autonomous Roads Corporation, carry out periodic monitoring with the total station [32].

In 1972, during its construction, the bridge suffered the collapse of two cantilevers. To understand the reasons, some in-depth studies were carried out [33].

In Figure 10, one can see the elevation of the bridge (Figure 10a), a photo of the collapsed cantilevers (Figure 10b) and some details at the breaking points (Figure 10c). Figure 11 shows the south side balanced cantilever with the post-tensioned segments (Figure 11a) and their cross section (Figure 11b); the red dot indicates the line scanned during the test. The height of the segments ranges from 2.00 to 7.80 m. Figures 10 and 11 are taken from [33].

The above described VZ1000 TLS was used, set up as a line scanner. To record the sampling time of each acquired point, the timestamp function has been activated. GPS time was acquired using the receiver of which TLS is provided.

Figure 12a shows the layout of the test. The TLS station point was chosen under a cantilever, close to a pile. A path parallel to the longitudinal axis of the bridge, at the bottom of the sidewalk, was acquired for 60 s, in order to avoid an excessive size of the file to be processed (Figure 12b). The scans were carried out with the bridge open to traffic. It was chosen a 110,000 points per second scanning speed.

Figure 9. Views of Cannavino bridge: (**a**) Panorama; (**b**) Zoom of the end of the central cantilevers; (**c**) deck surface.

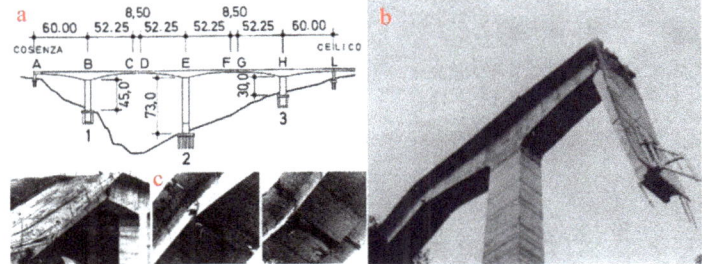

Figure 10. Drawing and photo after the collapse: (**a**) Quoted Elevation; (**b**) Collapsed Cantilevers; (**c**) Details of the broken zones.

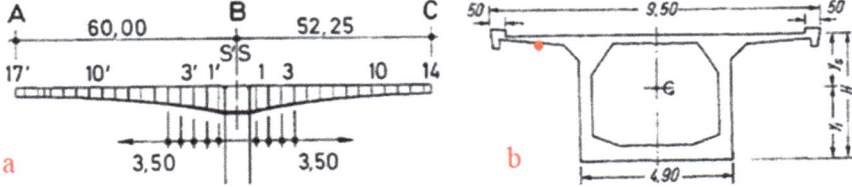

Figure 11. Draft of the south side balanced cantilever: (**a**) Post-tensioned segments; (**b**) Cross section. The red dot indicates the line scanned during the test.

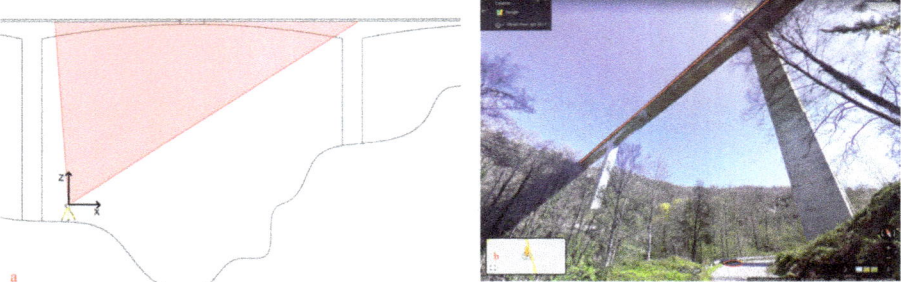

Figure 12. The Test on the Cannavino Bridge: (**a**) The Layout; (**b**) the position of TLS (red circle) and the scanned line (red line).

RiSCAN PRO® software was used to process the acquired data. Point coordinates and timestamps have been recorded in a text file. Subsequently the file was processed with the previously described Matlab® code and the single lines were extracted and corrected, by imposing the structural constraints. The splines were then calculated for each scanned line.

The video of the loads was acquired with the above described digital camera, equipped with a GP-1 unit, an accessory capable of providing Coordinated Universal Time (UTC) (Figure 13).

Figure 13. A frame of the video used to obtain position and estimate of the mobile loads.

Figure 14 shows a stretch of the splines obtained for six load combinations; the reference system of the TLS shown in Figure 12 was used for the coordinates in meters.

The green line is obtained for the maximum load visible in the frame of Figure 13. The cyan line refers to the unloading bridge, while the other lines refer to intermediate load conditions. As regards the position of the loads, they were obtained approximately taking into account the posts of the guardrails. Loads were estimated based on the type of vehicles present in each selected frame. A FEM analysis was performed using the aforementioned code [31], the structural data provided by [33] and the six load combinations.

The following remarks can be made: (a) towards the free end of the cantilever, the splines diverge (different colors correspond to different times) when the mobile loads increase; (b) the maximum distance is about 6 mm; (c) The FEM analysis provides 16% lower results, with a maximum displacement of 5 mm. This difference is understandable given the uncertainty in the evaluation of loads, the unavailability of the *as built* and the age of the structure. A comparison was made with the outcomes of the load tests carried out by the technicians of ANAS some years earlier with higher loads. The results obtained, taking into account the ratio between the loads in the two tests, show a difference of about 10%.

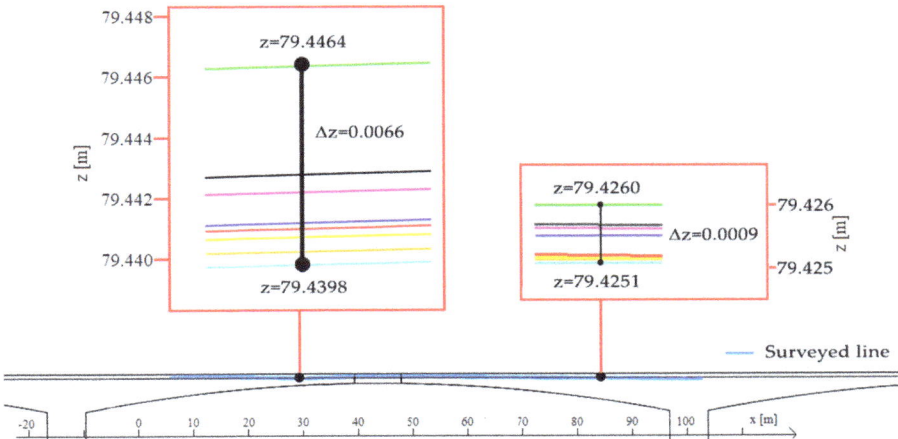

Figure 14. A stretch of the splines obtained for six load combinations. In the boxes, the enlargement of the splines in correspondence of the two points indicated by dots, near to the free end of the cantilever (left box) and near the fixed end (right box). The scale is arbitrary.

3.3. The Third Test: The Santiago Calatrava's San Francesco Bridge at Cosenza

Built to allow an access large and of artistic value to the city of Cosenza, the large structure of San Francesco Bridge (Figure 15) crosses the Crati river.

Figure 15. The Santiago Calatrava's San Francesco Bridge at Cosenza.

It is a cable-stayed bridge with a cantilever spar. The characterizing structural element is the single inclined pylon, which with its height of 95 m marks the surrounding landscape. The span, 200 m long, is counterbalanced by twenty twin cable lengths. Two large rods contrast the actions of the cables. A special feature of the bridge is the upper part of the pylon, where the cables are anchored so as to give the impression of a sail. The metal structure of the deck rests on artistically sculpted concrete abutments.

To represent work and the surrounding area of the city, a survey was conducted using a TLS [34]. The surveyed area extends to the historic center and covers the riverbed. (Figure 16).

The experiment was performed during the official load test carried out on 17 January 2018, a windy day; a series of trucks of known weight were placed on the deck (Figure 17b).

Figure 16. The 3D model of San Francesco Bridge.

Several instruments were used during the test. In addition to a digital level and a total station; some strain sensors were placed both on the pylon and in various positions on the deck. Furthermore, two inclination sensors provided inclination at different pylon heights in real time, thanks to a wireless connection [35].

A station point for VZ1000 TLS was chosen, aligned with the central axis of the bridge. The chosen location is in a stable area and is about 90 m from the anchorage of the rods. This position allows a vertical section of the pylon to be described with the laser beam of the instrument. In particular, the outer part of its cylindrical surface with elliptical section is detected. The rods hidden only a short stretch of the pylon. In this way, we can get the elastic line of the pylon for each line scan, during the runs of the trucks used as mobile loads.

Figure 17a shows the layout of the test. On the left side of the panel, you can see the TLS station. The laser trace is highlighted by a red line. Figure 18a shows a vertical section extracted from a line scan. Figure 18b shows the point cloud obtained for the pylon head. Overlaid you can see the upper part of the laser track, circled in blue in Figure 18a.

The official load test involved the sequential positioning of ten trucks side by side on two adjacent lanes (Figure 17b). The TLS acquisitions were made in two configurations: (a) with unloaded bridge, (b) during the load test.

The unloaded bridge acquisition was carried out from 11.33 am to 11.35 am local time (UTC + 1). The acquisition with mobile loads was carried out from 10:41 UTC (GPS time 1,200,220,888) to 10.51 UTC (GPS time 1,200,221,500). The sampling rate was 120 lines per second. This acquisition began before the first truck left and continued until before the departure of the fourth truck. The scans were then stopped due to the heavy rain, so the runs of the last vehicles were not acquired. The trucks were equipped with Leica GS15 GNSS receivers, mounted on a solid magnetic base that allows them to be fixed to the cab roof.

The recording rate was set to 5 Hz. The data collected by the permanent GNSS station of the University of Calabria were used to perform post processing.

Figure 17. The Test on the San Francesco Bridge: (**a**) The Layout with the TLS position (red circle) and the scanned line (red line); (**b**) Trucks used as loads; (**c**) The TLS, positioned after the alignment operation, preliminary to the test.

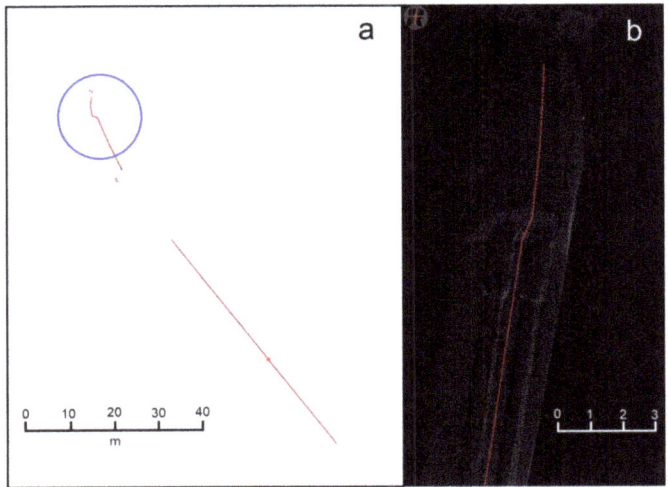

Figure 18. A single line scanned: (**a**) The section of the pylon; (**b**) The upper part of the scanned line superimposed to 3D model of the pylon.

Table 4 shows the GPS time for the actions performed during the load test.

Table 4. GPS time for the actions performed during the load test.

Event	GPS Time (s)
Start of TLS acquisitions	1,200,220,888
Start of first Truck	1,200,221,072
Stop of first Truck	1,200,221,098
Start of second Truck	1,200,221,186
Stop of second Truck	1,200,221,214
Start of third Truck	1,200,221,243
Stop of third Truck	1,200,221,268
Stop of TLS acquisitions	1,200,221,500

A comparison between the results obtained by the total station and the acquisitions of the TLS in line scanner mode showed a substantial agreement between the two techniques. Also the comparison with the acquisitions of inclination sensors confirmed the reliability of the methodology.

Figure 19 shows some test results. The magnifications correspond to the areas circled in blue in the thumbnails. The point cloud of the profile acquired with the unloaded bridge is in blue.

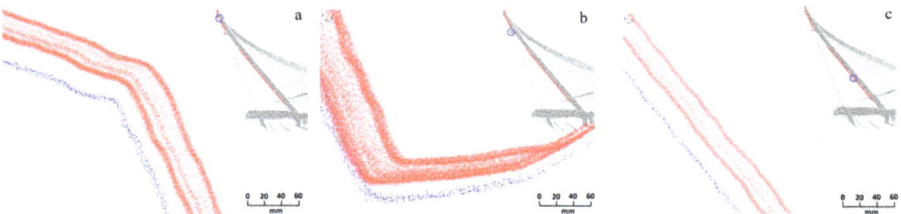

Figure 19. Enlargements of the point clouds obtained before the load test (blue points) and during the test (red points): (**a**) Upper zone of the pylon ($z = 81$ m); (**b**) Constraint device in which the rods can slide ($z = 69$ m); (**c**) lower part of the pylon ($z = 39$ m).

First of all we can see that the precision declared for the instrument and reported in Table 1 is confirmed. The cloud of blue points, in fact, is about 8 mm thick. Therefore the precision of the acquisitions is about ten times lower than the displacements we have to measure.

In the panel 19a we can see the displacements of the pylon head. The height, referred to the TLS centre, is about 81 m. The dense red point cloud on the left was obtained during the stop of the first truck. The red dense point clouds on the right were obtained during the short stop of the second truck and during the stop of the third one. At the end of the TLS acquisitions, three trucks were positioned on the bridge. Also this point cloud is 8 mm thick. Scattered points were acquired while the second and third trucks were running. Point clouds acquired during truck slowdown become denser. We observe an overall displacement of 41 mm. The first two mobile loads caused roughly half of the total displacement.

In the panel 19b one can see the points acquired in correspondence with the constraint device in which the rods can slide, 69 m above the TLS. In this case, we can see that the horizontal displacements begin with the run of the second truck. The analysis of this behavior is outside the scope of our study. The maximum displacements are slightly lower with respect to the upper part. Since these zones are close together, this behavior seems reasonable.

The points acquired in the lowest part of the pylon are shown in panel 19c. The height of this zone is 39 m respect to the TLS center. The displacements reach a maximum value of 26 mm.

Using the Matlab code described above, the individual lines from the overall file supplied by the TLS were extracted and a timestamp was associated to each line. The interpolating lines were then determined.

Figure 20 shows a stretch of four splines, obtained for different instants, in the upper zone of the pylon. The coordinates are referred to the center of the instrument, so x is the horizontal distance and z is the difference in height with respect to the center of the TLS. The magenta line (GPS TIME 1200220928) is related to the unloading bridge. The other lines relate to the parking of the first truck (blue line), the parking of the second truck alongside the first one (green line) and the parking of the first three trucks (yellow line). From the data collected it is possible to extract the movements of the pylon, at each height, as a function of time.

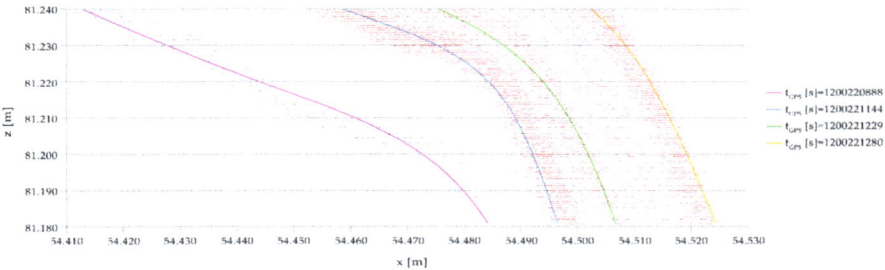

Figure 20. Enlargements of the point clouds obtained at a height of 80 m above the TLS. Superimposed are the interpolating lines of the 2D point cloud provided by the TLS in line scanner mode.

Figure 21 shows the horizontal movements of the pylon at three different heights. In panel (a) we observe the movements of a point at a height of 80 m with respect to the instrument, as a function of the GPS time. The displacements due to the moving loads are easily recognizable by analyzing the 10 samples moving average, drawn in red: during the run of the first truck the point moves up to 15 mm towards the midspan. The point does not move during the stop and undergoes a further displacement during the run of the second truck. After a brief pause, the third movement begins, ending with the halt of the third truck. The total displacement is about 41 mm. It is worth remarking that the graph reflects the times of the events reported in Table 4.

The comments made above can be repeated for the point at a height of 50 m, which movements are drawn in the panel (b). Displacements are reduced, as expected.

As regards the point at 20 m height, the various phases are less evident. A final displacement of about 16 mm can be observed.

Measurements were performed with a total station after positioning each of the three trucks. The station pointed targets placed on the pylon at various heights and its measurements were used for the official load test. The results fully agree with those obtained with our method. Also the inclination variations sampled by two sensors, positioned inside the pylon and connected wirelessly, are consistent with the splines extracted using the method described.

Thanks to the acquisitions made by the GNSS receivers positioned on the trucks, the displacements of the points of the pylon can finally be correlated to the position of the mobile loads. Since the TLS position was obtained thanks to its GPS receiver, the coordinates of the each mobile vehicle, for each GPS time, can be converted into local coordinates in the TLS reference frame.

The correlation is shown in Figure 22 in which the horizontal positions of a point of the upper part of pylon are represented as a function of GPS time. The point is 85 m above the TLS centre. The trend-line evidences runs and stops of the trucks. In the thumbnails, the GPS time and the instantaneous local coordinate x of the moving loads, represented by dots, are shown. The red, blue and green dots represent, respectively, the first, second and third trucks.

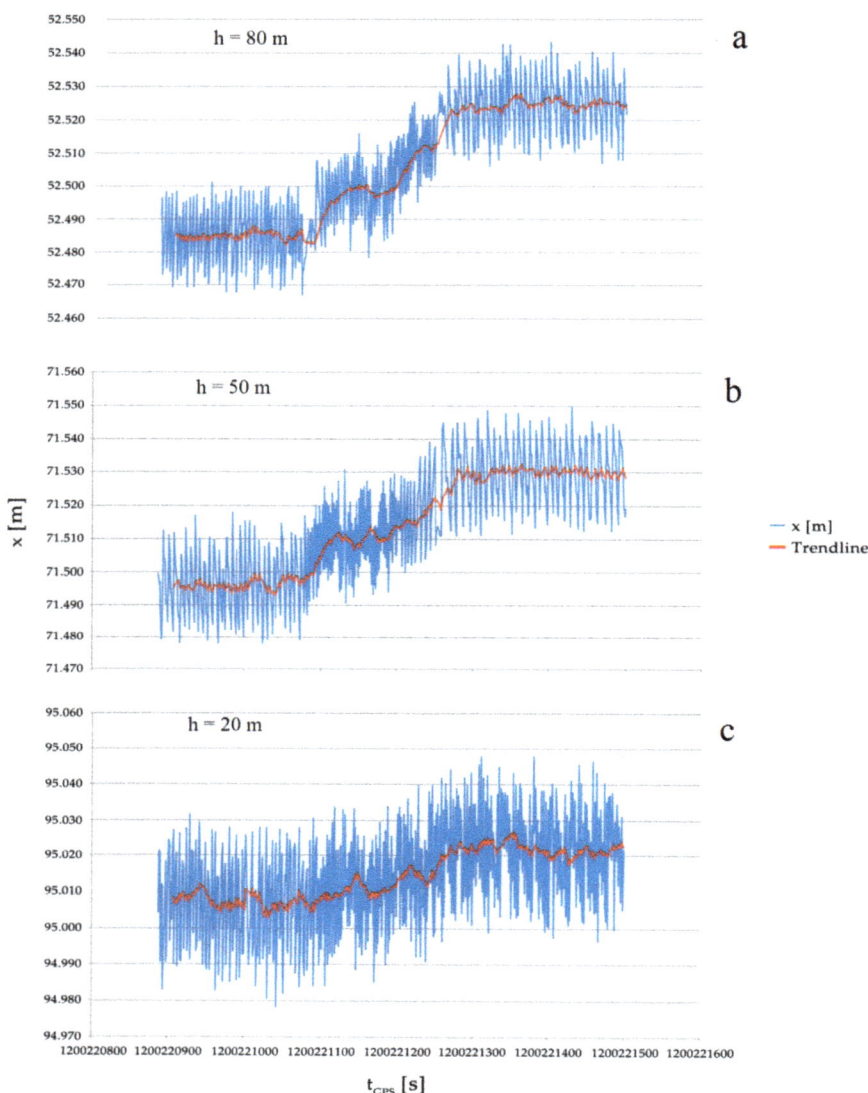

Figure 21. Horizontal displacements, as a function of the GPS time, for three points of the pylon during the test: (**a**) at 80 m above the TLS; (**b**) at 50 m above the TLS; (**c**) at 20 m above the TLS. Ordinates are the horizontal distances from the TLS.

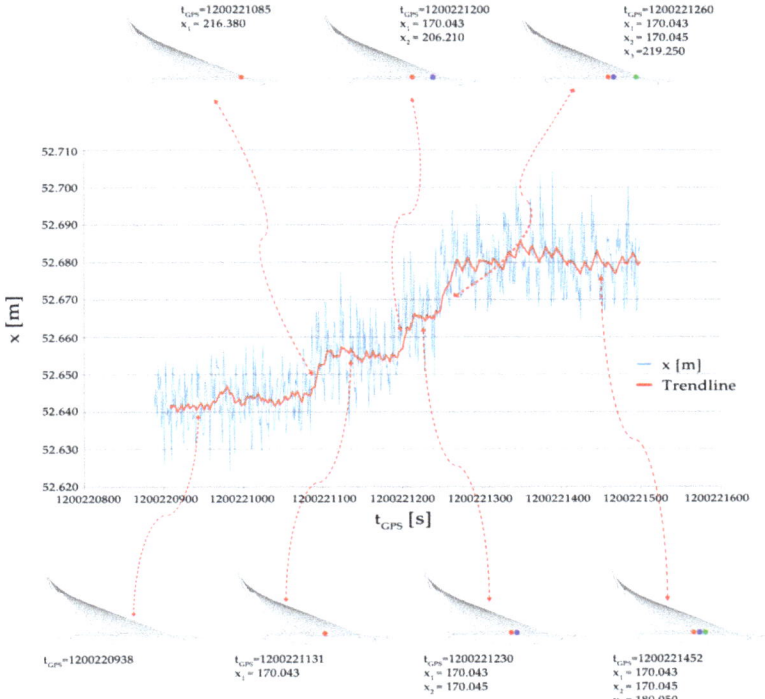

Figure 22. Horizontal displacements for a point of the pylon 85 m above the TLS during the test, correlated to the position of the mobile loads. The trend-line is a 10 sample moving average. In the thumbnails, GPS time and position of mobile loads (local coordinates) are reported.

4. Discussion

The three tests carried out allows us to make the following remarks.

The reliability of the method, if complete information on the materials and on the structural scheme is available, has been demonstrated with the first test. The displacements obtained under static loading conditions, approximated using a cubic polynomial, differ by approximately 2.5% from those obtained from the FEM analysis and by about 5% from the results provided by independent precise measurements.

The applicability of the method to structures more complex and with less information was verified with the second test, carried out on a bridge characterized by relaxation and creep phenomena, in normal traffic conditions. The difference between the results obtained with the method described and those of a FEM analysis is about 16%. Compared to the outcomes of the aforementioned measurements carried out by the technicians of ANAS, the difference decreases to about 10%. In any case, the results obtained are acceptable for an estimate of the behavior of the structures.

As regards the third test, the comparison with the results obtained with other techniques and instrumentations (total station and inclination sensors) confirmed the validity and precision of the methodology.

In all the tests, the use of interpolating lines allowed to get higher precision with respect to the unmodeled observations. The working hypotheses, therefore, proved to be reliable.

A peculiarity of the technique described is the possibility of obtaining the 3D model of the monitored structure and the dynamic linear survey with the same instrument and reference system (Figure 18).

Possible fields of application of the methodology presented are both road and railway infrastructures, tall structures and buildings. The method proves particularly valid for constructions characterized by remarkable spans and/or great heights.

A limit to its usability is given by the need to have a point to position the TLS, with a clear view of the whole or at least of most of the structure to be monitored.

The integration with different techniques opens up the prospect of interesting uses. In the immediate future, activities are planned to exploit the high acquisition rate of TLS, in order to determine the natural frequencies of structures. To this end, acquisitions will be performed simultaneously using TLS and Ground Based Radar Interferometry.

Author Contributions: Conceptualization, S.A.; methodology, S.A.; software S.A.; validation, S.A. and R.Z.; structural investigation, R.Z.; resources, R.Z.; data curation, S.A.; writing—original draft preparation, S.A.; writing—review and editing, S.A. and R.Z.; visualization, S.A. All authors have read and agreed to the published version of the manuscript.

Funding: This research received no external funding.

Conflicts of Interest: The authors declare no conflict of interest.

References

1. Lantsoght, E.O.L.; der Veen, C.; Boer, A.; Hordijk, D.A. State-of-the-art on load testing of concrete bridges. *Eng. Struct.* **2017**, *150*, 2312–2341. [CrossRef]
2. Zinno, R.; Artese, S.; Clausi, G.; Magarò, F.; Meduri, S.; Miceli, A.; Venneri, A. Structural Health Monitoring (SHM). In *The Internet of Things for Smart Urban Ecosystems*; Cicirelli, F., Guerrieri, A., Mastroianni, C., Spezzano, G., Vinci, A., Eds.; Springer International Publishing: Switzerland, Cham, 2019; pp. 225–249. [CrossRef]
3. Fan, W.; Qiao, P. Vibration-based damage identification methods: A review and comparative study. *Struct. Health Monit.* **2011**, *10*, 83–111. [CrossRef]
4. Malekjafarian, A.; McGetrick, P.J.; OBrien, E.J. A Review of Indirect Bridge Monitoring Using Passing Vehicles. *Shock Vib.* **2015**, *2015*, 286139. [CrossRef]
5. Carpinteri, A.; Lacidogna, G.; Pugno, N. Structural damage diagnosis and life-time assessment by Acoustic Emission monitoring. *Eng. Fract. Mech.* **2007**, *74*, 273–289. [CrossRef]
6. Davis, A.; Mirsayar, M.; Hartl, D. Structural health monitoring using embedded magnetic shape memory alloys for magnetic sensing. In Proceedings of the SPIE–International Society for Optics and Photonics, Nondestructive Characterization and Monitoring of Advanced Materials, Aerospace, Civil Infrastructure, and Transportation XIII, Bellingham, WA, USA, 4–7 March 2019; Gyekenyesi, A.L., Ed.; Volume 10971.
7. Morris, I.; Abdel-Jaber, H.; Glisic, B. Quantitative Attribute Analyses with Ground Penetrating Radar for Infrastructure Assessments and Structural Health Monitoring. *Sensors* **2019**, *19*, 1637. [CrossRef]
8. Artese, S.; Perrelli, M. Monitoring a Landslide with High Accuracy by Total Station: A DTM-Based Model to Correct for the Atmospheric Effects. *Geosciences* **2018**, *8*, 46. [CrossRef]
9. Artese, G.; Perrelli, M.; Artese, S.; Manieri, F. Geomatics activities for monitoring the large landslide of Maierato, Italy. *Appl. Geomat.* **2015**, *7*, 171. [CrossRef]
10. Lienhart, W.; Ehrhart, M.; Grick, M. High Frequent Total Station Measurements for the Monitoring of Bridge Vibrations. In Proceedings of the 3rd Joint International Symposium on Deformation Monitoring (JISDM), Vienna, Austria, 30 March–1 April 2016; Available online: https://www.fig.net/resources/proceedings/2016/ (accessed on 2 September 2019).
11. Yu, J.; Zhu, P.; Xu, B.; Meng, X. Experimental assessment of high sampling-rate robotic total station for monitoring bridge dynamic responses. *Measurement* **2017**, *104*, 60–69. [CrossRef]
12. Lovse, J.W.; Teskey, W.F.; Lachapelle, G.; Cannon, M.E. 7-Dynamic Deformation Monitoring of Tall Structure Using GPS Technology. *J. Surv. Eng.* **1995**, *121*, 35–40. [CrossRef]
13. Roberts, G.W.; Meng, X.; Dodson, A.H. Integrating a global positioning system and accelerometers to monitor the deflection of bridges. *J. Surv. Eng.* **2004**, *130*, 65–72. [CrossRef]
14. Yu, J.; Meng, X.; Shao, X.; Yan, B.; Yang, L. Identification of dynamic displacements and modal frequencies of a medium-span suspension bridge using multimode GNSS processing. *Eng. Struct.* **2014**, *81*, 432–443. [CrossRef]

15. Jo, B.W.; Lee, Y.S.; Jo, J.H.; Khan, R.M.A. Computer Vision-Based Bridge Displacement Measurements Using Rotation-Invariant Image Processing Technique. *Sustainability* **2018**, *10*, 1785. [CrossRef]
16. Baohua Shan, B.; Wang, L.; Huo, X.; Yuan, W.; Xue, Z. A Bridge Deflection Monitoring System Based on CCD. *Adv. Mater. Sci. Eng.* **2016**, *2016*, 4857373. [CrossRef]
17. Xu, Y.; Brownjohn, J.M.W. Review of machine-vision based methodologies for displacement measurement in civil structures. *J. Civil Struct. Health Monit.* **2018**, *8*, 91. [CrossRef]
18. Feng, D.; Feng, M.Q. Experimental validation of cost-effective vision-based structural health monitoring. *Mech. Syst. Signal Process.* **2017**, *88*, 199–211. [CrossRef]
19. Yoneyama, S.; Ueda, H. Bridge Deflection Measurement Using Digital Image Correlation with Camera Movement Correction. *Mater. Trans.* **2012**, *53*, 285–290. [CrossRef]
20. Pepe, M. Image-based methods for metric surveys of buildings using modern optical sensors and tools: From 2D approach to 3D and vice versa. *Int. J. Civil Eng. Technol.* **2018**, *9*, 729–745.
21. Yan, Y.; Hang, L.; Dongsheng, L.; Xingquan, M.; Jinping, O. Bridge Deflection Measurement Using Wireless MEMS Inclination Sensor Systems. *Int. J. Smart Sens. Intell. Syst.* **2013**, *6*, 38–57. [CrossRef]
22. Zhao, X.; Liu, H.; Yu, Y.; Xu, X.; Hu, W.; Li, M.; Ou, J. Bridge displacement monitoring method based on laser projection-sensing technology. *Sensors* **2015**, *15*, 8444–8463. [CrossRef]
23. Artese, S.; Achilli, V.; Zinno, R. Monitoring of Bridges by a Laser Pointer: Dynamic Measurement of Support Rotations and Elastic Line Displacements: Methodology and First Test. *Sensors* **2018**, *18*, 338. [CrossRef]
24. Jung, J.; Kim, D.-J.; Palanisamy Vadivel, S.K.; Yun, S.-H. Long-Term Deflection Monitoring for Bridges Using X and C-Band Time-Series SAR Interferometry. *Remote Sens.* **2019**, *11*, 1258. [CrossRef]
25. Di Pasquale, A.; Nico, G.; Pitullo, A.; Prezioso, G. Monitoring Strategies of Earth Dams by Ground-Based Radar Interferometry: How to Extract Useful Information for Seismic Risk Assessment. *Sensors* **2018**, *18*, 244. [CrossRef] [PubMed]
26. Gordon, S.J.; Lichti, D.D. Modeling Terrestrial Laser Scanner Data for Precise Structural Deformation Measurement. *J. Surv. Eng.* **2007**, *133*, 72–80. [CrossRef]
27. Truong-Hong, L.; Laefer, D.F. Using Terrestrial Laser Scanning for Dynamic Bridge Deflection Measurement. In Proceedings of the IABSE Istanbul Bridge Conference, Istanbul, Turkey, 11–13 August 2014; ISBN 978-605-64131-6-2.
28. Artese, S. Survey, diagnosis and monitoring of structures and land using geomatics techniques: Theoretical and experimental aspects. In *Geomatics Research 2016*; Vettore, A., Ed.; AUTeC: Caldogno, Italy, 2017; pp. 31–42. ISBN 978-88-905917-9-2.
29. Chen, W.; Lui, E. *Handbook of Structural Engineering*, 2nd ed.; CRC Press: Boca Raton, FL, USA, 2005; p. 1768. ISBN 9780849315695.
30. Gordon, S.J.; Lichti, D.D.; Chandler, I.; Stewart, M.P.; Franke, J. Precision Measurement of Structural Deformation using Terrestrial Laser Scanners. In Proceedings of the Optical 3D Methods, Zurich, Switzerland, 22–25 September 2003; Gruen, A., Kahmen, H., Eds.;
31. Madeo, A.; Casciaro, R.; Zagari, G.; Zinno, R.; Zucco, G. A mixed isostatic 16 dof quadrilateral membrane element with drilling rotations, based on Airy stresses. *Finite Elem. Anal. Des.* **2014**, *89*, 52–66. [CrossRef]
32. Stradeanas.it. Available online: https://www.stradeanas.it/it/calabria-strada-statale-107-l%E2%80%99anas-prosegue-il-monitoraggio-delle-condizioni-statiche-del-viadotto (accessed on 30 September 2019).
33. Wittfoht, H. Ursachen fur den Teil-Ensturz des "Viadotto Cannavino" bei Agro di Celico. *Beton Stahlbetonbau* **1983**, *78*, 29–36, ISSN 0005-9900. [CrossRef]
34. Artese, S. The Survey of the San Francesco Bridge by Santiago Calatrava in Cosenza, Italy. *Int. Arch. Photogramm. Remote Sens. Spatial Inf. Sci.* **2019**, *XLII-2/W9*, 33–37. [CrossRef]
35. Artese, G.; Perrelli, M.; Artese, S.; Meduri, S.; Brogno, N. POIS, a Low Cost Tilt and Position Sensor: Design and First Tests. *Sensors* **2015**, *15*, 10806–10824. [CrossRef]

 © 2020 by the authors. Licensee MDPI, Basel, Switzerland. This article is an open access article distributed under the terms and conditions of the Creative Commons Attribution (CC BY) license (http://creativecommons.org/licenses/by/4.0/).

Article

Study on the Method of Moving Load Identification Based on Strain Influence Line

Jing Yang [1], Peng Hou [1], Caiqian Yang [1,2,*] and Yang Zhang [3]

1. Key Laboratory of Concrete and Prestressed Concrete Structures of the Ministry of Education, School of Civil Engineering, Southeast University, Nanjing 210096, China; yangjingseu@seu.edu.cn (J.Y.); houpeng@seu.edu.cn (P.H.)
2. College of Civil Engineering & Mechanics, Xiangtan University, Xiangtan 411105, China
3. Jiangsu Suhuaiyan Expressway Management Co., Ltd., Huaian 223000, China; ceseunj@163.com
* Correspondence: ycqjxx@seu.edu.cn

Abstract: In order to improve the accuracy of load identification and study the influence of transverse distribution, a novel method was proposed for the moving load identification based on strain influence line and the load transverse distribution under consideration. The load identification theory based on strain influence line was derived, and the strain integral coefficient was proposed for the identification. A series of numerical simulations and experiments were carried out to verify the method. The numerical results showed that the method without considering the load transverse distribution was not suitable for solving the space problem, and the method with the load transverse distribution under consideration has a high identification accuracy and excellent anti-noise performance. The experimental results showed that the speed identification error was smaller than ±5%, and the vehicle speed had no obvious influence on the identification results of the vehicle weight. Moreover, the average identification error of the vehicle weight was smaller than ±10%, and the error of more than 90% of samples was smaller than ±5%.

Keywords: moving load identification; strain influence line; load transverse distribution; strain integral coefficient; identification error

1. Introduction

The pace of urban infrastructure construction was further increased with the development of the national economy, the bridge had become an indispensable structural form of transportation infrastructure. Therefore, the safety operation, long-term performance maintenance, and real-time state assessment are very important for bridges. The traditional safety inspection of bridge structures was mainly based on manual inspection. However, with the intensive and large-scale development of vehicles, the phenomenon of vehicle overloading was ubiquitous. Moreover, the safety load rating of the old bridge was relatively low, the results obtained by manual inspection may lag behind the development of the structural state, so the safety of bridges has drawn widespread concern in the society.

Therefore, it is important to install structural health monitoring (SHM) system on the bridge, which can monitor the working state and damage condition of bridge structures in a real-time manner. Li et al. [1] elaborated the efficiency and ascendancy of the proposed distributed fiber optic sensing system in SHM. Cardini et al. [2] presented an approach to use strain data from a multi-girder, composite steel bridge for long-term SHM. Brownjohn et al. [3] described the motivations for and recent history of SHM applications to various forms of civil infrastructure and provided case studies on specific types of structure. Wong et al. [4] studied the health monitoring of cable-supported bridges involving the integration of instrumentation, analytical and information technologies. Li et al. [5] described three commonly used fiber optic sensors, and presented an overview of current research and development in the field of SHM with civil engineering applications. In general, the SHM system can monitor the internal response of the structure, such as strain, displacement, and other parameters as well as the external effects of the structure, such as temperature, load, etc. [6,7]. Vehicle load is the most important external effect on the bridge, and the working state of the bridge can be effectively evaluated if the actual load acting on the structure is identified. Kim et al. [8] proposed that the evaluation of vehicle loads for bridge safety assessment may be adjusted according to the traffic conditions, such as the traffic volume, the proportion of heavy vehicles, and the consecutive vehicle traveling patterns. In addition, he presented a method for evaluating the reliability of an in-service highway bridge, and the bridge performance was evaluated by considering traffic conditions [9]. Ghosh et al. [10] presented a framework for joint seismic and live-load fragility assessment of highway bridges. Therefore, it is of great significance to the study of load identification.

In recent years, due to the rapid development of signal processing and computer processing technologies, some load identification methods and weighing techniques with a wider application and higher accuracy have been proposed and studied. Some scholars have studied static weighing techniques with excellent accuracy. Pinkaew et al. [11] used the least-squares method based on conventional regularization to identify the static gross weight of the vehicle. Han et al. [12] presented an adaptive algorithm to improve the efficiency of static weighing. Richardson et al. [13] systematically summarized a variety of weighing techniques. However, the disadvantages of static weighing technology are: It is troublesome to install, as well as costly and time-consuming. In order to solve these disadvantages, weigh-in-motion (WIM) techniques have been widely developed and applied since the 1980s. Among them, the pavement-based WIM system required installing sensors on the road, which lead to high installation and maintenance costs as well as a great impact on the traffic [14,15]. The further development of the bridge weigh-in-motion (B-WIM) systems has the advantage that the installation and maintenance process has little impact on the traffic [16]. However, its defect is that it needs to add additional equipment to assist the function, which increases the operating cost, and it is greatly affected by the outside, these have limited its application [17–19]. Several methods have been developed in recent years to identify the moving loads. Zhu et al. [20] identified moving loads on top of a continuous beam using measured vibration responses and orthogonal function approximation method, but the road surface roughness and the variation of the speed lead to a large error. Yang et al. [21] used the method of the BP neural network in bridge moving loads identification, and the influences of different activation function combinations and algorithms on identification results were discussed. It was found that the transfer function in different combinations has little effect on the results, but the different training methods have a great influence on the results. Wang et al. [22] presented a dynamic displacement influence line method for moving load identification on bridge, and the simulation of multi-axle moving train loads was carried out, which was identified with annealing genetic algorithm, but its practical performance is questionable. Some researchers used the strain response measured by strain sensors to identify the moving load. Chen et al. [23] presented a B-WIM system to measure the vehicle velocity, wheelbase, and axial and gross weight merely based on a single set of long-gauge fiber Bragg grating (FBG) sensors. Zhang et al. [24] established the correlations among the peak values of static macrostrain curves and vehicle loads based on the macrostrain influence line theory. Wang et al. [25] used strain-monitoring data and influence line theory to identify the moving train load parameters, including train speed, gross train weight, and axle weights. They were characterized by good accuracy and easy operation but these methods were greatly affected by noise. Yang et al. [26] presented a method for moving load identification based on the influence line theory and distributed optical fiber sensing technique, and the numerical results showed that the method had excellent resistance to noise. However, the method does not consider the influence of load transverse distribution, and the experiment results of the actual bridge showed that the method had a large identification error. Zuo et al. [27] proposed a vehicle weight identification method using the measured strain responses of the T-girders caused by the passing vehicle accordingly, and the load transverse distribution was considered. However, the study did not effectively compare the methods without considering the load transverse distribution, and the influence of speed on the load identification error was not considered.

In order to improve the accuracy of load identification and study the influence of transverse distribution, a moving load identification method was proposed based on strain influence line and the load transverse distribution under consideration. The feasibility and effect of this method were verified by numerical simulations and model bridge experiments.

2. Theoretical Background

2.1. Identification Theory of Influence Lines

According to the strain influence line theory [28] and material mechanics for a simply supported beam, as shown in Figure 1, the strain of the mid-span point C can be expressed as below:

$$\varepsilon_c = \begin{cases} \frac{Phx}{2EI} & 0 < x < \frac{L}{2} \\ \frac{PhL}{2EI}\left(1 - \frac{x}{L}\right) & \frac{L}{2} < x < L \end{cases} \quad (1)$$

where x is the distance between the moving load P and the beam end A, L is the beam span, h is the height of the neutral axis, I is the section inertia moment of point C, and E is the elastic modulus.

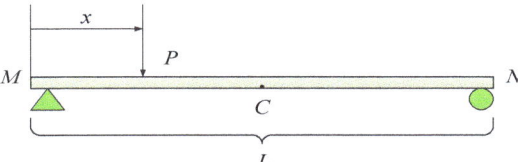

Figure 1. Moving load acting on the simply supported beam.

Generally, the moving load on the bridge is a multi-axle vehicle load, so the measured strain response can be seen as the superposition of multiple concentrated loads [29]. Figure 2 shows a three-axle vehicle load as a sample. The strain equation of mid-span section under vehicle load can be expressed as below:

$$\varepsilon(x) = \varepsilon_1(x - x_1) + \varepsilon_2(x - x_2) + \varepsilon_3(x - x_3) \quad (2)$$

where x is the distance between the vehicle's first axis and the left end of the bridge, x_1, x_2, and x_3 are the distances between each axle and the first axle, P_1, P_2, and P_3 are the axle loads, respectively.

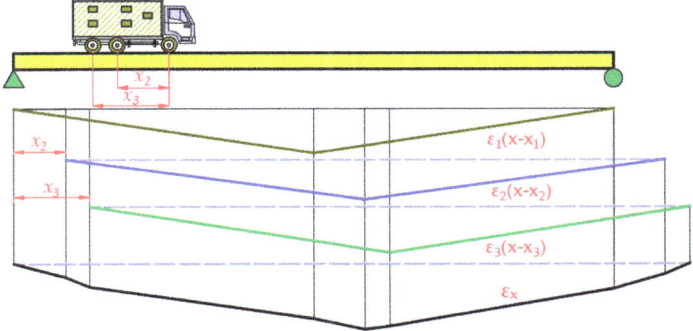

Figure 2. Mid-span strain influence line under moving load.

From the above case of three-axle vehicle, the strain response of the mid-span beam under the n-axle vehicle load can be expressed as below:

$$\varepsilon(x) = \sum_{i=1}^{n} \varepsilon_n(x - x_n) \tag{3}$$

By introducing Equation (3) into Equation (1), the following equation can be obtained:

$$\varepsilon_n(x - x_n) = P_n f(x - x_n) \tag{4}$$

in which:

$$f(x) = \begin{cases} \frac{hx}{2EI} & 0 < x < \frac{L}{2} \\ \frac{hL}{2EI}\left(1 - \frac{x}{L}\right) & \frac{L}{2} < x < L \end{cases} \tag{5}$$

When the multi-axle vehicle passes through the bridge, the area enclosed by the mid-span strain function and the x axis can be expressed as:

$$A = \int_{-\infty}^{+\infty} \varepsilon(x) dx \tag{6}$$

By introducing Equation (3) into Equation (6), the following equation can be obtained:

$$A = \int_{-\infty}^{+\infty} \sum_{i=1}^{n} P_n f(x - x_n) dx = \sum_{i=1}^{n} P_n \int_{-\infty}^{+\infty} f(x - x_n) dx \tag{7}$$

Consequently, the total weight P of the vehicle can be expressed as:

$$P = \sum_{i=1}^{n} P_n = \frac{A}{\int_{-\infty}^{+\infty} f(x - x_n) dx} = \frac{A}{\alpha} \tag{8}$$

in which:

$$\alpha = \int_{-\infty}^{+\infty} f(x - x_n) dx \tag{9}$$

where α is the mid-span strain integral coefficient, it is related to the envelope area of the strain influence line. The α can be calibrated by Equation (9) when a known vehicle load passes through the bridge. Then, the α can be used to identify the vehicle load.

2.2. Moving Load Identification Method Considering the Load Transverse Distribution

When the vehicle load acts on the bridge, the load is not only transmitted in the longitudinal direction, but also in the horizontal direction. Therefore, the force analysis of the bridge under the vehicle load is a space calculation problem. Then, the internal force analysis of the bridge section can be carried out through the influence surface. The influence surface of the bridge internal force can be expressed by a two-valued function $\eta(x, y)$, then the internal force value of section a can be expressed as $S = P \cdot \eta(x, y)$, in which S is the internal force value of the section, and P is the vehicle load. In addition, $\eta(x, y)$ can be separated into the product of two single-valued functions by the separation variable method. That is $\eta(x, y) = \eta_1(x) \cdot \eta_2(y)$, in which $\eta_1(x)$ is the internal force influence line of the beam section, and $\eta_2(y)$ is the change curve of the load ratio when the unit load acts in different positions along the horizontal direction. Then, the internal force value P' of the beam section can be expressed as $P' = P \cdot \eta_2(y)$, equivalent to assigned the load to the beam along the horizontal direction when the load P acts on point $a(x, y)$.

For the simply supported T-beam bridge, as shown in Figure 3, it is approximately assumed that $S = P \cdot \eta(x, y) \approx P \cdot \eta_1(x) \cdot \eta_2(y)$, which neglects the spatial effect of the bridge and turns it into a plane problem. When a moving load acts on the bridge deck and its position changes with the x coordinate but y coordinate is constant, then the $P \cdot \eta_2(y)$ is constant too. That is the direction of the load transverse distribution coefficient along the beam span does not change. Therefore, each beam can be analyzed individually when analyzing the internal force influence line of the beam section, and the equivalent load of each beam can be obtained according to the load transverse distribution coefficient.

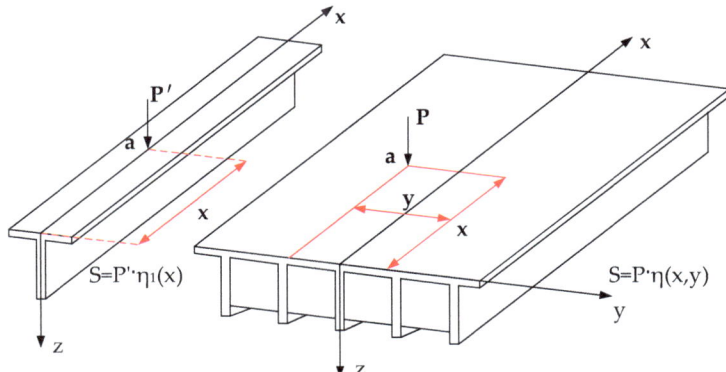

Figure 3. The internal force calculation under vehicle load.

Take the two-axle vehicle as an example to analyze the internal force (as shown in Figure 4). The axle weight is P_{11}, P_{12}, P_{21}, and P_{22}, respectively, and its action position is (x_1, y_1), (x_1, y_2), (x_2, y_1), (x_2, y_2), respectively. The y coordinate values of the four wheel loads are constant when the vehicle travels parallel to the x coordinate on the bridge, that is, the transverse distribution coefficient of each wheel load is constant. When analyzing the internal force of a single beam, the equivalent load (P_1^n, P_2^n) acting on it can be obtained by the following equation:

$$\begin{cases} P_1^n = P_{11} \cdot \eta_2(y_1) + P_{12} \cdot \eta_2(y_2) \\ P_2^n = P_{21} \cdot \eta_2(y_1) + P_{22} \cdot \eta_2(y_2) \end{cases} \quad (n = 1, 2, 3, 4, 5) \tag{10}$$

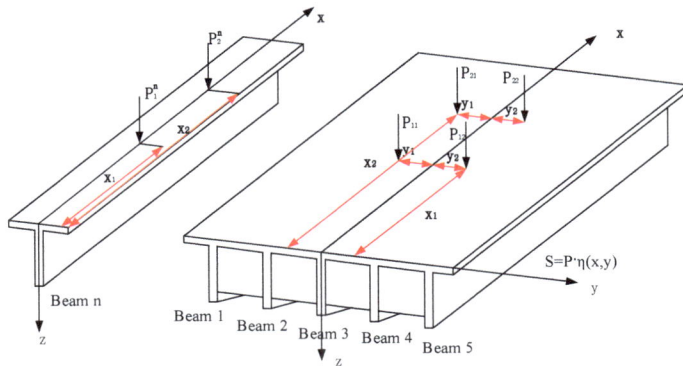

Figure 4. Wheel load transverse distribution on the bridge.

The total weight of the vehicle can be expressed as:

$$P = P_{11} + P_{12} + P_{21} + P_{22} = \sum_{n=1}^{5} (P_1^n + P_2^n) \tag{11}$$

According to the influence line theory, when the strain integral coefficient of one beam section is known, the total weight of the moving load can be calculated through the monitored strain integral value on the section.

Assuming that the above vehicle loads drive parallel to the x coordinate from one end to the other end of the beam, the measured strain integral values of the mid-span section of each beam bottom are A_2, A_3, A_4 and A_5, respectively. In addition, it is assumed that the mid-span strain integral coefficients of each beam bottom are α_1, α_2, α_3, α_4, and α_5, respectively. The following equation can be obtained:

$$P = \sum_{n=1}^{5}(P_1^n + P_2^n) = \sum_{n=1}^{5} \frac{A_n}{\alpha_n} \tag{12}$$

From Equation (12), it can be seen that the mid-span strain integral coefficient of each beam bottom must be obtained first in order to get the total weight P of the vehicle. The P, A_1, A_2, A_3, A_4, and A_5 in Equation (12) can be obtained by test. Therefore, the essence of calculating the strain integral coefficient is to solve a five-element linear equation. Keeping the vehicle weight constant but changing the driving position (five different values of y_1 and y_2), the equation group can be obtained as below:

$$\begin{cases} \frac{A_{11}}{\alpha_1} + \frac{A_{12}}{\alpha_2} + \frac{A_{13}}{\alpha_3} + \frac{A_{14}}{\alpha_4} + \frac{A_{15}}{\alpha_5} = P \\ \frac{A_{21}}{\alpha_1} + \frac{A_{22}}{\alpha_2} + \frac{A_{23}}{\alpha_3} + \frac{A_{24}}{\alpha_4} + \frac{A_{25}}{\alpha_5} = P \\ \frac{A_{31}}{\alpha_1} + \frac{A_{32}}{\alpha_2} + \frac{A_{33}}{\alpha_3} + \frac{A_{34}}{\alpha_4} + \frac{A_{35}}{\alpha_5} = P \\ \frac{A_{41}}{\alpha_1} + \frac{A_{42}}{\alpha_2} + \frac{A_{43}}{\alpha_3} + \frac{A_{44}}{\alpha_4} + \frac{A_{45}}{\alpha_5} = P \\ \frac{A_{51}}{\alpha_1} + \frac{A_{52}}{\alpha_2} + \frac{A_{53}}{\alpha_3} + \frac{A_{54}}{\alpha_4} + \frac{A_{55}}{\alpha_5} = P \end{cases} \tag{13}$$

In order to obtain the strain integral coefficient, its reciprocal can be calculated first:

$$\left\{\frac{1}{\alpha}\right\} = \{A\}^{-1}\{P\} \tag{14}$$

After obtaining the strain integral coefficient by the above method, the vehicle load identification can be carried out subsequently. Moreover, it should be noted that the theoretical derivation above is aimed at the simply supported T-beam bridge, but the method is still applicable to the similar bridge types, such as box girder bridges.

3. Numerical Simulation

3.1. Model Building

In order to verify the effectiveness of the above method, a numerical analysis model of T-beam bridge was established, as shown in Figure 5, the model beam is 3 m in length, 1.175 m in width, 0.21 m in height, the section size is shown in Figure 5a,b. The density of the mode material is 1170 kg/m^3, the Poisson's ratio is 0.35. The elastic modulus of 1# beam is $E_1 = 3.25 \times 10^4$ MPa, and the elastic modulus of 2#~5# beams are $E_2 = 1.02\, E_1$, $E_3 = 1.05\, E_1$, $E_4 = 1.07\, E_1$, $E_5 = 1.1\, E_1$, respectively.

Assuming that the vehicle load acting on the bridge was represented by four time-varying forces, its equation can be expressed as below:

$$\begin{cases} P_{11} = P_{12} = P(0.2 + 0.025\sin(6.67\pi t)) \\ P_{21} = P_{22} = P(0.3 + 0.025\sin(6.67\pi t)) \end{cases} \tag{15}$$

where, P_{11} and P_{12} represented the front wheel loads, P_{21} and P_{22} represented the rear wheel loads, P was the total weight of the vehicle. The vehicle wheelbase was 300 mm, the wheel-track was 180 mm, and the vehicle speed was 1 m/s, as shown in Figure 5c.

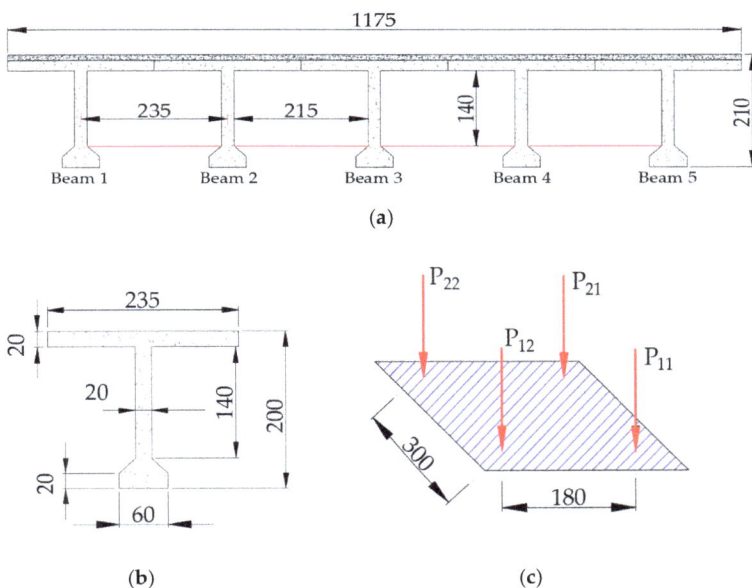

Figure 5. (a) Bridge model, (b) bridge section size, (c) vehicle load (unit: mm).

The vehicle load P was divided into three grades, 10 kg, 20 kg, and 30 kg, respectively. The load grade of 10 kg was used to calibrate the strain integral coefficient, and the other load grades were used to test. As shown in Figure 6, the bridge model was divided into three lanes, and the vehicle acted on the left, middle, and right positions of each lane. Thus, the vehicle load position was divided into nine conditions.

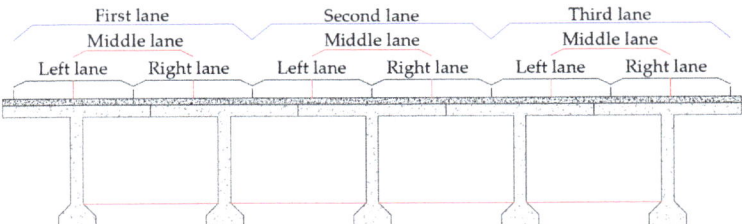

Figure 6. Vehicle load position.

3.2. Simulation Results Analysis

3.2.1. Analysis of the Identification Results without Considering the Load Transverse Distribution

For the identification method without considering the load transverse distribution, it is only necessary to know the mid-span strain integral coefficient α of a single beam. Firstly, the following simulation conditions were carried out: (1) Vehicle driving in the middle of the first lane with 10 kg weight at 1 m/s, (2) vehicle driving in the middle of the second lane with 10 kg weight at 1m/s, (3) vehicle driving in the middle of the third lane with 10 kg weight at 1m/s. According to each working conditions, the corresponding mid-span strain influence line (1#, 3#, 5# beam) were obtained, as shown in Figure 7, then the corresponding strain integral coefficient was obtained in Table 1.

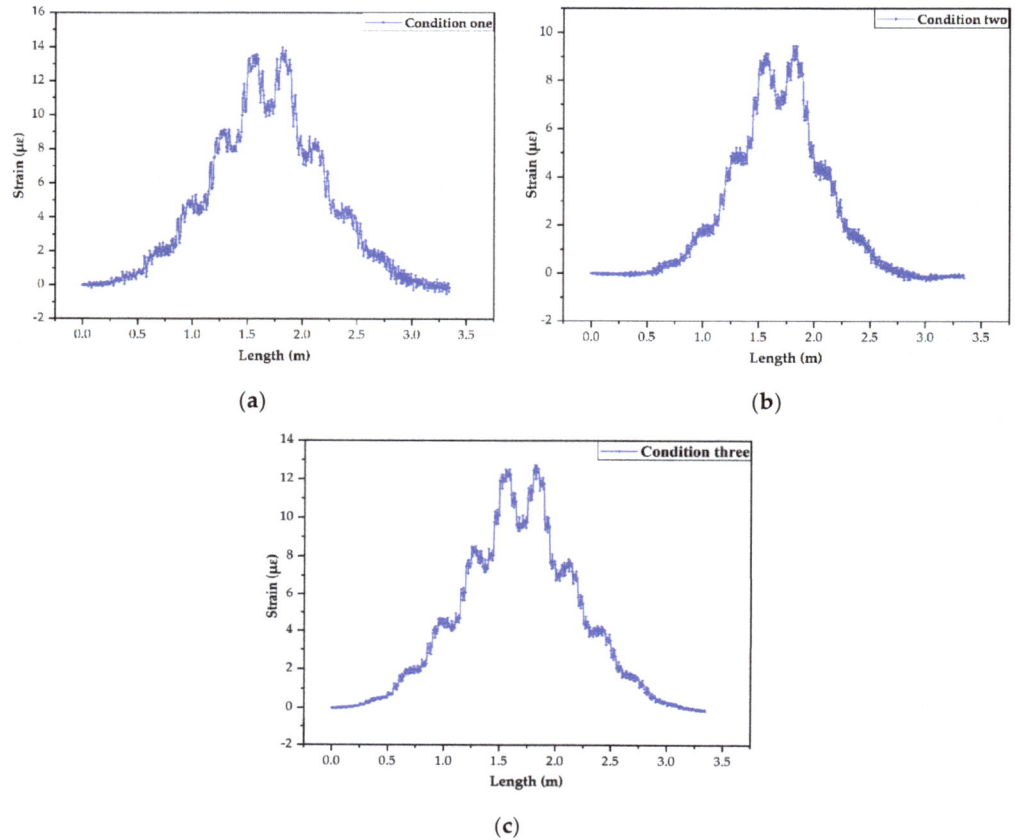

Figure 7. The middle-span strain influence line of (**a**) 1# beam, (**b**) 3# beam, (**c**) 5# beam.

Table 1. The strain integral coefficient.

Vehicle Driving Position	The Middle of the First Lane	The Middle of the Second Lane	The Middle of the Third Lane
Beam number	1#	3#	5#
Total weight (kg)	10	10	10
The integral value of strain influence line (10^{-6} m)	14.46	7.57	13.48
Strain integral coefficient (10^{-8})	14.75	7.73	13.75

Figure 8 shows the load identification results obtained according to the strain integral coefficient. Taking the strain integral coefficient of 3# beam as an example, it can be seen that the identification error was smaller than 10% when the vehicle load drove in the second lane. Especially when the vehicle load drove in the middle of the second lane, the error was almost zero. However, the identification error was large when the vehicle drove in the first and third lane. Therefore, it was not suitable for load identification. For the load identification results obtained according to the strain integral coefficient of 1# beam and 5# beam, the identification error was close to zero when the vehicle load drove in the middle of the first and third lane. However, the identification accuracy was still poor when the vehicle load drove in the left or the right line. In addition, the farther away from the vehicle position of coefficient calibration, the worse the identification accuracy was. In summary, the load identification accuracy was closely related to the driving position of the vehicle load when the influence of load transverse distribution was not considered. The identification accuracy was relatively high when it was close to the vehicle position of coefficient calibration, conversely, the identification accuracy was poor. Therefore, it was no longer suitable for load identification.

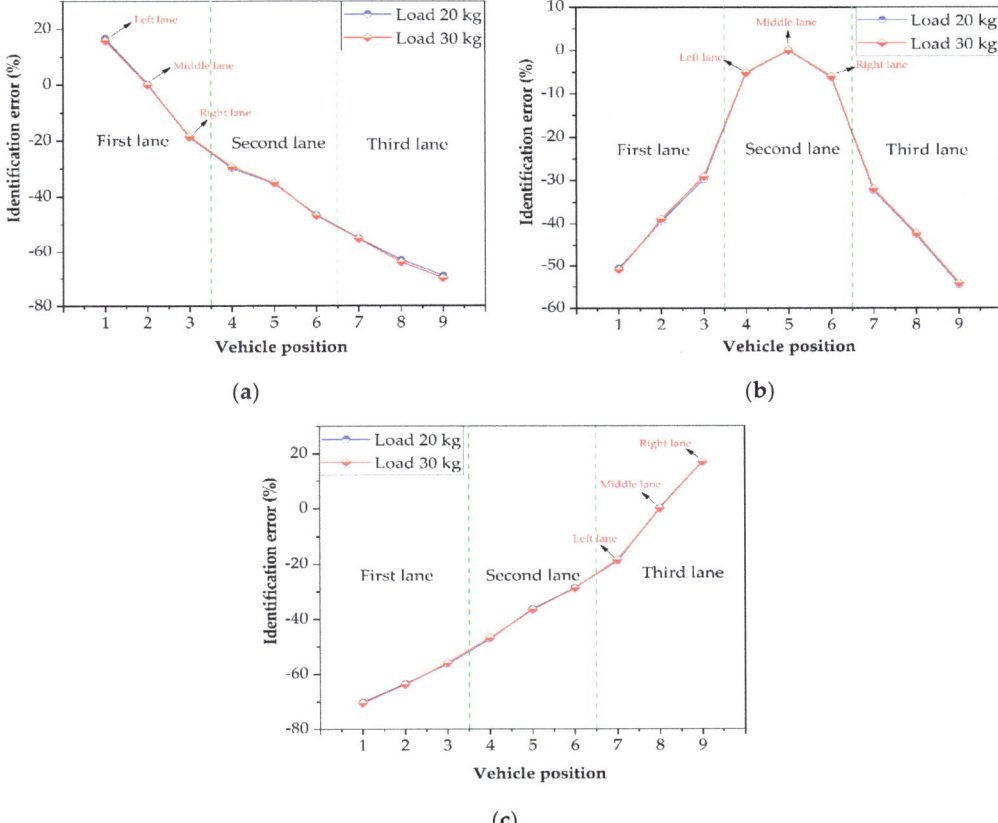

Figure 8. Identification results according to the strain integral coefficient of (**a**) 1# beam, (**b**) 3# beam, (**c**) 5# beam.

3.2.2. Analysis of the Identification Results Considering the Load Transverse Distribution

For the identification method considering the load transverse distribution, it was necessary to obtain the strain integral coefficient of each beam by Equation (14). Firstly, the following simulation conditions were carried out with the vehicle load of 10 kg: (1) Vehicle driving in the left of the first lane, (2) vehicle driving in the right of the first lane, (3) vehicle driving in the middle of the second lane, (4) vehicle driving in the left of the third lane, (5) vehicle driving in the right of the third lane. According to Equations (13) and (14), the strain integral coefficient of each beam was obtained (as shown in Table 2). It can be seen that the strain integral coefficient of each beam bottom was basically proportional to the reciprocal of the stiffness, and the reason for the error was that the load identification method considering the load transverse distribution, which ignores the influence of spatial effect and diaphragm in the theoretical derivation. The obtained strain integral coefficient was used to identify the load of the test sample, and the results are shown in Figure 9. It can be seen that the identification accuracy was very high no matter where the vehicle was, and the error was close to zero. Therefore, compared with the identification method without considering the load transverse distribution, the identification method considering the load transverse distribution has obvious advantages.

Table 2. The mid-span strain integral coefficient of each beam bottom.

	1# Beam	2# Beam	3# Beam	4# Beam	5# Beam
Strain integral coefficient (10^{-8})	29.17	28.84	27.93	27.70	26.61
Ratio to integral coefficient of 1# beam	1	0.989	0.957	0.949	0.912
Ratio to reciprocal 1# beam stiffness	1	0.980	0.952	0.936	0.910

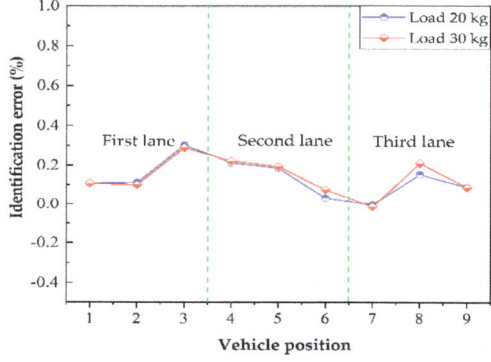

Figure 9. Identification results considering the load transverse distribution.

3.2.3. Analysis of the Anti-Noise Performance

Two kinds of noise (5% and 10%) were added to the numerical simulation to verify the anti-noise performance of the method. The strain values of the test sample were extracted, and 5% and 10% of the noise were added as condition 1 and condition 2, and the noise can be expressed as:

$$\varepsilon'(x) = \varepsilon(x) + \beta \cdot rand(n) \cdot var(x) \tag{16}$$

where $\varepsilon'(x)$ is the strain output after the added noise, $\varepsilon(x)$ is the original strain input, β is the noise level, rand is short for random and $rand(n)$ is a set of values with the mean is 0 and the variance is 1, var is short for variance and $var(x)$ is the variance of the original strain input.

When the vehicle load (20 kg) drives in the middle of the second lane, the mid-span strain time history with different kinds of noise of 3# beam bottom is shown in Figure 10. It can be seen that there is only a slight fluctuation of the strain output when the noise level is 5%, which can better simulate the environmental noise. The strain output has an obvious difference for the original value when the noise level reaches 10%, both of these two working conditions are representative. The load identification results with different levels of noise are shown in Figure 11. It can be seen that the load identification error with different noise levels was slightly larger compared with the no-noise condition. The load identification errors were nearly the same when the noise levels were 5% and 10%. In addition, the overall error was smaller than 0.5%, which showed a good identification accuracy. Therefore, the method can keep the identification accuracy under different kinds of noise, and the noise reduction for the following analysis processing can be ignored.

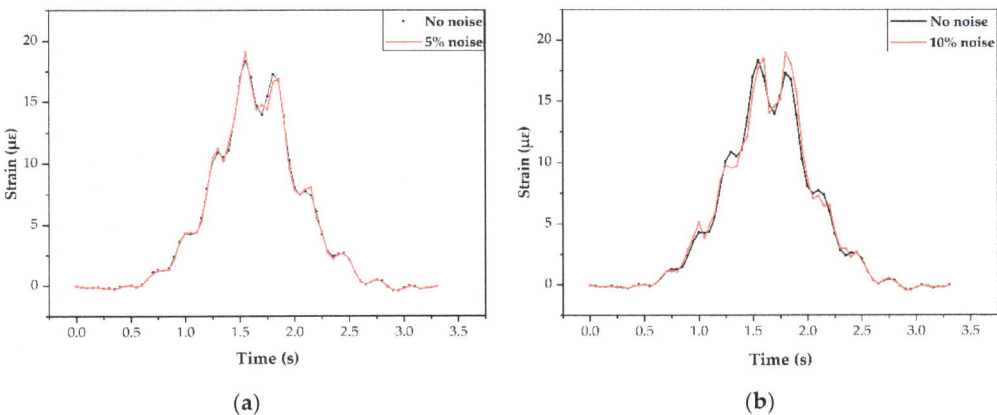

Figure 10. The mid-span strain time history of 3# beam with (**a**) 5% noise, (**b**) 10% noise.

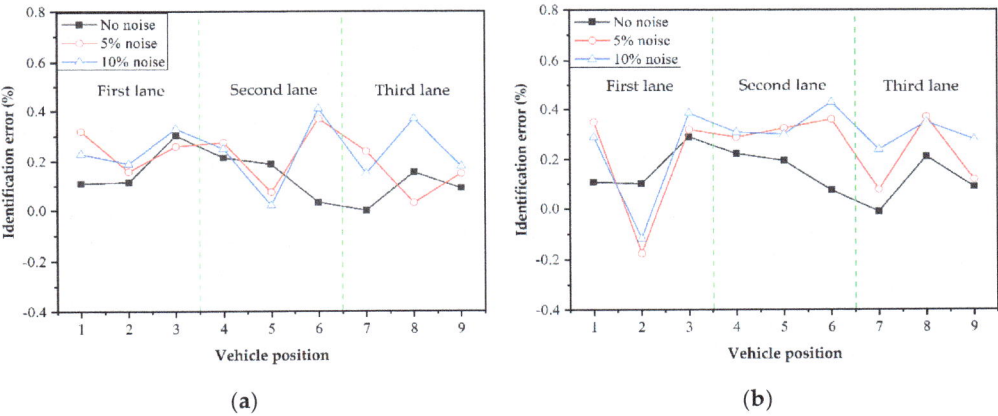

Figure 11. Identification error under different levels of noise with the load of: (**a**) 20 kg, (**b**) 30 kg.

4. Verification by Experiment

4.1. Experimental Setup

To test the feasibility of the proposed method, a series of bridge model experiments were conducted. The experimental platform was made of an acceleration platform, test bridge, and deceleration platform, which is shown in Figure 12a. The model bridge was made according to the size of the numerical simulation bridge, as shown in Figure 5, the material of the model bridge is polymethyl methacrylate (as shown in Figure 12b), and its density is 1170 kg/m^3, the Poisson's ratio is 0.35, and the elastic modulus is 3.25×10^4 MPa.

The experimental vehicle models were divided into two-axle and three-axle vehicles, as shown in Figure 13, and the way to change the vehicle weight was to add counterweight in the vehicle. In addition, the long-gauge FBG strain sensors were used to collect and analyze the data in the experiment [1]. Based on the influence line method considering the load transverse distribution, five FBG strain sensors were arranged in the mid-span of each beam bottom, as shown in Figure 12b.

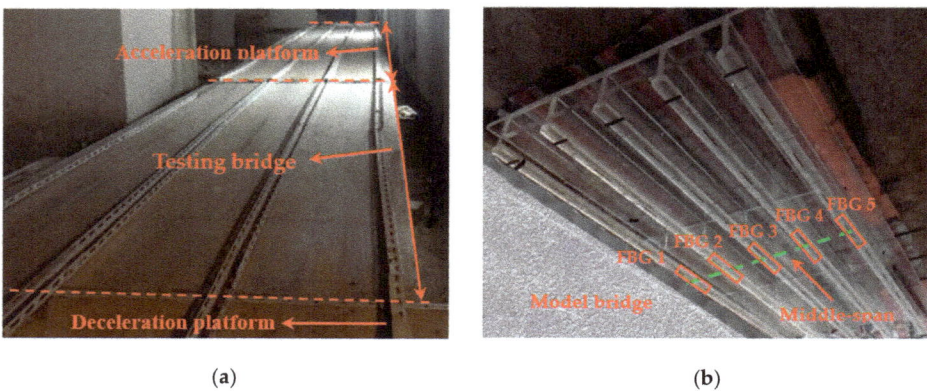

Figure 12. (a) Experimental platform, (b) model bridge.

Figure 13. Vehicle model in the experiment: (a) Two-axle vehicle, (b) three-axle vehicle.

4.2. Analysis of Experiment Results

The MOI's S130 model acquisition instrument was used to collect the data of FBG sensors in the experiment, and the measured data were used for load identification. When the vehicle drives in the second lane with the speed of 1.33 m/s, the typical long-gauge strain time history curve of each beam bottom is shown in Figure 14.

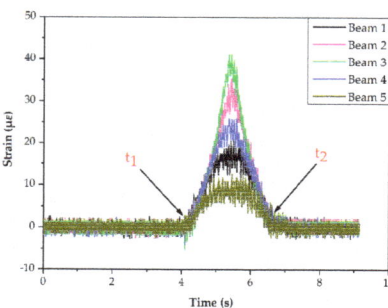

Figure 14. Measured strain time history curve.

4.2.1. Speed Identification

Firstly, the vehicle speed was identified. Assuming that the vehicle crosses the bridge with a constant speed, the integral Equation (17) of strain influence line can be obtained by modifying Equation (6):

$$A = \int_{-\infty}^{+\infty} \varepsilon(x)dx = \int_{-\infty}^{+\infty} \varepsilon(t)vdt = v\int_{-\infty}^{+\infty} \varepsilon(t)dt \qquad (17)$$

As shown in Figure 14, it corresponds to the starting point t_1 of the wave peak when the front axle of the vehicle contacts the bridge, and it corresponds to the ending point t_2 when the rear axle of the vehicle leaves the bridge. Then the vehicle speed V can be calculated according to the time difference and the driving distance, as shown in Equation (18):

$$v = \frac{d}{\Delta t} = \frac{L+x}{t_2 - t_1} \qquad (18)$$

where, x is the vehicle wheelbase, L is the bridge length. In this experiment, the vehicle weight was 16.95 kg, and the vehicle speed was divided into nine levels by changing the speed of traction motor, as shown in Table 3. Meanwhile, each experimental condition was repeated three times. According to the above method, the strain data of 3# beam bottom were used to identify the vehicle speed, and the results are shown in Figure 15. It can be seen that the average relative errors of the speed identification were smaller than ±4%, and they were within an acceptable range, which shows the great performance of the method.

Table 3. Speed levels.

Vehicle Speed (m/s)	0.86	1.33	1.81	2.19	2.59	3.06	3.53	4.01	4.39

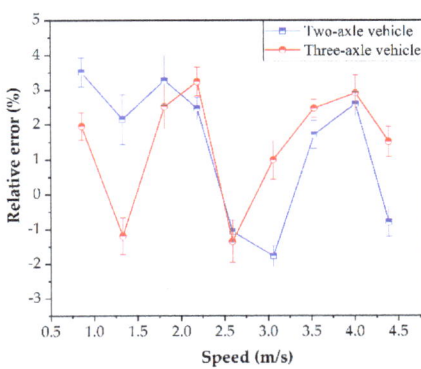

Figure 15. Relative error of the speed identification.

4.2.2. Influence of Speed on Load Identification

Based on the vehicle speed obtained from inversion, the strain time history curve was converted into strain influence line. Then, the load identification was carried out by the method considering the load transverse distribution. The samples with the weight of 28 kg and speed of 0.86 m/s were selected to calibrate the strain integral coefficient. It should be noted that the driving path of the vehicle was limited to three lanes, so the five equations that were shown in Equation (13) cannot be obtained. However, the optimal solution of the strain integral coefficient can be obtained by using three equations, as shown in Table 4.

Table 4. The mid-span strain integral coefficient of each beam bottom.

	1# Beam	2# Beam	3# Beam	4# Beam	5# Beam
The strain integral coefficient (10^{-7})	8.11	8.33	8.04	7.23	7.08

According to Equation (17), the integral value of the influence line is independent of the vehicle speed, but the identification accuracy of the speed has an effect on the integral value. The actual speed and inversion speed were used to identify the vehicle weight of the above 18 samples, which was mentioned in Section 4.2.1, and the results are shown in Figure 16. It can be seen that the speed had no obvious influence on vehicle weight identification, and the average identification error was smaller than ±5%. In addition, it should be noted that the results obtained by using the actual speed to calculate the vehicle weight were more accurate than those obtained by using the inversion speed. Moreover, the average identification error was smaller than 2% when using the actual speed to identify the vehicle weight.

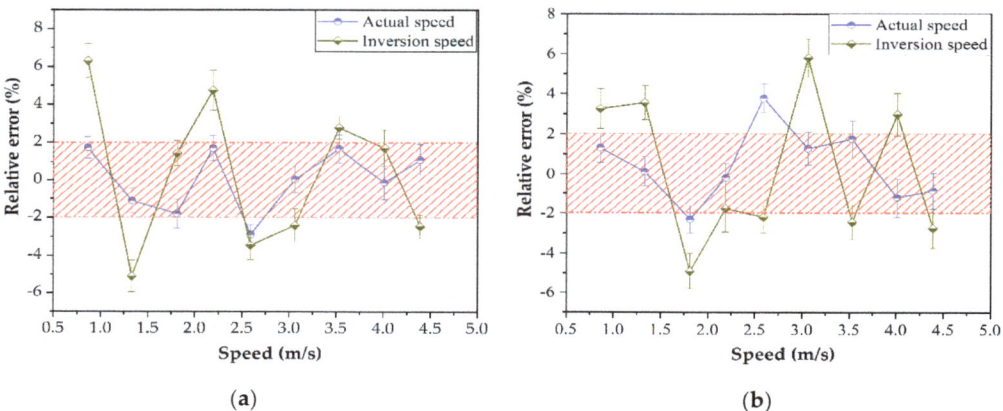

Figure 16. Influence of vehicle speed on load identification: (a) Two-axle vehicle, (b) three-axle vehicle.

4.2.3. The Identification Results of Vehicle Weight

As the vehicle speed has no obvious influence on the identification results of the vehicle weight, the vehicle speed was set as 1.33 m/s in the following analysis. The two-axle vehicle and three-axle vehicle were divided into four grades of weight in the experiment. Each grade of vehicle weight was tested in three lanes, and the load identification results were shown in Figure 17. It can be seen that no matter which lane the vehicle drives, the vehicle weight identification error of each sample can be controlled within ±10%. The error of more than 90% of samples was smaller than ±5%, and the error was relatively larger compared with the simulation results, which was within an acceptable range. In addition, the error fluctuation of load identification was small, and the variance was smaller than 2%. Compared with reference [26], the method proposed in this paper greatly improves the identification accuracy. Therefore, it can be considered that the load identification method considering the load transverse distribution was effective. The error sources should be analyzed in the following aspects: (1) The vehicle weight identification is based on the speed identification, so the error of vehicle speed will affect the result of vehicle weight identification, (2) although there are lane restrictions in the experiment, the vehicle's trajectory is not always in a straight line along the bridge span direction, and its trajectory is relatively random.

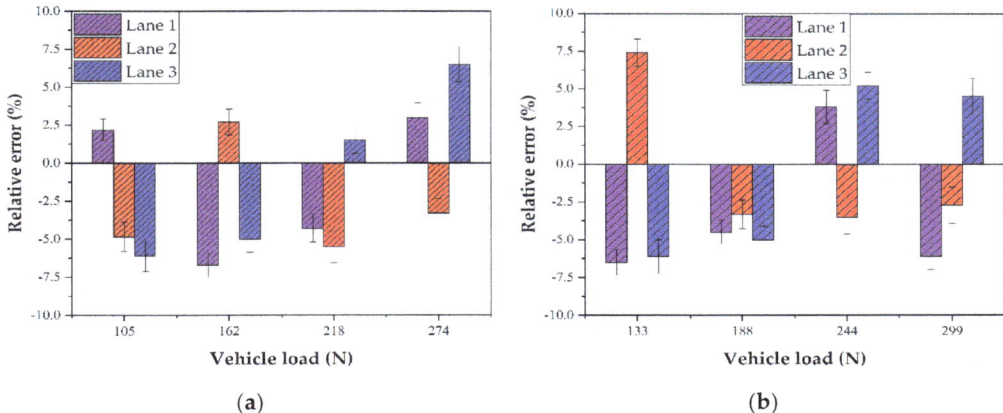

Figure 17. The relative error of vehicle weight identification: (**a**) Two-axle vehicle, (**b**) three-axle vehicle.

5. Conclusions

In this paper, considering the load transverse distribution, a novel moving load identification method was proposed based on distributed strain sensing technique and influence line. The load identification accuracy and anti-noise performance were greatly improved, and it was more universal for a variety of bridge types. In addition, a series of numerical simulations and experiments were conducted to verify the proposed method. The main conclusions are as follows:

1. Through the verification of numerical simulation, the method without considering the load transverse distribution was not suitable for solving the space problem, and the method considering the load transverse distribution has a high identification accuracy and excellent performance of anti-noise performance.
2. The results of the model test showed that the average relative error of the speed identification was smaller than ±4%, which shows the great performance of the method.
3. The speed has no obvious influence on the vehicle weight identification, and the average identification error was smaller than ±5%. In addition, it should be noted that the results obtained by using the actual speed to calculate the vehicle weight were more accurate than those obtained by using the inversion speed.
4. The relative error of the vehicle weight identification was smaller than ±10%, and the error of more than 90% of samples was smaller than ±5%.

Author Contributions: J.Y. proposed the topic of this study and designed the experiments; P.H. and Y.Z. analyzed the experimental date; C.Y. guided the research and provided advice. The paper was finally written by J.Y. and confirmed by all the authors. All authors have read and agreed to the published version of the manuscript.

Funding: Innovative Venture Technology Investment Project of Hunan Province (2018GK5028). National Natural Science Foundation of China (52078122).

Institutional Review Board Statement: Not applicable.

Informed Consent Statement: Not applicable.

Data Availability Statement: Not applicable.

Acknowledgments: The authors gratefully appreciate the financial support of Innovative Venture Technology Investment Project of Hunan Province (2018GK5028), and National Natural Science Foundation of China (52078122).

Conflicts of Interest: The authors declare no conflict of interest.

References

1. Li, S.Z.; Wu, Z.S. Development of distributed long-gage fiber optic sensing system for structural health monitoring. *Struct. Health Monit.* **2007**, *6*, 133–143. [CrossRef]
2. Cardini, A.J.; DeWolf, J.T. Long-term structural health monitoring of a multi-girder steel composite bridge using strain data. *Struct. Health Monit.* **2009**, *8*, 47–58. [CrossRef]
3. Brownjohn, J.M.W. Structural health monitoring of civil infrastructure. *Philos. Trans. R. Soc. Lond. Ser. A* **2007**, *365*, 589–622. [CrossRef] [PubMed]
4. Wong, K.Y. Instrumentation and health monitoring of cable-supported bridges. *Struct. Control Health Monit.* **2004**, *11*, 91–124. [CrossRef]
5. Li, H.N.; Li, D.S.; Song, G.B. Recent applications of fiber optic sensors to health monitoring in civil engineering. *Eng. Struct.* **2004**, *26*, 1647–1657. [CrossRef]
6. Karoumi, R.; Wiberg, J.; Liljencrantz, A. Monitoring traffic loads and dynamic effects using an instrumented railway bridge. *Eng. Struct.* **2005**, *27*, 1813–1819. [CrossRef]
7. Sekuła, K.; Kołakowski, P. Piezo-based weigh-in-motion system for the railway transport. *Struct. Control Health Monit.* **2012**, *19*, 199–215. [CrossRef]
8. Kim, S.H.; Heo, W.H.; You, D. Vehicle loads for assessing the required load capacity considering the traffic environment. *Appl. Sci.* **2017**, *7*, 365. [CrossRef]
9. Kim, S.H.; Choi, J.G.; Ham, S.M. Reliability evaluation of a PSC highway bridge based on resistance capacity degradation due to a corrosive environment. *Appl. Sci.* **2016**, *6*, 423. [CrossRef]
10. Ghosh, J.; Caprani, C.C.; Padgett, J.E. Influence of traffic loading on the seismic reliability assessment of highway bridge structures. *J. Bridge. Eng.* **2014**, *19*, 04013009. [CrossRef]
11. Pinkaew, T. Identification of vehicle axle loads from bridge responses using updated static component technique. *Eng. Struct.* **2006**, *28*, 1599–1608. [CrossRef]
12. Han, L.D.; Ko, S.S.; Gu, Z.; Jeong, M.K. Adaptive weigh-inmotion algorithms for truck weight enforcement. *Transp. Res.* **2012**, *24*, 256–269.
13. Richardson, J.; Jones, S.; Brown, A.; O'Brien, E.; Hajialzadeh, D. On the use of bridge weigh-in-motion for overweight truck enforcement. *Int. J. Heavy Veh. Syst.* **2014**, *21*, 83–104. [CrossRef]
14. Yu, Y.; Cai, C.S.; Deng, L. State-of-the-art review on bridge weigh-in-motion technology. *Adv. Struct. Eng.* **2016**, *19*, 1514–1530. [CrossRef]
15. Bajwa, R.; Coleri, E.; Rajagopal, R.; Varaiya, P.; Flores, C. Development of a cost-effective wireless vibration weigh-in-motion system to estimate axle weights of trucks. *Comput. Aided Civ. Inf. Eng.* **2017**, *32*, 443–457. [CrossRef]
16. Zolghadri, N.; Halling, M.; Johnson, N.; Barr, P. Field verification of simplified bridge weigh-in-motion techniques. *J. Bridge. Eng.* **2016**, *21*, 04016063. [CrossRef]
17. Ojio, T.; Carey, C.; Obrien, E.; Doherty, C.; Taylor, S.E. Contactless bridge weigh-in-motion. *J. Bridge. Eng.* **2016**, *21*, 04016032. [CrossRef]
18. Zhao, H.; Uddin, N.; O'Brien, E.; Hao, X.S.; Zhu, P. Identification of vehicular axle weights with a bridge weigh-in-motion system considering transverse distribution of wheel loads. *J. Bridge. Eng.* **2014**, *19*, 04013008. [CrossRef]
19. Helmi, K.; Taylor, T.; Ansari, F. Shear force based method and application for real-time monitoring of moving vehicle weights on bridges. *J. Intell. Mater. Syst. Struct.* **2015**, *26*, 505–516. [CrossRef]
20. Zhu, X.Q.; Law, S.S. Practical aspects in moving load identification. *J. Sound Vib.* **2002**, *258*, 123–146. [CrossRef]
21. Yang, H.; Yan, W.; He, H. Parameters Identification of Moving Load Using ANN and Dynamic Strain. *Shock Vib.* **2016**, *2016*, 1–13. [CrossRef]
22. Wang, Y.; Qu, W.L. Moving train loads identification on a continuous steel truss girder by using dynamic displacement influence line method. *Int. J. Steel Struct.* **2011**, *11*, 109–115. [CrossRef]
23. Chen, S.Z.; Wu, G.; Feng, D.C.; Zhang, L. Development of a bridge weigh-in-motion system based on long-gauge fiber Bragg grating sensors. *J. Bridge. Eng.* **2018**, *23*, 04018063. [CrossRef]
24. Zhang, L.; Wu, G.; Li, H.; Chen, S. Synchronous Identification of Damage and Vehicle Load on Simply Supported Bridges Based on Long-Gauge Fiber Bragg Grating Sensors. *J. Perform. Constr. Fac.* **2020**, *34*, 04019097. [CrossRef]
25. Wang, H.; Zhu, Q.; Li, J.; Mao, J.; Hu, S.; Zhao, X. Identification of moving train loads on railway bridge based on strain monitoring. *Smart Struct. Syst.* **2019**, *23*, 263–278.
26. Yang, C.Q.; Yang, D.; He, Y.; Wu, Z.S.; Xia, Y.F.; Zhang, Y.F. Moving load identification of small and medium-sized bridges based on distributed optical fiber sensing. *Int. J. Struct. Stab. Dyn.* **2016**, *16*, 1640021. [CrossRef]
27. Zuo, X.H.; He, W.Y.; Ren, W.X. Vehicle weight identification for a bridge with multi-T-girders based on load transverse distribution coefficient. *Adv. Struct. Eng.* **2019**, *22*, 3435–3443. [CrossRef]
28. Ojio, T.; Yamada, K. Bridge weigh-in-motion systems using stringers of plate girder bridges. In Proceedings of the Third International Conference on Weigh-in-Motion (ICWIM3) Iowa State University, Ames, Orlando, FL, USA, 13–15 May 2002.
29. O'Connor, C.; Chan, T.H.T. Dynamic wheel loads from bridge strains. *J. Struct. Eng.* **1988**, *114*, 1703–1723. [CrossRef]

Review

Health Monitoring of Civil Infrastructures by Subspace System Identification Method: An Overview

Hoofar Shokravi [1,*], Hooman Shokravi [2], Norhisham Bakhary [1,3], Seyed Saeid Rahimian Koloor [4] and Michal Petrů [4]

1. Faculty of Civil Engineering, Universiti Teknologi Malaysia, Skudai 81310, Malaysia; norhisham@utm.my
2. Department of Civil Engineering, Islamic Azad University, Tabriz 5157944533, Iran; hooman.shokravi@gmail.com
3. Institute of Noise and Vibration, Universiti Teknologi Malaysia, City Campus, Jalan Semarak, Kuala Lumpur 54100, Malaysia
4. Institute for Nanomaterials, Advanced Technologies and Innovation (CXI), Technical University of Liberec (TUL), Studentska 2, 461 17 Liberec, Czech Republic; s.s.r.koloor@gmail.com (S.S.R.K.); michal.petru@tul.cz (M.P.)
* Correspondence: hf.shokravi@gmail.com

Received: 27 March 2020; Accepted: 13 April 2020; Published: 17 April 2020

Abstract: Structural health monitoring (SHM) is the main contributor of the future's smart city to deal with the need for safety, lower maintenance costs, and reliable condition assessment of structures. Among the algorithms used for SHM to identify the system parameters of structures, subspace system identification (SSI) is a reliable method in the time-domain that takes advantages of using extended observability matrices. Considerable numbers of studies have specifically concentrated on practical applications of SSI in recent years. To the best of author's knowledge, no study has been undertaken to review and investigate the application of SSI in the monitoring of civil engineering structures. This paper aims to review studies that have used the SSI algorithm for the damage identification and modal analysis of structures. The fundamental focus is on data-driven and covariance-driven SSI algorithms. In this review, we consider the subspace algorithm to resolve the problem of a real-world application for SHM. With regard to performance, a comparison between SSI and other methods is provided in order to investigate its advantages and disadvantages. The applied methods of SHM in civil engineering structures are categorized into three classes, from simple one-dimensional (1D) to very complex structures, and the detectability of the SSI for different damage scenarios are reported. Finally, the available software incorporating SSI as their system identification technique are investigated.

Keywords: structural health monitoring (SHM); vibration-based damage detection; system identification; subspace system identification (SSI)

1. Introduction

In the recent years, there has been a considerable interest in "Smart City" concept and the monitoring, controlling, and preservation of the health state of critical infrastructures, like roads, buildings, bridges, and tunnels. Structural health is required to be diagnosed at every moment during a structure life cycle in order to provide high quality services. Civil engineering structures are designed for the lifetime of the occupants or facilities and their failure might lead to catastrophic consequences in terms of human life and economic assets [1]. Structural health monitoring (SHM) is one of the main stakeholders of the said Smart City concept [2]. SHM is an effective solution contributing to the need for safety, lowering the maintenance costs and reliable condition assessment of structures [3,4].

SHM is an interdisciplinary subject that incorporates knowledge and experiences from synergetic technologies in civil, mechanical, control, and computer engineering to deal with the health assessment of structures. The health monitoring of structures has been the subject of many studies for the past three decades [5,6]. The development of a reliable SHM method for civil structures is a challenging task due to ambient-induced uncertainty and the associated complexity measures. Four analysis levels of damage detection are applied in the context of SHM that include: (1) detection, (2) localization, (3) quantification, and (4) prediction of the remaining life, whereas the first three levels are more explicitly reported in literature [7,8].

Based on the acting load, SHM methods can be divided into two classes of static and dynamic-based methods [9]. The methods that use vibration characteristics of structures to assess the health state of structures, so-called vibration-based damage detection (VDD) [10,11]. The key premise of VDD is to estimate the modal parameters of a structure while using the analytical model that was constructed by system identification methods [12,13]. Static-based damage detection (SDD) methods rely on measuring the change in static response of structure, such as load bearing capacity, strain, deflection, and stiffness. Posenato et al. and Wu et al. [14] used strain data for damage detection of structure. Chen et al. [15] proposed a method to take advantages of stay cable force measurements and structural temperatures for damage detection. Yu et al. [16] used deflection data for damage identification in structures. Zhu et al. [17] introduced a temperature-driven method while using strain information of structures for anomaly detection.

Weigh in motion (WIM) is a widely used vehicle classification method for the health monitoring of structures. Bridge weigh in motion (BWIM) is a type of WIM technology that can identify traffic data, including speed, number of axles, axles spacing, and gross and axle weight of the passing vehicles using a series of conventional strain gauges. BWIM is particularly suitable for short-term measurements of traffic data, as it can be easily installed and detached from the bridge. Cardini and Dewolf [18] applied BWIM through using strain gauges to gain information on the quantity and weights of the trucks crossing highway bridges. Cantero et al. [19] proposed a BWIM-based damage identification method through introducing the concept of 'Virtual Axle' for deriving a damage indicator. Gonzalez and Karoumi [20] proposed a model-free damage detection method using deck accelerations response and BWIM. Kalyankar and Uddin [21] developed a three dimensional finite element model to estimate multi-vehicles–bridge interaction in a BWIM. The environmental factors are a weak point of vibration- and static-based damage detection [22]. However, in some cases, the temperature based methods shown higher sensitivity when compared to vibration based methods.

Estimations of the modal parameters in SHM are generally performed by system identification methods [23]. System identification is a mathematical procedure for establishing an analytical model based on experimental data. System identification is a mature field in SHM to extract modal parameters in VDD methods [24]. System identification methods in SHM can be classified into three categories based on their domains, including: time-domain (TD), frequency-domain (FD), and time/frequency domain (TFD) [25]. TD methods are more attractive for monitoring of civil structures, owing to the direct use of vibration signals. TD methods are generally classified within three groups of subspace system identification (SSI), natural excitation technique (NExT), and auto-regressive moving average (ARMA) [26]. The premise of the NExT method is that the response signals of a structure for ambient excitation and free-vibration have the same analytical form [27]. ARMA-based methods are popular statistical strategies for VDD of civil engineering structures. The auto regressive (AR) part ARMA models a linear function for the response time-history and the moving average (MA) section determines the moving average of the measurement response. The SSI algorithm presents a harmonious combination of algebraic, mathematical, statistical, and geometrical tools to identify the system parameters. SSI takes advantages of LS, angles between subspaces, QR decomposition, singular value decomposition (SVD), Kalman filter, and stochastic realization theory to deal with the problem of modal parameter identification. Subspace-based methods that are used for the parameter identification of civil structures are mainly one of the following two methods of data-driven subspace system identification (SSI-DATA) and covariance-driven subspace system identification (SSI-COV) approaches [28]. The ARMA model has the most similarity to the SSI-COV model, as both methods use the correlation function of the vibration measurement in their preprocessing stage.

Recently, a large number of subspace-based methods have been applied in VDD. However, the previously conducted surveys have not kept pace with the changing environment and diversity in this field. Therefore, there is a need for a review focusing on the most important recent studies conducted in the considered area. The presented review attempts to address the available studies that employed subspace-based techniques in VDD of civil structures. This paper provides an overview of the background and new findings in SSI with a focus on both theory and practice. In addition, it describes some contributions toward the development and application of the SSI algorithm in recent years.

This review study is organized, as follows. In Section 2, SHM methods are outlined and the strength and drawbacks of each class are highlighted. Section 3 evaluated vibration-based damage detection (VDD) methods. The focus is on subspace system identification in the three later sections (Sections 4–6). Section 7 introduces some commercially available software that use subspace as their main constituent. Future research directions and conclusions are provided in Sections 8 and 9, respectively.

2. Structural Health Monitoring (SHM)

An SHM system implements strategies for the damage detection of structures [29–31]. Currently, SHM is known as a well-established tool for the diagnosis of damages in civil engineering communities and it is employed in a number of different structures, such as buildings, bridges, and dams. Structural data are collected from several points through installed sensors and they are analyzed to evaluate the health of a structure. Figure 1 shows the categorization of SHM methods with a focus on addressing key subspace-based algorithms. SHM methods are divided into local methods, which mainly rely on non-destructive evaluation (NDE) strategies and global methods. Based on the incorporated domain, the VDD strategies could be categorized as TD, FD and TFD methods. The TD methods are generally from one of the auto-regressive moving average (ARMA), natural excitation technique (NExT), or subspace system identification (SSI) families. The SSI methods are an important class of algorithm and they can be divided into three categories of canonical variate analysis (CVA), numerical algorithms for state-space subspace system identification (N4SID), and multivariable output error state-space (MOESP). N4SID is of the most favored SSI algorithm in SHM due to its capability to cope with output-only data (stochastic, unknown input). Two classes of N4SID algorithms are commonly practiced in SHM of civil engineering structures, namely the SSI-DATA method and the SSI-COV method.

NDE is a local SHM method used to perform constrained random tests to diagnose the state and severity of the possible defects. Some NDE methods are concerned with measuring defects in steel components, whereas others are designed for concrete substructures. Several NDE methods are available to identify defects in steel structures such as the ultrasonic test (UT), radiographic test (RT) [32], and eddy current test (ET) [33]. UT is an acoustic NDE method that uses ultrasonic waves passing through a structure for detecting defects. The phased array ultrasonic test (PAUT) method is a more reliable type of UT that uses a greater number of arrays in order to reliably simulate a specimen's profile [33]. A variety of methods is applied for NDE of concrete components. These methods range from the very simple strength evaluation methods, such as using rebound hammer [34], to more complex methods, such as impact-echo [35] and radiography testing [36]. Even though local SHM methods yield excellent performance for detection and localization of damages, they have some limitations and drawbacks. The main disadvantage of these techniques is that the evaluation process cannot be implemented without any prior knowledge of the approximate damage location. Moreover, in many cases, access to below of the test area is an essential requirement that is not always affordable or practical [37]. Further detailed information on local SHM methods in civil engineering structures are provided in [38].

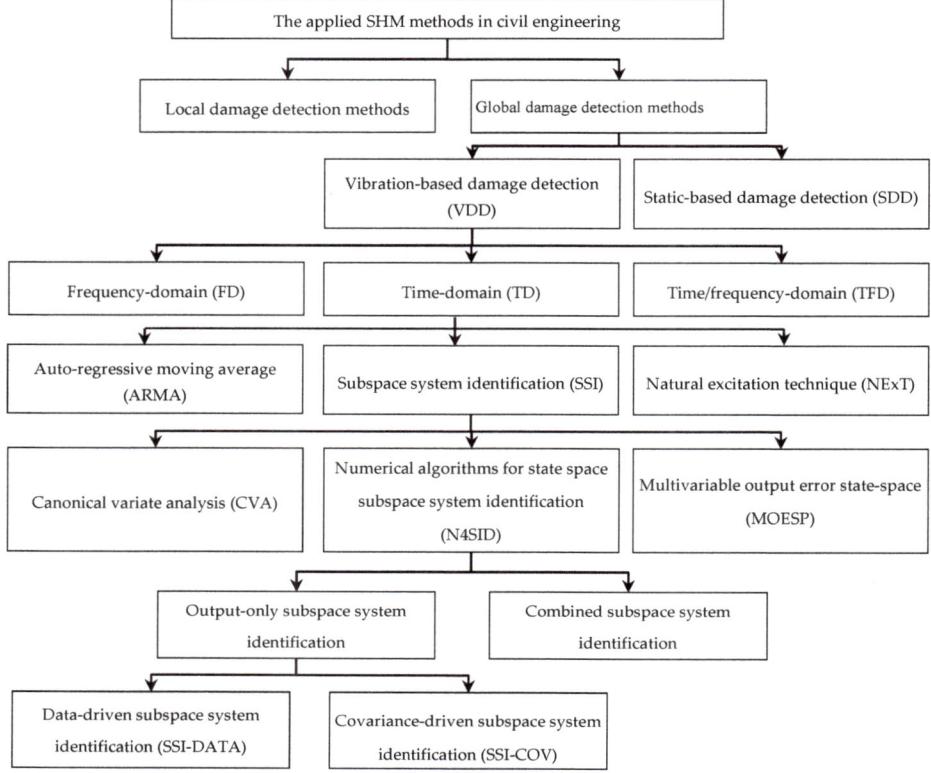

Figure 1. Classification of the structural health monitoring (SHM) methods with focus on subspace system identification (SSI) algorithm.

Researchers have proposed using global damage detection methods to identify damages in structures in order to overcome the previously mentioned limitations. Global methods are very effective choices to overcome the limitation of local SHM methods in civil engineering. In global methods, there is no limitation regarding the location of damage or even access to and preparation of the damaged area. These methods can localize and estimate the extent of damages while using the global characteristics of structures. The SDD methods require large measurement datasets for a reliable damage detection process. The targeted structure must generally be removed from its normal service in order to implement loading tests in SDD methods. Therefore, the SDD methods are not appropriate for continuous monitoring applications. When compared to VDD, the number of studies that employed SDD techniques are limited [39,40]. SDD methods are beyond the scope of this review paper and, for the sake of brevity, will not be covered in this study. A full review of the SDD has been well-documented elsewhere (for example [41]). VDD methods will be studied in the proceeding subsection.

3. Vibration-Based Damage Detection (VDD)

VDD is considered to be the most popular methodology in global SHM. VDD methods rely on changes in dynamic properties as an indicator of damage existence. These methods exploit observable variations in modal parameters, such as resonant frequency, damping, and mode shape or their derivative as indicators of change in physical properties of a structure. A thorough review of the system identification methods in VDD and modal analysis is provided in Song et al. [42] and Reynders [43].

3.1. Frequency-Domain (FD) Methods

An FD representation of a signal is extracted from a TD response signal using a Fourier transform. A Fourier transform describes a harmonic function by a linear combination of complex exponentials. A number of enhanced algorithms have been developed in order to improve the computational efficiency of Fourier transform, which are collectively termed as a fast Fourier transforms (FFT). FFT is generally used for deriving frequency response functions (FRF), and it plays the key role for many FD damage detection approaches [44]. FD methods in VDD can be classified into three main categories, which include [26]:

1. peak picking (PP) methods;
2. complex mode indication function (CMIF) methods; and,
3. least squares complex frequency-domain (LSCF) methods.

PP methods, also referred to as basic frequency-domain (BFD), are the most typical approaches in modal testing and they initially rely on power spectral analysis and Fourier transforms. PP methods are qualitative in nature and founded upon the fact that FRF reaches an extreme approximately around the natural frequency. Furthermore, these methods could be accompanied with a half power bandwidth approach to extract damping ratio [45]. Frequency domain decomposition (FDD) is an accurate and simple technique for system identification that is widely used in modal analysis. FDD has been developed based on spectral density decomposition. The obtained spectra are a reduced form of a dynamic response for individual modes [46]. The enhanced frequency domain decomposition (EFDD) method is an extension of FDD for estimating reliable modal parameters [47].

CMIF methods can also be considered to be an extension of PP techniques. They have been widely used for the output-only identification of system parameters. CMIF is developed by performing SVD on a normal FRF matrix at each spectral line [48]. CMIF is combined with other algorithms to be used as a standalone model, such as the enhanced frequency response function (eFRF) and enhanced mode indicator function (EMIF). eFRF is the subsequent development of the CMIF method and it is used to estimate the frequencies that are associated with a particular peak in the CMIF [49]. The eFRF is rooted in the concept of physical coordinate transformation to enhance the estimation of modal parameters. EMIF could be considered as an extension of the CMIF/eFRF, which estimates modal parameters in several modes at one time. The distinctive property of this method is due to the fixity in the number of natural frequencies based on the peaks of CMIF plots.

LSCF is a fast and accurate method for estimating modal parameters. Originally, LSCF was applied to extract initial values in the maximum likelihood method. LSCF performs reliably due to its clear stabilization diagram [50]. The polyreference least-squares complex frequency-domain method (PolyMAX) is the polyreference version of the LSCF that takes advantage of the right matrix-fraction model. The main benefit of this method is that the closely spaced modal frequencies can be separated from each other [51]. El-Kafafy and Peeters [52] introduced the poly-reference least squares complex frequency-domain (pLSCF) for modal analysis. A two-step scheme is proposed to enhance the damping estimates. The proposed method can improve the processing time and accuracy of the modal identification, particularly for damping estimates.

FD methods are fast and accurate, but they suffer from some limitations in the frequency resolution of the estimated spectral data [53]. Conventional FD methods are not accurate and reliable for the analysis of non-linear and non-stationary signals. The resolution of the identified system parameters in low-frequency ranges or fewer numbers of incorporated modes is poor in these methods [54]. Moreover, the estimated damping coefficients are not accurate in the non-parametric FD methods [55]. The strong demand in the field of system identification to achieve higher accuracy and extract more information from the vibration responses led to the development of TD methods.

3.2. Time-Domain (TD) Methods

TD techniques rely on the fact that the vibrational properties of structures can be captured through the time-history response of a dynamic system. Hence, the extracted response in a healthy state is different from that of one that is in a damaged state. Figure 2 shows the schematic architecture of TD methods that were used for the identification of dynamic systems. Different numerical techniques, for example, FFT, SVD, least squares (LS), QR decomposition, Eigen-vector decomposition (EVD) [56], and statistical methods, were used to develop these algorithms. Observer/Kalman filter identification (OKID), NExT, and random decrement (RD) are the most common TD methods for extracting the FRF when there is no access to the input data. The input signal can be estimated using an auto-correlation or cross-correlation function [57].

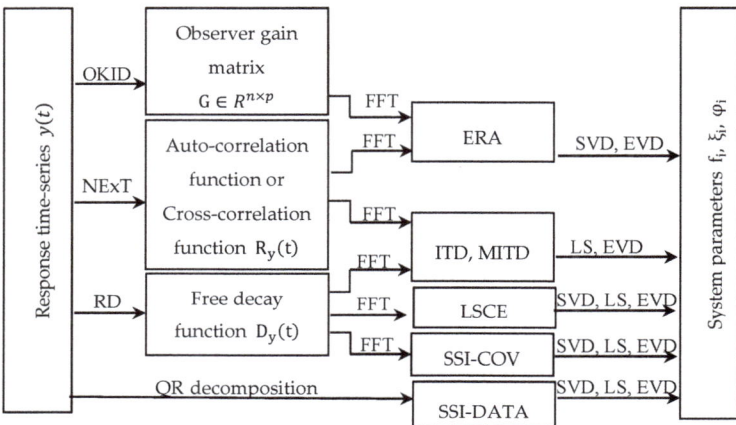

Figure 2. Schematic model for some of output-only time-domain modal identification methods.

The TD techniques in the health monitoring of civil structures can be classified into three widely known categories, which include [26]:

1. NExT methods;
2. ARMA methods; and,
3. SSI methods.

These methods are adopted for the identification of system parameters in civil engineering structures by introducing output-only extensions. NExT and SSI-COV methods are generally indirect where the SSI-DATA and ARMA methods are direct in nature, as can be seen in Figure 2. The TD methods are more appropriate for continuous monitoring compared to modal analysis methods due to direct use of response signals. Furthermore, the information extracted using TD methods is more complete than that of FD methods, particularly when a large number of modes or a large frequency range exist. A further explanation of the TD class of the identification methods are provided in the following subsections.

3.2.1. Natural Excitation Technique (NExT) Methods

James et al. [58] first proposed the NExT for the modal analysis of systems exposed to noise uncertainty. The key idea behind the NExT scheme is that the cross-correlation of the response signal from random excitation can be considered as a summation of decaying sinusoids. The complex exponential (CE) model is based on Prony's method and it was the first modal estimation method used as single-input single-output (SISO). The CE algorithm fits the curves of analytical impulse response into the experimental impulse response data [9]. The CE algorithm has been extended to the single-input multiple-output (SIMO) version by applying the LS technique. The algorithm was named the least-squares complex exponential (LSCE) [59]. The polyreference complex exponential (PRCE) is an extension of LSCE in the form of the multiple-input multiple-output (MIMO) identification of modal parameters [60].

Juang et al. [61] first proposed the Eigensystem realization algorithm (ERA) method for modal identification in the field of aerospace engineering. This method uses state-space forms together with Markov parameters in order to extract the coefficients of a dynamic system. Juang et al. [62] introduced an extension of ERA while using data correlations (ERA/DC). In the proposed method, the ERA realization-based approach is combined with a state-space correlation fit. The extracted modal parameters in this method are less sensitive to noise corruption and less prone to bias error. The improved polyreference complex exponential (IPCE) [63] is another extension to ERA and it is specifically designed for enhancing the reliability of PRCE and reducing the influence of random noise in modal identification. The IPCE technique uses correlation filtering as a pre-processing step to reduce the noise effects on measured data and minimize system order.

The Ibrahim time-domain (ITD) was of the first SIMO algorithms for estimating eigenvalues and eigenvectors in one-step. The ITD technique is reported not to be appropriate for heavy damped systems or systems with low natural frequencies [64]. The multiple references Ibrahim time-domain (MRITD) is an enhanced extension of ITD for MIMO modal analysis. The method is a high-resolution modal decomposition approach that is based on eigenanalysis [65]. OKID was originally a companion of the ERA, being denoted as OKID-ERA. The method suggested establishing a non-recursive LS observer to relate input and output data [66]. In recent years, the OKID has been introduced as a separate class of algorithms that could be combined with other models in a pairwise basis. Output-only ERA-OKID [67] and output-only observer/Kalman filter identification (O3KID) [68] methods are two recent versions of OKID for experimental modal analysis of civil engineering structures. The RD method is a TD approach for modal analysis through transforming system responses into random decrement functions [69]. RD functions are considered to be the free-vibration responses of a system. With the assumption of a zero-mean stationary Gaussian stochastic process, RD functions are proportional to correlation functions. The proposed method uses the concept of averaging to extract the random decrement signatures (RDS) of structures. Brincker [70] presented a general overview of the random decrement application in the modal analysis of structures. NExT methods provide a reliable tool for the modal identification of civil engineering structures. These methods have been implemented on several real world structures. However, the obtained results for damping ratios were less accurate that those that were obtained with other counterparts, such as the SSI and FDD algorithms [55].

3.2.2. Auto-Regressive Moving Average (ARMA) Methods

Auto-regressive moving average (ARMA)-based methods outperformed the purely statistical methods. The auto regressive (AR) part aims to model linear function time-history and the moving average (MA) part aims to determine the moving average of the time-series [71]. The general structure of the AR-based models depends on which of the A(q), B(q), C(q), D(q), and F(q) polynomials are used in the model. Equation (1) shows the general structure of AR-based models.

$$A(q)y(t) = \frac{B(q)}{F(q)}u(t) + \frac{C(q)}{D(q)}e(t) \tag{1}$$

Table 1 shows the most common ARMA models that were used in the system identification of civil engineering structures.

Table 1. Common auto-regressive moving average (ARMA) models shown with their associated polynomials.

Used Polynomials	Name of Model
$A(q), B(q)$	ARX
$A(q), B(q), C(q)$	Auto-Regressive Moving Average with eXogenous variable (ARMAX)
$A(q), C(q)$	ARMA
$A(q), B(q), D(q)$	Auto-Regressive Auto-Regressive with eXogenous input (ARARX)

Several studies have been conducted while using autoregressive-based methods for the identification of damage in structures. Table 2 presents a review of applied ARMA methods in SHM and modal analysis with a focus on the test methods, damage features, and the incorporated pattern recognition techniques.

Table 2. Autoregressive-based methods applied in the structural health monitoring (SHM) of civil engineering structures.

Method/Reference	Test Model	Damage Feature	Pattern Recognition
AR [72]	Progressive failure test of retrofitted reinforced concrete bridge column	AR coefficients	X-bar control chart
AR-ARX [73]	Laboratory-scale 8- DOF mass-spring system	Residual error between prediction and reference signal	Multi-paradigm statistical pattern recognition
ARMAV [74]	Numerical model of a simply supported beam and a real-size steel-quake structure	Changes in natural frequency	Confidence interval and normal distribution of random variables
AR-ARX [75]	A simulated three-story frame structure	Residual errors	Sequential Probability Ratio Test (SPRT)
VFAR [76]	Numerical simulations of a six-story shear building and a three-span pre-stressed box-girder bridge	Natural frequency and damping ratio	-
VBAR [77]	Numerical model of a lumped mass system and experimental model of a steel cantilever beam	Natural frequency	-
ARMAV [78]	Steel-quake structure	Natural frequencies and damping ratio	-
TV-ARMAV [79]	Supported beam with a moving mass	Natural frequencies and mode shapes and damping ratio	-
ARMA [80]	Analytical and experimental results of the ASCE benchmark structure	First three AR components	Multi-paradigm statistical signal processing approach combined with the pattern classification
ARMA [81]	IASC–ASCE four-story experimental benchmark structure, Z24 bridge and the Malaysia–Singapore Second Link Bridge	Normalized AR coefficient	T-test statistical control chart
ARMA [82]	Simulated ASCE benchmark structure	GMM	Mahalanobis criterion function
ARIMA [83]	Data from Malaysia-Singapore Second Link bridge	Coefficients of the ARIMA model	Statistical outlier detection
TARMA [84]	Laboratory-scale steel beam with a moving mass	Natural frequencies	-

The structures range from a very simple model of a mass spring system to more elaborate real world structures, as shown in Table 2. In many cases, the coefficients of the incorporated AR-based algorithm are directly used as damage features. For modal analysis, the extracted system parameters are utilized for the evaluation of dynamic behavior in a structure. Using modal parameters as damage features, the variation of the extracted parameters for the reference and actual state could be considered for the condition assessment of a structure. The obtained results for modal parameters are more robust when compared to other TD methods. However, higher scatterings in damping ratios are observed and the reported error is much more than other TD methods, such as SSI.

3.2.3. Subspace System Identification (SSI) Methods

Parametric TD methods provide a powerful and versatile mathematical framework for analyzing dynamic systems. Among all system identification methods, subspace-based techniques are the most remarkable achievement in the field of control and system identification. Meanwhile, many research studies on system identification have been concentrated on subspace methods in recent years. The subspace approach is a suitable technique for estimating the state-space model of a dynamic system. The SSI is a numerically reliable algorithm and it estimates models with good quality, particularly for multivariable systems [85,86]. The state-space form of the equation of motion can be written as Equation (2).

$$\begin{cases} x(t+1) = Ax(t) + Bu(t) + w(t) \\ y(t) = Cx(t) + Du(t) + v(t) \end{cases} \quad (2)$$

where $x(t+1)$, $u(t+1)$, $x(t)$, and $u(t)$ are state vectors and scalars at time instant of $t+1$ and t, respectively. y_k is output vector, A, B, C, and D are system, control, output, and feedback matrices, respectively. $w(t)$ and $v(t)$ are measurement and process noise, respectively. Most subspace algorithms reported in the literature are closely related to the LS-based methods [87]. In the first step, an oblique projection is calculated and it is pre- and post-multiplied by appropriate weight matrices to infer the system order and state sequence. In the second step, a geometrical projection is adapted in order to eliminate the dependence of the SSI algorithm on future output. In the third step, LS is deployed to drive the A and C matrices. Finally, the Kalman predictor is used to estimate the system model by inferring the Kalman gain K of the state-space model. In a general sense, the most researched subspace methods in the field of system identification can be classified within the following three main categories [87]:

- CVA methods;
- MOESP methods; and,
- N4SID methods.

Larimore and Wallace [88] proposed CVA methods that are based on Markov parameters for TD system identification. The study continues with the same principles as the pioneering activities of Akaike [89] in a statistical setting. SVD is used as a tool to extract the incorporated canonical variates. Verhaegen [90] proposed the MOESP method for the identification of the multivariable state-space model from noisy input-output data. The MOESP subspace algorithm is known for two characteristics, those of the reduced-size Hankel matrix and the extended observability matrix. The method was not applicable for stochastic systems. Van Overschee and De Moor [91] unified proposed subspace schemes into a pragmatic approach, referred to as N4SID. The algorithm was analytically robust and reliable due to the use of SVD and QR decomposition. Based on the way, the subspace algorithm deals with the measurement time history data; they can be divided into the two categories of SSI-DATA and SSI-COV. In the next two subsections, a review on application of the SSI-DATA and SSI-COV algorithms in the SHM of civil engineering structures is outlined.

Data-Driven Stochastic Subspace Identification Method (SSI-DATA)

SSI-DATA is a method for identifying modal parameters by the direct use of measured response time-history [92]. Overschee et al. [93] introduced a subspace algorithm using power spectrum data. The state-space coefficients were derived using inverse discrete Fourier transform. The computational complexity in this method is higher when compared to that of direct subspace methods or FDD.

Table 3 shows some examples of the methods that use SSI-DATA for damage detection and modal analysis.

Table 3. Some examples of schemes that use data-driven subspace system identification (SSI-DATA) in their damage identification process.

Method/Reference	Damage Feature	Pattern Recognition	Test Model	Specification
SSI-DATA [94]	Eigensolutions	Statistical local approach	Analytical model	Application of a combined SSI
SSI-DATA [95]	Residual error	Statistical process control	Aircraft model and real bridge structure	Kalman model is extracted to represent the reference state of the structure.
Null-space based algorithm [96]	Residue matrix of orthonormality between subspaces	Statistical process control	Aircraft model and steel lighting device	Subspace analysis of the Hankel matrices is used for extracting features.
CSI/ref [97]	Modal parameters	-	Z24 bridge benchmark structure	Stabilization diagram is adopted for post processing of modal data.
McKelvey FD SSI [98]	Modal parameters	-	Uncontrolled cantilever plate	Natural frequency was predicted with an average error of 3.2% and damping ratio had an average error of 2.8%
KPCA subspace based damage detection [99]	Subspace angles	Through mapping the Subspace angles	A beam with non-linear stiffness	KPCA performs damage detection problem without modal identification.
CC-SSI [100]	Modal parameters	-	Ariane 5 launch vehicle	Clear results even in the case of non-stationary data obtained using this algorithm
RSSI [101]	Modal parameters	-	3-story steel frame and 2-story reinforced concrete frame	In this method the SVD algorithm is replaced by an advanced algorithm to update LQ decomposition.
Improved SSI [102]	Modal parameters	-	A numerical example of 7 DOF and an experimental model of Chaotianmen bridge	Less computing time due to not having QR decomposition CH matrix is constructed as a replacement for Hankel matrix. Spurious modes are removed using the model similarity index
SSA [103]	Subspace-based and nullspace-based damage indices	SVD for change-point detection	6-story steel frame	SSA algorithm uses SVD for change-point detection
Fast CC-SSI [104]	Modal parameters	-	Operational data from a ship	Fast multi-order computation
ECCA-based SSI algorithm [105]	Modal parameters	-	FE model and experimental wind tunnel bridge model	Enhanced results are achieved for weakly excited modes and noisy response signals

Covariance-Driven Stochastic Subspace Identification (SSI-COV) Method

SSI-COV is a parametric output-only method that is used for modal analysis. The method use the vibration response time-history to extract the state-space model of a dynamic system. SSI-COV is a two-step procedure that utilizes the correlation function of vibration time-history. The system order is the only user defined parameter in SSI and it must be carefully chosen to obtain meaningful results. In these methods, the applied excitation is considered as white noise and it is equal to the covariance of the measured response time-history.

Table 4 shows some SSI-COV algorithms that are applied for the damage detection of civil engineering structures. The damage indices in these methods are either modal parameters or a novel indicator for detecting changes that are caused by damage. In the previous two subsections, the application of SSI-DATA and SSI-COV algorithms for the damage detection of civil engineering structures was presented. In the following section, a comparison between the SSI algorithm and other key system identification methods in TD and FD is presented.

Table 4. Some examples of the schemes that use covariance-driven subspace system identification (SSI-COV) in their damage identification process.

Method/Reference	Damage Feature	Pattern Recognition	Test Model	Specification
SSI-COV [106]	Subspace residual	Sensitivity analysis	Numerical analysis	Damage localization using the FE model
EMD-SSI-COV [107]	Intrinsic mode functions	Sifting process	Beichuan arch bridge	Capable of dealing with non-stationary signals
BSSI-COV [108]	Curvature evolution index	Stabilization diagram	Numerical simulation of an 7-DOF MSD model	Suitable for identifying weakly excited modes
SSI-COV [109]	Natural frequencies	Fuzzy clustering	Z24 Bridge and Republic Plaza Office Tower	No need for mode shape data
SSI-COV [110]	Residual of the observability null-space	Averaging operation	FE model of a bridge deck and laboratory test of a clamped beam within a climatic chamber	Temperature effect is evaluated using an averaging operation
SSI-COV [111]	Subspace residual	χ_2 test	Workbench and instrumented beam	Handling the temperature effect is a target in this paper
SSI-COV [112,113]	State space matrices	Mahalanobis and Euclidean distance decision functions	Two simulated beam model, a laboratory-scale reinforced concrete beam and a full-scale arch bridge structure	The method was capable of locating damage in beam structures.
Recursive SSI-COV-IV [114]	Modal parameters	-	Numerical models of a Single Degree of Freedom (SDOF) structure and an ASCE benchmark steel frame structure	Model identification was conducted for a system with time-varying measurement noise
RD–SSIcov [115]	Modal parameters	Modal strain energy method	Numerical model and a small scale wind turbine tower	The RD method was selected in this study for its noise reduction capabilities.
SSI-COV [116]	Modal parameters	Local optimization methods of NMA, GA, HS and PSO	Numerical examples and a laboratory model of cantilever beams	The model is appropriate to handle incomplete measurement data and truncated mode shapes
SSI-COV [117]	Curvature change, novelty index and PCA-SSA (for localization) modal updating and stiffness reduction (for quantification)	Novelty analysis	Laboratory-scale bridge foundation scouring	Detection, localization and quantification was conducted in a unified and online recursive subspace algorithm
SSI-COV [118]	Eigenfrequencies, damping ratios and mode shapes	Stabilization diagram	Experimental models of a damaged pre-stressed concrete bridge and a mid-rise building	Uncertainty quantification and effect of different setups in modal parameters was evaluated in this study

4. Comparison between SSI and Other System Identification Algorithms

The SSI-DATA is a direct method and it does not need any data pre-processing to calculate correlation functions or spectrum analysis. On the other hand, the subspace identification algorithm uses QR factorization, SVD decomposition, and LS robust numerical techniques in the analysis process [119]. Several comparative studies are presented below in order to evaluate the advantages and drawbacks of the subspace identification algorithm when compared to other time and frequency domain damage identification methods in SHM. Table 5 presents a comparison between SSI method and other system identification algorithms.

Table 5. Comparison between subspace system identification (SSI) method and other system identification algorithms.

Reference	Identification Method	Advantages and Drawbacks
Magalhães et al. [55]	SSI PP	• Generally, there was a good agreement between the obtained modal frequencies and mode shape in all algorithms, however the damping ratio shows the most significant scatter. • The PP method is an appropriate technique for modal identification of closely spaced modes. • SSI and FDD is a more appropriate choice for modal identification where closely spaced frequencies are not of the same type (e.g., bending or torsion).
	FDD	• PP and FDD are not suitable tools for identification of the modal parameters in cable-stayed bridges. Further analysis must be undertaken to successfully discriminate the vibrational influence of the cables from the global bridge frequencies. • The FDD and SSI yielded unreliable results for the second pair of closely spaced frequencies due to the coincidence of the modes and coupling effect in ambient excitation. • Estimation of damping ratios with wind-induced excitation has shown a significant level of scatter. Further research on ambient vibration testing is needed to approach a reliable identification of damping coefficients. • All the above-mentioned algorithms require human judgment during implementation. The user defined parameters for the SSI, FDD and PP are system order, singular values, and global modes, respectively.
Magalhães et al. [120]	SSI-COV p-LSCF	• Performance of both SSI-COV and p-LSCF was found to be good in the identification of natural frequencies and mode shapes. • It is demonstrated that p-LSCF can provide better estimation for the daily variations of damping coefficients compared to SSI. • All of the modal identification methods perform well when the responses are exposed to noise.
Wang et al. [121]	SSI ERA ARMA	• Model order is an important parameter in extracting correct modal parameters in all methods. • Significant errors are observed in the damping coefficient when system order varies. • Damping coefficient is sensitive to noise pollution, specifically to the first mode.
	ITD	• More stable results for modal identification were obtained by the SSI method in numerical simulations. However, the ERA method outperforms other algorithms for field testing. • ARMA shows the worst results for both field and numerical experiments.
Moaveni et al. [122]	MNExT-ERA SSI-DATA	Comparing output-only methods of SSI-DATA, MNExT-ERA and EFDD, it can be inferred that: • The identified natural frequencies using these methods display good agreement in each damage scenario. • The identified damping coefficient shows larger variability but still in a reasonable range.
	EFDD	• Lower damping ratios are obtained through FDD than the results identified using SSI-DATA and MNExT-ERA. • The identified mode shapes using SSI-DATA were confirmed to be the most reliable among all methods.
	DSI OKID-ERA	Comparing input-output methods of DSI, OKID-ERA, and GRA, the following were observed: • The identified mode-shapes in all of the above-mentioned algorithms are close to the actual mode shape of the structure.
	GRA	• In general, the calculated mode shapes by the input-output methods are more accurate than those obtained from output-only methods. • The mode-shape parameters obtained from the GRA appear to be in good agreement with the actual mode shapes of the structure.
Cunha et al. [123]	SSI-COV FDD	• The same result in terms of damage detection was obtained using both the SSI-COV and FDD procedures. • In terms of modal parameters, perfect results were extracted through using both methods.

Table 5. *Cont.*

Reference	Identification Method	Advantages and Drawbacks
Liu et al. [54]	SSI and ERA	• Both the ERA and SSI methods are accurate and provide stable results for modal parameters. The methods are extremely sensitive to the size of the Hankel matrix.
	PolyMAX	• The work burden of the algorithms is large and usually needs to perform multiple analyses. • The PolyMAX and PolyLSCF methods are fast but the operator needs to have a high qualification. • The algorithm is sensitive to the size of the Hankel matrix and sometimes needs to perform multiple analyses.
	PPM	• PPM is a simple and less demanding algorithm. • The algorithm can identify modal parameters in the pre-defined frequency range.
	PZM	• PZM and PolyLSCF are basically similar and the obtained results can complement each other. • PZM uses a non-power spectrum for modal frequencies and damping.
	EFDD and FSDD	• EFDD and FSDD are fast and easy for modal parameter identification under ambient excitation. • The estimation precisions in these methods depend on the accuracy of the power spectral density. • Both methods are sensitive to noise and leakage error.
	WT	• The accuracy of the WT is relatively high. • The WT is not conducive to the real-time monitoring due to analysis load.
Kim and Lynch [53]	SSI	• Similar performance was confirmed for the SSI and FDD methods.
	FDD	• The FDD method slightly suffered from resolution problems due to the limited number of data points.
Ceravolo and Abbiati [124]	SSI-DATA AR	• All three output-only methods of SSI-DATA, AR and ERA/RDS show high robustness to modal frequency estimation in non-stationary conditions.
	ERA/RDS	• By using the same methods, damping estimates were very scattered. • The estimation error of modal frequency in all methods remained less than 1.5%. • The error in estimating damping ratio ranged from 15% (in SSI) to 50% (in AR).

Table 5 shows the results of a comparison between several TD, FD, and time-frequency domain methods for damage detection and modal analysis of civil engineering structures. The obtained results confirm the reliable performance of the TD methods. It was shown that SSI is a powerful tool for modal identification, in which closely spaced frequencies are not of the same type (e.g., bending or torsion) [55]. It was reported that the identified mode shapes using SSI-DATA were the most reliable when compared to MNExT-ERA or EFDD [122]. Moreover, fewer errors were obtained in estimating the damping ratio using the SSI algorithm as compared to ARMA [124]. However, SSI has some disadvantages, such as requiring human judgment of system order for implementation [55]. On the other hand, the work burden of the SSI algorithm is large and it usually needs to perform multiple analyses [54].

5. Challenges of SSI in Practical Application

Research on the application of subspace methods for the damage detection of civil structures emerged in the mid-1990s. Most of the methods used in SHM presume a parametric model of a dynamic system in order to characterize structural behavior under an applied excitation load. However, civil engineering structures still have many challenges to achieve a robust SHM model. The size of civil structures does not permit a large number of sensors to be mounted on a structure. Moreover, forced-vibration is not considered to be practical due to interruption in the serviceability of structures. On the other hand, civil structures are complex in terms of geometry and their material properties involve a large range of uncertainty due to operational and environmental factors [125]. In Table 6, the problems that are faced with practical implementation of SHM systems using the SSI algorithm and the researches to resolve the associated problems are presented.

Table 6 reviews the challenges of implementing the subspace algorithm in real-world applications. In practical applications of SHM, the response signal of structures is generally in the form of a non-Gaussian random signal. In such conditions, deterministic techniques result in unreliable system models. On the other hand, data loss or corruption caused by failure or loss of sensing, transmission, or storage devices during their normal use is a concern for a reliable damage detection scheme. Consequently, appropriate procedures must be considered in order to deal with the uncertainty that is caused by such instrumental failures. The accuracy of an identification algorithm is due not only to its insensitivity to environmental variation and instrumental failure, but also to the inherent performance of the estimation scheme. Studies have been conducted to improve the performance of the SSI method and enhance the modal identification process itself. The inherent performance of an estimation scheme to deal with problems, such as the short-length of a signal, non-stationarity measurement data, system non-linearity, leakage error, or different measurement setups remains a challenge for a reliable SHM. In some cases, resolving the problem and increasing the accuracy demands exhaustive expert assistance and time-consuming computation burdens. Gluing the non-simultaneously measured set-ups of sensor data is another controversial issue that needs to be considered before applying any identification platform. The aforementioned drawbacks are the topic of ongoing research in the field of SHM.

Table 6. The challenges of SSI in practical applications.

Reference	Method	Concern
Peeters and De Roeck [126]	SSI	Dealing with the sensor data of different measurement setups.
Pridham and Wilson [127]	Correlation-driven subspace algorithm	The effect of sample size and the dimension of the Hankel matrix on the estimated damping coefficient
Pan [128]	-	Discriminates the effects of epistemic uncertainty
Benveniste and Mevel [129]	SSI	Consistency of the subspace algorithm against non-stationarities in the measurement data
Reynders et al. [130]	SSI	Removing bias errors in ambient vibrations
Brasiliano et al. [131]	SSI-COV and SSI-DATA	Effect of non-structural elements in the vibrational behavior of bridge structures.
Alıcıoğlu & Luş [132]	SSI-COV and SSI-DATA techniques through BR and CVA.	The effect of structural complexity and ambient conditions on modal identification.
Marchesiello et al. [133]	ST-SSI	Time-variant identification
Balmès et al. [134]	Extended SSI-COV-based	Handling temperature effect in the identified modal parameters.
Wang et al. [135]	SSI together with ARX	Damage detection of Hammerstein systems or non-linearity introduced to a linear dynamic system through piece-wise constant inputs
Magalhães et al. [136]	EFDD and SSI-COV	Quality of the extracted damping ratios considering the proximity of natural frequencies, non-proportional damping and accuracy of the identification algorithms
Döhler et al. [137]	SSI-COV	Multiple non-simultaneously recorded measurement setups.
Carden and Mita [138]	A combined SSI-DATA method and first order perturbation technique	The challenges faced in extracting the accurate confidence interval of modal parameters in civil engineering structures.
Döhler and Mevel [139]	SSI- a modular and scalable approach	Merging sensor data by applying a modular approach.
Döhler et al. [140]	First-order SSI method perturbation analysis	Evaluation of the statistical uncertainty in a multi-setup configuration
Loh and Chen [141]	Hybrid RSSI-COV, NLPCA and AANN	Distinguishing the damage abnormality from those caused by environmental and operational variations
Brehm et al. [142]	Power spectral amplitudes and FE model	Determination of the optimum position of the reference sensor
Cara et al. [143]	Kalman filter SSI algorithm	Contribution of the modal parameters in each mode to the recorded measurement data.
Ashari and Mevel [144]	SSI-based algorithm	Injecting auxiliary input to extract the unexcited modes
Tondreau and Deraemaeker, [145]	Monte-Carlo and SSI	Introducing uncertainty into an FE model
Rainieri and Fabbrocino [146]	SSI	Influence of the system order and number of block rows in identification accuracy
Cho et al. [147]	SDSI	Implementation of VDD using wireless sensor networks
Markovsky [148]	SSI	Influence of missing measurement data in dynamic system
Banfi & Carassale [149]	SSI, SRA and FDD	Effect of uncertainty in determining modal parameters
Spiridonakos et al. [150]	PCE	Humidity and temperature effects
Huynh et al. [151]	A combined SSI algorithm and short-time Fourier transform	Structural analysis of a cable-stayed bridge under typhoons with various wind speeds.
Ren et al. [152]	Improved SSI algorithm	Eliminating spurious modes caused by non-white noise

6. Application of Subspace Identification in Civil Engineering Structures

Currently, stochastic subspace methods are widely accepted tools in civil engineering communities. Large number of SHM methods is operating in structures that are subjected to dynamic vehicular, seismic, wind, or impact loading. In this study, the application of the SSI algorithm in civil engineering structures is investigated in the following three categories:

1. beams and two dimensional (2D) frames;
2. three dimensional (3D) frames and building structures; and,
3. bridges and other structures.

Beams and two-dimensional (2D) models attempt to explain the performance of the SHM algorithm while using numerical FE or laboratory-scale models [153]. Many of the studies on these structures are generally academic in nature. Three-dimensional (3D) frames and building structures are generally used to investigate the practical aspects of SHM. Bridges are spectacular due to their specific loading conditions when compared to building-type structures [154]. Vehicular and pedestrian loading, together with exposure to environmental variations due to temperature and wind precipitation, have been the focus of the analysis of bridge structures in many pioneering studies of SHM [155]. Tables 7–9 show some application of SSI algorithms for damage detection of 2D frames, 3D frames, and buildings and bridges, and other structures, respectively.

Table 7. Application of SSI algorithms for damage detection of beams and two-dimensional (2D) frames.

Reference	Applied Damage Detection Technique	Test Structure	Damage Configuration	Damage Identification Results
Vanlanduit et al. [156]	Subspace theory and RSVD	Laboratory-scale aluminum beam	(1) 30% and 50% sew cut in the middle (2) A beam with fatigue crack under several geometrical and damping conditions.	Reliable results were obtained by applying the introduced RSVD method.
Naseralavi et al. [157]	Theory of subspaces and Kernel parallelization	Three simulations of 1D and 2D case studies	Reduction in Young's modulus ranging from 10% to 30% correlated with the extent of the damage	Most single and multiple damage scenarios
Döhler and Hille [158]	Subspace-based	(1) Numerical simulations (2) A 1D steel frame structure	The crack-like damage simulated by the loosened 3 to 7 number of bolts, which correlates with a reduction of bending stiffness from 3% to 30%, respectively.	All damage cases were clearly detected under variable excitation using the new test method
Banan and Mehdi-Pour, [159]	Concept of subspace rotation and Monte Carlo simulation	A simulation model of a braced-frame structure	Single damage case with a 25%, 50%, and 95% reduction in Young's Modulus	Damages were detected at noise ratios less than 5% but for a noise ratio of 20%, the results were unreliable for all damage rates.

Table 8. Application of SSI algorithms for damage detection of three-dimensional (3D) frames and buildings.

Reference	Applied Damage Detection Technique	Test Structure	Damage Configuration	Damage Identification Results
Siegert et al. [160]	Statistical subspace-based algorithm	Laboratory-scale composite bridge deck	40% reduction of the Young modulus	Algorithm can successfully detect the damages of the structure investigated
Li and Chang [114]	SSI and IV	Numerical model of a four-story, ASCE benchmark steel frame	A damage case of 30% reduction in the stiffness of the first story	Application of the technique tracking damages in the presence of measurement noise was successful.
Huang et al. [161]	SSI and DLV	Numerical model and laboratory-scale of a five-story steel frame	Damage scenarios with single and two damage case inflicted on the first second, third and fourth stories' columns	(1) Poor results were recorded for multiple damage cases. (2) The obtained result for the ill-conditioned counterpart was effective.
Kim and Lynch [162]	SSI algorithms	Laboratory scale of a single-bay six-story steel frame structure	Six damage cases comprised of sew cuttings of 3, 6 and 9 cm of the columns on the first and second floors	All damage in the structure was easy to identify using the proposed algorithm even in the presence of noise.
Wang et al. [163]	SSI and DLV	Five-story shear frame structure with diagonal bracings.	Damage conditions simulated by partially removing some of the bracings	The proposed algorithm for damage localization was successful
Moaveni et al. [164]	Deterministic-stochastic subspace identification	Laboratory-scale, three-story, two-bay, infilled RC	Progressive damage induced by shaking table experiencing scaled earthquake records.	Natural frequencies decrease with increase in damage level during the test experiment
Weng et al. [165]	SI and model updating	Six-story steel frame structure and a two-story RC frame.	(1) Loosening the connection bolts in single and multiple damage case in 2nd, 4th and 6th stories. (2) by progressive damage through shaking table in a RC frame.	The method was capable of detecting, localizing and quantifying the damage in both steel and concrete frames.
Karami and Akbarabadi [166]	Active control strategy and SSI	Numerical model of a 5 story and a 20 story shear building	Reduction in the stories' lateral stiffness ranged from 10 to 50%	The algorithm can detect damages accurately.
Belleri et al. [167]	Combined deterministic-stochastic subspace identification	Real-size three-story half-scale precast concrete building	Base excitation leading to loss of stiffness, cracks, and failure of the connections	Damage reduces natural frequencies and increases the damping ratios
Shinagawa and Mita [168]	Subspace identification	Full-scale four-story steel building	Collapse test.	Strong correlation between the extracted features with structural damage was observed.
Zhou et al. [169]	Subspace-based methods	A six-story residential building	More than 20 earthquake aftershocks recorded after instrumentation	Results of damage detection were not very reliable for earthquake induced excitation
Yoshimoto et al. [170]	MIMO SSI method and sub-structuring	An existing 7-story base-isolated building.	Stiffness reduction in the stories	The estimated damage of the stories correlates with the reduction ratio.

Table 9. Application of SSI algorithms for damage detection of bridges and other structures.

Reference	Applied Damage Detection Technique	Test Structure	Damage Configuration	Damage Identification Results
Allahdadian et al. [171]	SSDD and χ^2 test	Reibersdorf Bridge	Damage simulated in girder, column, bearing, deck and cap beam with the ratios varied between 20%, 40% and 80%.	Unable to detect minor damage with a ratio of 20%
Dohler et al. [172]	Statistics-based subspace and χ^2 test	S101 Bridge	Progressive damage test including cutting column, column settlement, horizontal crack in column, settling of bridge deck, and cutting pre-stress tendons.	(1) All steps of the column settlement and uplifting were clearly visible (2) Cutting of the pre-stressing tendons was not detectable.
Deraemaeker et al. [173]	Automated SSI and Shewhart-T control charts	Numerical model of a 3 span bridge	Four damage scenarios simulated by stiffness reduction in three different locations.	Proposed method was capable of detecting damage scenarios
Hu et al. [174]	SSI-COV and PCA algorithm	Pedro e Ines footbridge	Several realistic damage scenarios of 5%, 10%, 15%, 20% and 30% reduction of model spring constants	The method was capable of detecting all damage scenarios
Loh and Chao [175]	SSI-COV and MSSA	Laboratory-scale bridge	Displacement and lowering of a pier due to scouring	Localization and quantification were successful
Mevel and Goursat [176]	SSI	Z24 Bridge	Pier settlement of 20mm and 80 mm.	The efficiency of the proposed method was confirmed
Lin et al. [113]	Statistics-based SSI	(1) Laboratory-sale pre-stressed RC beams (2) a full-scale RC arch bridge	Gradually releasing the pre-stressed tendon from 17% to 73%	Algorithm was successful in detecting damage
Kullaa [177]	Subspace and stabilization diagram and control chart	Z24 Bridge	Pier settlements of 40mm and 95mm	Multivariate control charts have better performance in damage detection compared to univariate methods
Reynders, and De Roeck [178]	SSI algorithm and KPCA	Z24	Pier settlement, tilt of foundation, concrete spalling, landslide of abutment, failure of concrete hinge, failure of anchor heads and rupture of tendons were introduced to the structure	The proposed algorithm was capable of detecting imperfection in most cases
Nguyen et al. [179]	NSA and EPCA	Numerical model of Champangshiehl Bridge	Several intentionally created cuttings in pre-stressed tendons	The obtained result was encouraging.

In most structures, damage occurs in the form of a reduction in the cross-section of structural members. Partial reductions in 2D structures are usually detectable with a high level of precision. The robust applicability of the proposed methods to solve the problem of detecting local damage in real and complicated civil engineering structures has not been validated, even though these methods work relatively well in simple structures. Three-dimensional frames and building structures are generally complex and they pose challenges for both practitioners and researchers. Damage in building structures is usually in the form of a partial reduction in the cross-sectional area in column elements. In some cases, damage is made by opening bolts in a beam-column connection. However, beam damages are less important and they require higher detectability resolution. Bridge structures are a very important element of transportation. An in-service bridge is subject to loads, such as traffic, temperature variation, wind loading, and deterioration, under aggressive environments. Applying SHM to bridge structures poses significant challenges due to the specific types of loading and complexity of the structure. As a general conclusion, it could be derived that damage detection strategies that use modal frequencies, mode shapes or mode shape curvatures as their damage sensitive features [180] are only efficient for the detection of global damage and are not generally sensitive enough to detect changes in the local elements of structures [181].

7. Application of Subspace Identification in Civil Engineering Structures

In general, an individual program or a combination of software packages implement the process of damage detection [182,183]. A structural monitoring program is considered an algorithm for analyzing response signals, extracting damage features and deploying pattern recognition paradigms, ultimately leading to damage identification [184]. Subspace methods have been used as the central part of many of the structural monitoring programs used in industry [185,186]. In this subsection, the industrial software packages used in modal identification and SHM, which have adapted SSI as their core identification process, are further investigated.

Table 10 provides some of the commercially available software that use subspace identification algorithms as their standard technique for SHM and modal analysis. Most of the available algorithms are generally used for modal analysis. The SHM algorithms are composed of (i) identification, (ii) feature extraction, and (iii) pattern recognition steps and the implementation of a unified algorithm for huge diversity of each category is quite challenging and rewarding.

Table 10. SSI-based commercially available software for SHM and modal analysis.

Reference	Software	Analysis Tool	Environment	Purpose
Döhler et al. [182]	ARTeMIS	CC-SSI	-	Modal identification
Y. Zhou et al. [183]	ModalVIEW	SSI, RPF and LSCF	LabVIEW	Modal identification
Hu et al. [184]	SMI	EFDD	LabVIEW	Modal identification
Hu et al. [184]	CSMI	PP, FDD and SSI	LabVIEW	Modal identification
Goursat & Mevel [186]	COSMAD	SSI-COV	Scilab	Modal identification
Chang et al. [185]	SMIT	SSI, auto-regressive-based methods and realization-based algorithms	MATLAB	SHM

8. Future Research Directions

In the future, researchers should focus on the identification of local damages by improving the accuracy and noise-robustness of damage identification algorithms [187,188]. Furthermore, they must think about introducing a novel platform for the implementation of commercialized SHM software that is versatile enough to deal with the diversity of techniques in a damage detection system. Providing a user-friendly platform for the implementation of the SHM algorithm will improve the general usage of SHM software in solving real-life engineering problems. Furthermore, the extensible architecture would enhance the applicability of the software by enabling users to modify the existing base code and add their own extensions. A modular and flexible architecture enables a wide variety of reported methods to deal with within an integrated framework. Enhancing the SSI properties to deal with real-world applications, such as noise inclusion, short length data, and gluing sensor data, will enhance accuracy to provide more reliable damage detection results.

9. Conclusions

This paper presented a review of the recent advances in subspace-based SHM methods. SSI is a powerful choice for modal identification, particularly when dealing with closely spaced frequencies or noisy response signals. The identified mode shapes using SSI-DATA were confirmed to be the most reliable when compared to MNExT-ERA and EFDD. Moreover, fewer errors were obtained in estimating the damping ratio using the SSI algorithm as compared to ARMA. However, there are some disadvantages in using SSI, such as requiring human judgment for system order and large work burden of the SSI algorithm.

In most structures, damage is imposed in the form of a reduction in the cross-section of structural members. A review on applied models used for verification of damage detection algorithms shows that damages in the 2D structures are usually detectable with a high level of precision. Three-dimensional frames and building structures are generally complex and pose challenges for both practitioners and researchers. Even though lots of proposed SHM algorithms work relatively well in simple structures, the robust application of these methods to solve the problem of local damage detection in real and complicated civil engineering structures has not been validated. The obtained results demonstrate that damage detection strategies that use vibration parameters of a dynamic system are only efficient for the detection of global damages and they are not generally sensitive enough to detect local damages in structures.

Most commercially available software is generally designed for modal analysis and cannot be directly used for the SHM of structures. The commercial modal analysis toolboxes introduced are mainly implemented while using LabVIEW, which adapts a limited number of system identification methods, for example, SSI, LSCF, FDD, and SSI. The identification of modal parameters generally starts with extracting the response signal of structures and it results in the modal parameters of natural frequency, mode shapes, and damping ratio. However, SHM algorithms are generally complicated due to the wide variety of methods reported and their inherent structures. Hence, the implementation for the large diversity found in each category within a unified algorithm is quite a challenging problem.

Author Contributions: Resources, H.S. (Hoofar Shokravi), H.S. (Hooman Shokravi), N.B., S.S.R.K. and M.P.; investigation, H.S. (Hoofar Shokravi); writing—original draft preparation, H.S. (Hoofar Shokravi) and H.S. (Hooman Shokravi); writing—review and editing, H.S. (Hoofar Shokravi), H.S. (Hooman Shokravi), N.B. and S.S.R.K.; visualization, H.S. (Hoofar Shokravi), H.S. (Hooman Shokravi), N.B., S.S.R.K. and M.P.; supervision, N.B.; project administration, H.S. (Hoofar Shokravi), H.S. (Hooman Shokravi), N.B., S.S.R.K. and M.P.; funding acquisition, N.B., S.S.R.K. and M.P.; All authors have read and agreed to the published version of the manuscript.

Funding: This research was funded by Ministry of Higher Education, Malaysia, and Universiti Teknologi Malaysia (UTM) for their financial support through the Fundamental Research Grant Scheme (grant number: 4F800) and Higher Institution Centre of Excellent grant (grant number: 4J224) The APC was funded by Ministry of Education, Youth, and Sports of the Czech Republic and the European Union (European Structural and Investment Funds Operational Program Research, Development, and Education) in the framework of the project "Modular platform for autonomous chassis of specialized electric vehicles for freight and equipment transportation", Reg. No. CZ.02.1.01/0.0/0.0/16_025/0007293.

Acknowledgments: The authors would like to thank the Ministry of Higher Education, Malaysia, and Universiti Teknologi Malaysia (UTM) for their financial support through the Fundamental Research Grant Scheme (4F800) and Higher Institution Centre of Excellent grant (4J224), Ministry of Education, Youth, and Sports of the Czech Republic and the European Union (European Structural and Investment Funds Operational Program Research, Development, and Education) in the framework of the project "Modular platform for autonomous chassis of specialized electric vehicles for freight and equipment transportation", Reg. No. CZ.02.1.01/0.0/0.0/16_025/0007293.

Conflicts of Interest: The authors declare no conflict of interest.

Nomenclature

AR	Auto regressive
ARMA	Auto-regressive moving average
ARMAX	Auto-Regressive Moving Average with eXogenous variable
ARX	Auto-Regressiv with eXogenous input
BFD	Basic frequency-domain
CE	Complex exponential
CMIF	Complex mode indication function
CVA	Canonical variate analysis
EFDD	Enhanced frequency domain decomposition
eFRF	Enhanced frequency response function
ERA	Eigensystem realization algorithm
ERA/DC	ERA using data correlations
ERA-OKID-OO	Output-only ERA-OKID

ET	Eddy current test
EVD	Eigen-vector decomposition
FD	Frequency-domain
FE	Finite element
FFT	Fast Fourier transforms
FRF	Frequency response functions
IPCE	Improved polyreference complex exponential
ITD	Ibrahim time-domain
LS	Least-squares
LSCE	Least-squares complex exponential
MA	Moving average
MIMO	Multiple-input multiple-output
MOESP	Multivariable output error state-space
MRITD	Multiple references Ibrahim time-domain
N4SID	Numerical algorithms for state-space subspace system identification
NDE	Non-destructive evaluation
NExT	Natural excitation technique
O3KID	Output-only observer/Kalman filter identification
OKID	Observer/Kalman filter identification
PAUT	Phased array ultrasonic test
pLSCF	Poly-reference least squares complex frequency-domain
PolyMAX	Polyreference least-squares complex frequency-domain method
PP	Peak picking
PRCE	Polyreference complex exponential
RD	Random decrement
RDS	Random decrement signatures
RT	Radiographic test
SDD	Static-based damage detection
SHM	Structural health monitoring
SIMO	Single-input multiple-output
SISO	Single-input single-output
SSI	Subspace system identification
SSI-COV	Covariance-driven subspace system identification
SSI-DATA	Data-driven subspace system identification
SVD	Singular value decomposition
TARMA	Time-dependent auto-regressive moving average
TD	Time-domain
TFD	Time/frequency domain
UT	Ultrasonic test
VDD	Vibration-based damage detection

References

1. Zhao, H.; Ding, Y.; Nagarajaiah, S.; Li, A. Longitudinal Displacement Behavior and Girder End Reliability of a Jointless Steel-Truss Arch Railway Bridge during Operation. *Appl. Sci.* **2019**, *9*, 2222. [CrossRef]
2. Artese, S.; Nico, G. TLS and GB-RAR Measurements of Vibration Frequencies and Oscillation Amplitudes of Tall Structures: An Application to Wind Towers. *Appl. Sci.* **2020**, *10*, 2237. [CrossRef]
3. Zhou, L.; Guo, J.; Wen, X.; Ma, J.; Yang, F.; Wang, C.; Zhang, D. Monitoring and Analysis of Dynamic Characteristics of Super High-rise Buildings using GB-RAR: A Case Study of the WGC under Construction, China. *Appl. Sci.* **2020**, *10*, 808. [CrossRef]
4. Zinno, R.; Artese, S.; Clausi, G.; Magarò, F.; Meduri, S.; Miceli, A.; Venneri, A. *Structural Health Monitoring (SHM). Internet Things Smart Urban Ecosyst*; Springer: Cham, Switzerland, 2019; pp. 225–249.
5. Li, H.-N.; Ren, L.; Jia, Z.-G.; Yi, T.-H.; Li, D.-S. State-of-the-art in structural health monitoring of large and complex civil infrastructures. *J. Civ. Struct. Heal. Monit.* **2016**, *6*, 3–16. [CrossRef]

6. Kurka, P.R.G.; Cambraia, H.N. Application of a multivariable input–output subspace identification technique in structural analysis. *J. Sound Vib.* **2008**, *312*, 461–475. [CrossRef]
7. Rytter, A. Vibration Based Inspection of Civil Engineering Structures. Ph.D. Thesis, Aalborg Univ Denmark, Aalborg, Denmark, 1993.
8. Doebling, S.W.; Farrar, C.R.; Prime, M.B. A summary review of vibration-based damage identification methods. *Shock Vib. Dig.* **1998**, *30*, 91–105. [CrossRef]
9. Qiao, L.; Esmaeily, A. An Overview of Signal-Based Damage Detection Methods. *Appl. Mech. Mater.* **2011**, *94–96*, 834–851. [CrossRef]
10. Artese, S.; Zinno, R. TLS for Dynamic Measurement of the Elastic Line of Bridges. *Appl. Sci.* **2020**, *10*, 1182. [CrossRef]
11. Fan, W.; Qiao, P. Vibration-based damage identification methods: A review and comparative study. *Struct. Heal. Monit.* **2011**, *10*, 83–111. [CrossRef]
12. Ozer, E.; Feng, Q.M. Structural Reliability Estimation with Participatory Sensing and Mobile Cyber-Physical Structural Health Monitoring Systems. *Appl. Sci.* **2019**, *9*, 2840. [CrossRef]
13. Ljung, L. *System Identification*; Wiley Online Library: Hoboken, NJ, USA, 1999.
14. Wu, B.; Wu, G.; Yang, C.; He, Y. Damage identification method for continuous girder bridges based on spatially-distributed long-gauge strain sensing under moving loads. *Mech. Syst. Signal Process.* **2018**, *104*, 415–435. [CrossRef]
15. Chen, C.-C.; Wu, W.-H.; Liu, C.-Y.; Lai, G. Damage detection of a cable-stayed bridge based on the variation of stay cable forces eliminating environmental temperature effects. *Smart Struct. Syst.* **2016**, *17*, 859–880. [CrossRef]
16. Yu, Z.; Xia, H.; Goicolea, J.M.; Xia, C. Bridge damage identification from moving load induced deflection based on wavelet transform and Lipschitz exponent. *Int. J. Struct. Stab. Dyn.* **2016**, *16*, 1550003. [CrossRef]
17. Zhu, Y.; Ni, Y.-Q.; Jin, H.; Inaudi, D.; Laory, I. A temperature-driven MPCA method for structural anomaly detection. *Eng. Struct.* **2019**, *190*, 447–458. [CrossRef]
18. Cardini, A.J.; Dewolf, J.T. Implementation of a long-term bridge weigh-in-motion system for a steel girder bridge in the interstate highway system. *J. Bridge Eng.* **2009**, *14*, 418–423. [CrossRef]
19. Cantero, D.; Karoumi, R.; González, A. The Virtual Axle concept for detection of localised damage using Bridge Weigh-in-Motion data. *Eng. Struct.* **2015**, *89*, 26–36. [CrossRef]
20. Gonzalez, I.; Karoumi, R. BWIM aided damage detection in bridges using machine learning. *J. Civ. Struct. Heal. Monit.* **2015**, *5*, 715–725. [CrossRef]
21. Kalyankar, R.; Uddin, N. Axle detection on prestressed concrete bridge using bridge weigh-in-motion system. *J. Civ. Struct. Heal. Monit.* **2017**, *7*, 191–205. [CrossRef]
22. Xu, X.; Ren, Y.; Huang, Q.; Fan, Z.Y.; Tong, Z.J.; Chang, W.J.; Liu, B. Anomaly detection for large span bridges during operational phase using structural health monitoring data. *Smart Mater. Struct.* **2020**, *29*, 45029. [CrossRef]
23. Kroll, A.; Schulte, H. Benchmark problems for nonlinear system identification and control using soft computing methods: Need and overview. *Appl. Soft Comput.* **2014**, *25*, 496–513. [CrossRef]
24. Staszewski, W.J. Identification of non-linear systems using multi-scale ridges and skeletons of the wavelet transform. *J. Sound Vib.* **1998**, *214*, 639–658. [CrossRef]
25. Qiao, L. Structural Damage Detection Using Signal-Based Pattern Recognition. Ph.D. Thesis, Kansas State University, Manhattan, KS, USA, 2009.
26. Karbhari, V.M.; Guan, H. Sikorsky, C. Operational Modal Analysis for Vibration-based Structural Health Monitoring of Civil Structures. In *Structural Health Monitoring of Civil Infrastructure Systems*; Woodhead Publishing: Sawston/Cambridge, UK, 2009; pp. 213–259.
27. Runtemund, K.; Cottone, G.; Müller, G. Treatment of arbitrarily autocorrelated load functions in the scope of parameter identification. *Comput. Struct.* **2013**, *126*, 29–40. [CrossRef]
28. Pavlov, G.K. Design of Health Monitoring System to Detect Tower Oscilations. Master's Thesis, Technical University of Denmark, Lyngby, Denmark, 2008.
29. Shokravi, H.; Shokravi, H.; Bakhary, N.; Koloor, S.S.R.; Petru, M. Application of the Subspace-based Methods in Health Monitoring of the Civil Structures: A Systematic Review and Meta-analysis. *Appl. Sci.* **2020**. (under review).

30. Sohn, H.; Farrar, C.R.; Hemez, F.M.; Shunk, D.D.; Stinemates, D.W.; Nadler, B.R.; Czarnecki, J.J. *A Review of Structural Health Monitoring Literature: 1996–2001*; Los Alamos National Laboratory: Los Alamos, NM, USA, 2003.
31. Shokravi, H.; Shokravi, H.; Bakhary, N.; Koloor, S.S.R.; Petru, M. A Comparative Study of the Data-driven Stochastic Subspace Methods for Health Monitoring of Structures: A Bridge Case Study. *Appl. Sci.* **2020**. (under review).
32. Senthilnathan, K.; Hiremath, C.P.; Naik, N.K.; Guha, A.; Tewari, A. Microstructural damage dependent stiffness prediction of unidirectional CFRP composite under cyclic loading. *Compos. Part A Appl. Sci. Manuf.* **2017**, *100*, 118–127. [CrossRef]
33. Yu, Y.; Zou, Y.; Al Hosani, M.; Tian, G. Conductivity Invariance Phenomenon of Eddy Current NDT: Investigation, Verification, and Application. *IEEE Trans. Magn.* **2017**, *53*, 1–7. [CrossRef]
34. Alwash, M.; Breysse, D.; Sbartaï, Z.M.; Szilágyi, K.; Borosnyói, A. Factors affecting the reliability of assessing the concrete strength by rebound hammer and cores. *Constr. Build. Mater.* **2017**, *140*, 354–363. [CrossRef]
35. Epp, T.; Cha, Y.-J. Wavelet Transform-Based Damage Detection in Reinforced Concrete Using an Air-Coupled Impact-Echo Method. In *Structural Health Monitoring & Damage Detection*; Springer: Cham, Switzerland, 2017; Volume 7, pp. 23–25.
36. Davis, A.G.; Ansari, F.; Gaynor, R.D.; Lozen, K.M.; Rowe, T.J.; Caratin, H.; Heidbrink, F.D.; Malhotra, V.M.; Simons, B.P.; Carino, N.J.; et al. *Nondestructive Test Methods for Evaluation of Concrete in Structures*; American Concrete Institute, ACI: Farmington Hills, MI, USA, 1998.
37. Kaiser, H.; Karbhari, V.M. Non-destructive testing techniques for FRP rehabilitated concrete. I: A critical review. *Int. J. Mater. Prod. Technol.* **2004**, *21*, 349–384. [CrossRef]
38. Ettouney, M.M.; Alampalli, S. *Infrastructure Health in Civil Engineering: Theory and Components*; CRC Press: Boca Raton, FL, USA, 2016; Volume 1.
39. Terlaje, A.; Gould, P.; Dyke, S. An Algorithm and Methodology for Static Response Based Damage Detection in Structural Systems. *IEEE Trans. Signal Process.* **1994**, *8*, 2146–2157.
40. Eun, H.C.; Park, S.Y.; Lee, M.S. Static-Based Damage Detection Using Measured Strain and Deflection Data. In *Applied Mechanics and Materials*; Trans Tech Publications Ltd.: Stafa-Zurich, Switzerland, 2013; Volume 256, pp. 1097–1100.
41. Gul, M.; Catbas, F.N. Structural health monitoring and damage assessment using a novel time series analysis methodology with sensor clustering. *J. Sound Vib.* **2011**, *330*, 1196–1210. [CrossRef]
42. Song, G.; Wang, C.; Wang, B. Structural Health Monitoring (SHM) of Civil Structures. *Appl. Sci.* **2017**, *7*, 789. [CrossRef]
43. Reynders, E. System identification methods for (operational) modal analysis: Review and comparison. *Arch. Comput. Methods Eng.* **2012**, *19*, 51–124. [CrossRef]
44. Ingle, V.; Proakis, J. *Digital Signal Processing Using MATLAB*; Cengage Learning: Boston, MA, USA, 2011.
45. Bendat, J.S.; Piersol, A.G. *Engineering Applications of Correlation and Spectral Analysis*; Wiley-Interscience: New York, NY, USA, 1980.
46. Brincker, R.; Zhang, L.; Andersen, P. Modal identification of output-only systems using frequency domain decomposition. *Smart Mater. Struct.* **2001**, *10*, 441. [CrossRef]
47. Pioldi, F.; Rizzi, E. A refined Frequency Domain Decomposition tool for structural modal monitoring in earthquake engineering. *Earthq. Eng. Eng. Vib.* **2017**, *16*, 627–648. [CrossRef]
48. Khalilinia, H.; Zhang, L.; Venkatasubramanian, V. Fast frequency-domain decomposition for ambient oscillation monitoring. *IEEE Trans. Power Deliv.* **2015**, *30*, 1631–1633. [CrossRef]
49. Allemang, R.J.; Brown, D.L. A complete review of the complex mode indicator function (CMIF) with applications. In Proceedings of the ISMA International Conference on Noise and Vibration Engineering, Katholieke Universiteit Leuven, Leuven, Belgium, 18–20 September 2006; Volume 38, pp. 36–44.
50. Guillaume, P.; Verboven, P.; Vanlanduit, S. Frequency-domain maximum likelihood identification of modal parameters with confidence intervals. In Proceedings of the International Seminar on Modal Analysis, Katholieke Universiteit Leuven, Leuven, Belgium, 16 September 1998; Volume 1, pp. 359–366.
51. Peeters, B.; Van der Auweraer, H.; Guillaume, P.; Leuridan, J. The PolyMAX frequency-domain method: A new standard for modal parameter estimation? *Shock Vib.* **2004**, *11*, 395–409. [CrossRef]

52. El-Kafafy, M.; Guillaume, P.; Peeters, B.; Marra, F.; Coppotelli, G. Advanced frequency-domain modal analysis for dealing with measurement noise and parameter uncertainty. In *Topics in Modal Analysis I*; Springer: New York, NY, USA, 2012; Volume 5, pp. 179–199.
53. Kim, J.; Lynch, J.P. Comparison study of output-only subspace and frequency-domain methods for system identification of base excited civil engineering structures. In *Civil Engineering Topics*; Springer: New York, NY, USA, 2011; Volume 4, pp. 305–312.
54. Liu, C.W.; Wu, J.Z.; Zhang, Y.G. Review and prospect on modal parameter identification of spatial lattice structure based on ambient excitation. In *Applied Mechanics and Materials*; Trans Tech Publications Ltd.: Stafa-Zurich, Switzerland, 2011; Volume 94, pp. 1271–1277.
55. Magalhães, F.; Caetano, E.; Cunha, Á. Challenges in the application of stochastic modal identification methods to a cable-stayed bridge. *J. Bridge Eng.* **2007**, *12*, 746–754. [CrossRef]
56. Redif, S.; Weiss, S.; McWhirter, J.G. Relevance of polynomial matrix decompositions to broadband blind signal separation. *Signal Process.* **2017**, *134*, 76–86. [CrossRef]
57. Cunha, A.; Caetano, E.; Magalhaes, F.; Moutinho, C. Recent perspectives in dynamic testing and monitoring of bridges. *Struct. Control Heal. Monit.* **2013**, *20*, 853–877. [CrossRef]
58. James, G.H.; Carne, T.G.; Lauffer, J.P. *The Natural Excitation Technique (NExT) for Modal Parameter Extraction from Operating Wind Turbines*; Sandia National Labs: Albuquerque, NM, USA, 1993.
59. Brown, D.L.; Allemang, R.J.; Zimmerman, R.; Mergeay, M. Parameter estimation techniques for modal analysis. SAE Technical paper. *SAE Trans.* **1979**, *13*, 176–186.
60. Vold, H.; Kundrat, J.; Rocklin, G.T.; Russell, R. *A Multi-Input Modal Estimation Algorithm for Mini-Computers*; SAE Technical Paper; SAE: Warrendale, PA, USA, 1982.
61. Juang, J.-N.; Pappa, R.S. An eigensystem realization algorithm for modal parameter identification and model reduction. *J. Guid. Control. Dyn.* **1985**, *8*, 620–627. [CrossRef]
62. Juang, J.-N.; Cooper, J.E.; Wright, J.R. An eigensystem realization algorithm using data correlations (ERA/DC) for modal parameter identification. *Control Adv. Technol.* **1988**, *4*, 5–14.
63. Zhang, L.; Yao, Y.; Lu, M. An improved time domain polyreference method for modal identification. *Mech. Syst. Signal Process.* **1987**, *1*, 399–413. [CrossRef]
64. Zhang, L. An overview of major developments and issues in modal identification. In Proceedings of the 22nd International Modal Analysis Conference (IMAC), Detroit, MI, USA, 12 August 2004; pp. 1–8.
65. Fukuzono, K. Investigation of Multiple-Reference Ibrahim Time Domain Modal Parameter Estimation Technique. Ph. D. Thesis, University of Cincinnati, Cincinnati, OH, USA, 1986.
66. Juang, J.-N. *Applied System Identification*; Prentice Hall: Englewood Cliffs, NJ, USA, 1994.
67. Chang, M.; Pakzad, S.N. Observer Kalman Filter Identification for Output-Only Systems Using Interactive Structural Modal Identification Toolsuite. *J. Bridge Eng.* **2013**, *19*, 4014002. [CrossRef]
68. Vicario, F.; Phan, M.Q.; Betti, R.; Longman, R.W. Output-only observer/Kalman filter identification (O3KID). *Struct. Control Heal. Monit.* **2014**, *22*, 847–872. [CrossRef]
69. Cole, Jr. *On-line Failure Detection and Damping Measurement of Aerospace Structures by Random Decrement Signatures*; NASA: Washington, DC, USA, 1973; Volume 37.
70. Brincker, R.; Rodrigues, J.; Brincker, R. Application of the random decrement technique in operational modal analysis. In Proceedings of the 1st International Operational Modal Analysis Conference (IOMAC), Aalborg Universitet, Aalborg, Denmark, 26–27 April 2005.
71. Ljung, L. *System Identification: Theory for the User, PTR Prentice Hall Information and System Sciences Series*; Prentice Hall: Upper Saddle River, NJ, USA, 1999.
72. Sohn, H.; Czarnecki, J.A.; Farrar, C.R. Structural health monitoring using statistical process control. *J. Struct. Eng.* **2000**, *126*, 1356–1363. [CrossRef]
73. Sohn, H.; Farrar, C.R. Damage diagnosis using time series analysis of vibration signals. *Smart Mater. Struct.* **2011**, *10*, 446–451. [CrossRef]
74. Bodeux, J.B.; Golinval, J.C. Application of ARMAV models to the identification and damage detection of mechanical and civil engineering structures. *Smart Mater. Struct.* **2001**, *10*, 479–489. [CrossRef]
75. Fasel, T.R.; Gregg, S.W.; Johnson, T.J.; Farrar, C.R.; Sohn, H. Experimental modal analysis and damage detection in a simulated three story building. In Proceedings of the 20th International Modal Analysis Conference, Los Angeles, CA, USA, 4–7 February 2002; pp. 122–135.

76. Huang, C.S. Structural identification from ambient vibration measurement using the multivariate AR model. *J. Sound Vib.* **2001**, *241*, 337–359. [CrossRef]
77. Hung, C.F.; Ko, W.J. Identification of modal parameters from measured output data using vector backward autoregressive model. *J. Sound Vib.* **2002**, *256*, 249–270. [CrossRef]
78. Bodeux, J.B.; Golinval, J.C. Modal identification and damage detection using the data-driven stochastic subspace and ARMAV methods. *Mech. Syst. Signal Process.* **2003**, *17*, 83–89. [CrossRef]
79. Bertha, M.; Golinval, J.-C. Identification of non-stationary dynamical systems using multivariate ARMA models. *Mech. Syst. Signal Process.* **2017**, *88*, 166–179. [CrossRef]
80. Nair, K.K.; Kiremidjian, A.S.; Law, K.H. Time series-based damage detection and localization algorithm with application to the ASCE benchmark structure. *J. Sound Vib.* **2006**, *291*, 349–368. [CrossRef]
81. Carden, E.P.; Brownjohn, J.M.W. ARMA modelled time-series classification for structural health monitoring of civil infrastructure. *Mech. Syst. Signal Process.* **2008**, *22*, 295–314. [CrossRef]
82. Nair, K.K.; Kiremidjian, A.S. Time series based structural damage detection algorithm using Gaussian mixtures modeling. *J. Dyn. Syst. Meas. Control* **2007**, *129*, 285. [CrossRef]
83. Omenzetter, P.; Brownjohn, J. Application of time series analysis for bridge monitoring. *Smart Mater. Struct.* **2006**, *15*, 129–138. [CrossRef]
84. Spiridonakos, M.D.; Poulimenos, A.G.; Fassois, S.D. Output-only identification and dynamic analysis of time-varying mechanical structures under random excitation: A comparative assessment of parametric methods. *J. Sound Vib.* **2010**, *329*, 768–785. [CrossRef]
85. Gil, P.; Santos, F.; Palma, L.; Cardoso, A. Recursive subspace system identification for parametric fault detection in nonlinear systems. *Appl. Soft Comput.* **2015**, *37*, 444–455. [CrossRef]
86. Brincker, R.; Andersen, P. Understanding stochastic subspace identification. In Proceedings of the 24th IMAC, St. Louis, MO, USA, 30 January–2 February 2006; p. 126.
87. Gomez, H.C. System Identification of Highway Bridges using Long-Term Vibration Monitoring Data. Ph.D. Thesis, University of California, Irvine, CA, USA, 2012.
88. Larimore, W.E. Canonical variate analysis in identification, filtering, and adaptive control. In Proceedings of the 29th IEEE Conference on Decision and Control, Honolulu, HI, USA, 5–7 December 1990; pp. 596–604.
89. Akaike, H. Markovian representation of stochastic processes by canonical variables. *SIAM J. Control* **1975**, *13*, 162–173. [CrossRef]
90. Verhaegen, M. Identification of the deterministic part of MIMO state space models given in innovations form from input-output data. *Automatica* **1994**, *30*, 61–74. [CrossRef]
91. Van Overschee, P.; De Moor, B.L. *Subspace Identification for Linear Systems: Theory-Implementation-Applications*; Springer Science & Business Media: Berlin, Germany, 2012.
92. Shokravi, H.; Bakhary, N.H. Comparative analysis of different weight matrices in subspace system identification for structural health monitoring. *IOP Conf. Ser. Mater. Sci. Eng.* **2017**, *271*, 12092. [CrossRef]
93. Van Overschee, P.; De Moor, B.; Dehandschutter, W.; Swevers, J. A subspace algorithm for the identification of discrete time frequency domain power spectra. *Automatica* **1997**, *33*, 2147–2157. [CrossRef]
94. Basseville, M.; Abdelghani, M.; Benveniste, A. Subspace-based fault detection algorithms for vibration monitoring. *Automatica* **2000**, *36*, 101–109. [CrossRef]
95. Yan, A.-M.M.; De Boe, P.; Golinval, J.-C.C. Structural damage diagnosis by Kalman model based on stochastic subspace identification. *Struct. Heal. Monit.* **2004**, *3*, 103–119. [CrossRef]
96. Yan, A.-M.; Golinval, J.-C. Null subspace-based damage detection of structures using vibration measurements. *Mech. Syst. Signal Process.* **2006**, *20*, 611–626. [CrossRef]
97. Reynders, E.; De Roeck, G. Reference-based combined deterministic–stochastic subspace identification for experimental and operational modal analysis. *Mech. Syst. Signal Process.* **2008**, *22*, 617–637. [CrossRef]
98. Urgessa, G.S. Vibration properties of beams using frequency-domain system identification methods. *J. Vib. Control* **2010**, *17*, 1287–1294. [CrossRef]
99. Nguyen, V.H.; Golinval, J.-C. Fault detection based on kernel principal component analysis. *Eng. Struct.* **2010**, *32*, 3683–3691. [CrossRef]
100. Goursat, M.; Döhler, M.; Mevel, L.; Andersen, P. Crystal clear SSI for operational modal analysis of aerospace vehicles. In *Structural Dynamics*; Springer: New York, NY, USA, 2011; Volume 3, pp. 1421–1430.

101. Loh, C.H.; Weng, J.H.; Liu, Y.C.; Lin, P.Y.; Huang, S.K. Structural damage diagnosis based on on-line recursive stochastic subspace identification. *Smart Mater. Struct.* **2011**, *20*, 34–55. [CrossRef]
102. Zhang, G.; Tang, B.; Tang, G. An improved stochastic subspace identification for operational modal analysis. *Measurement* **2012**, *45*, 1246–1256. [CrossRef]
103. Chao, S.-H.; Loh, C.-H.; Weng, J.-H. Application of higher order SVD to vibration-based system identification and damage detection. In *Sensors and Smart Structures Technologies for Civil, Mechanical, and Aerospace Systems*; International Society for Optics and Photonics: Bellingham, WA, USA, 2012; Volume 8345, p. 834525. [CrossRef]
104. Döhler, M.; Andersen, P.; Mevel, L. Operational modal analysis using a fast stochastic subspace identification method. In *Topics in Modal Analysis I*; Springer: New York, NY, USA, 2012; Volume 5, pp. 19–24.
105. Hong, A.L.; Ubertini, F.; Betti, R. New Stochastic Subspace Approach for System Identification and Its Application to Long-Span Bridges. *J. Eng. Mech.* **2013**, *139*, 724–736. [CrossRef]
106. Basseville, M.; Mevel, L.; Goursat, M. Statistical model-based damage detection and localization: Subspace-based residuals and damage-to-noise sensitivity ratios. *J. Sound Vib.* **2004**, *275*, 769–794. [CrossRef]
107. Yu, D.-J.; Ren, W.-X. EMD-based stochastic subspace identification of structures from operational vibration measurements. *Eng. Struct.* **2005**, *27*, 1741–1751. [CrossRef]
108. Zhang, Z.; Fan, J.; Hua, H. Simulation and experiment of a blind subspace identification method. *J. Sound Vib.* **2008**, *311*, 941–952. [CrossRef]
109. Carden, E.P.; Brownjohn, J.M.W. Fuzzy Clustering of Stability Diagrams for Vibration-Based Structural Health Monitoring. *Comput. Civ. Infrastruct. Eng.* **2008**, *23*, 360–372. [CrossRef]
110. Balmès, E.; Basseville, M.; Bourquin, F.; Mevel, L.; Nasser, H.; Treyssède, F. Merging sensor data from multiple temperature scenarios for vibration monitoring of civil structures. *Struct. Heal. Monit.* **2008**, *7*, 129–142. [CrossRef]
111. Basseville, M.; Bourquin, F.; Mevel, L.; Nasser, H.; Treyssède, F. Handling the temperature effect in vibration monitoring: Two subspace-based analytical approaches. *J. Eng. Mech.* **2010**, *136*, 367. [CrossRef]
112. Ren, W.X.; Lin, Y.Q.; Fang, S.E. Structural damage detection based on stochastic subspace identification and statistical pattern recognition: I. Theory. *Smart Mater. Struct.* **2011**, *20*, 115009. [CrossRef]
113. Lin, Y.Q.; Ren, W.X.; Fang, S.E. Structural damage detection based on stochastic subspace identification and statistical pattern recognition: II. Experimental validation under varying temperature. *Smart Mater. Struct.* **2011**, *20*, 115010. [CrossRef]
114. Li, Z.; Chang, C.C. Tracking of structural dynamic characteristics using recursive stochastic subspace identification and instrumental variable technique. *J. Eng. Mech.* **2011**, *138*, 591–600. [CrossRef]
115. Loendersloot, R.; Schiphorst, F.B.A.; Basten, T.G.H.; Tinga, T. *Application of SHM Using an Autonomous Sensor Network*; DEStech Publications, Inc: Lancaster, PA, USA, 2013.
116. Miguel, L.F.F.; Lopez, R.H.; Miguel, L.F.F. A hybrid approach for damage detection of structures under operational conditions. *J. Sound Vib.* **2013**, *332*, 4241–4260. [CrossRef]
117. Chao, S.-H.; Loh, C.-H.; Tseng, M.-H. Structural damage assessment using output-only measurement: Localization and quantification. *J. Intell. Mater. Syst. Struct.* **2013**, *25*, 1097–1106. [CrossRef]
118. Reynders, E.; Maes, K.; Lombaert, G.; De Roeck, G. Uncertainty quantification in operational modal analysis with stochastic subspace identification: Validation and applications. *Mech. Syst. Signal Process.* **2016**, *66*, 13–30. [CrossRef]
119. Ozcelik, O.; Misir, I.S.; Amaddeo, C.; Yucel, U.; Durmazgezer, E. Modal Identification Results of Quasi-statically Tested RC Frames at Different Damage Levels. In *Topics in Modal Analysis*; Springer: Cham, Switzerland, 2015; Volume 10, pp. 215–226.
120. Magalhães, F.; Reynders, E.; Cunha, Á.; De Roeck, G. Online automatic identification of modal parameters of a bridge using the p-LSCF method. In Proceedings of the IOMAC, Ancona, Italy, 4–6 May 2009.
121. Wang, S.Q.; Zhang, Y.T.; Feng, Y.X. Comparative study of output-based modal identification methods using measured signals from an offshore platform. In *ASME 2010 29th International Conference on Ocean, Offshore and Arctic Engineering*; American Society of Mechanical Engineers: New York, NY, USA, 2010; pp. 561–567.
122. Moaveni, B.; He, X.; Conte, J.P.; Restrepo, J.I.; Panagiotou, M. System identification study of a 7-story full-scale building slice tested on the UCSD-NEES shake table. *J. Struct. Eng.* **2010**, *137*, 705–717. [CrossRef]
123. Cunha, A.; Caetano, E.; Ribeiro, P.; Müller, G. Vibration-based SHM of a centenary bridge: A comparative study between two different automated OMA techniques. *Preservation* **2011**, *1*, 12.

124. Ceravolo, R.; Abbiati, G. Time domain identification of structures: Comparative analysis of output-only methods. *J. Eng. Mech.* **2012**, *139*, 537–544. [CrossRef]
125. Sohn, H. A Bayesian Probabilistic Approach to Damage Detection for Civil Structures. *Dep. Civ. Environ. Eng.* **1998**.
126. Peeters, B.; De Roeck, G. Reference-based stochastic subspace identification for output-only modal analysis. *Mech. Syst. Signal Process.* **1999**, *13*, 855–878. [CrossRef]
127. Pridham, B.A.; Wilson, J.C. A study of damping errors in correlation-driven stochastic realizations using short data sets. *Probabilistic Eng. Mech.* **2003**, *18*, 61–77. [CrossRef]
128. Pan, Q. System identification of constructed civil engineering structures and uncertainty. Ph.D. Thesis, Drexel University, Philadelphia, PA, USA, 2007.
129. Benveniste, A.; Mevel, L. Nonstationary consistency of subspace methods. *Autom. Control IEEE Trans.* **2007**, *52*, 974–984. [CrossRef]
130. Reynders, E.; Pintelon, R.; De Roeck, G. Uncertainty bounds on modal parameters obtained from stochastic subspace identification. *Mech. Syst. Signal Process.* **2008**, *22*, 948–969. [CrossRef]
131. Brasiliano, A.; Doz, G.; Brito, J.L.; Pimentel, R. Role of non-metallic components on the dynamic behavior of composite footbridges. In Proceedings of the Third International Conference–Footbridges, Porto, Portugal, 2–4 July 2008; pp. 501–522.
132. Alıcıoğlu, B.; Luş, H. Ambient vibration analysis with subspace methods and automated mode selection: Case studies. *J. Struct. Eng.* **2008**, *134*, 1016–1029. [CrossRef]
133. Marchesiello, S.; Bedaoui, S.; Garibaldi, L.; Argoul, P. Time-dependent identification of a bridge-like structure with crossing loads. *Mech. Syst. Signal Process.* **2009**, *23*, 2019–2028. [CrossRef]
134. Balmès, É.; Basseville, M.; Mevel, L.; Nasser, H. Handling the temperature effect in vibration monitoring of civil structures: A combined subspace-based and nuisance rejection approach. *Control Eng. Pract.* **2009**, *17*, 80–87.
135. Wang, J.; Sano, A.; Chen, T.; Huang, B. Identification of Hammerstein systems without explicit parameterisation of non-linearity. *Int. J. Control* **2009**, *82*, 937–952. [CrossRef]
136. Magalhães, F.; Cunha, Á.; Caetano, E.; Brincker, R. Damping estimation using free decays and ambient vibration tests. *Mech. Syst. Signal Process.* **2010**, *24*, 1274–1290. [CrossRef]
137. Döhler, M.; Reynders, E.; Magalhaes, F.; Mevel, L.; De Roeck, G.; Cunha, A. Pre-and post-identification merging for multi-setup OMA with covariance-driven SSI. In *Dynamics of Bridges*; Springer: New York, NY, USA, 2011; Volume 5, pp. 57–70.
138. Carden, E.P.; Mita, A. Challenges in developing confidence intervals on modal parameters estimated for large civil infrastructure with stochastic subspace identification. *Struct. Control Heal. Monit.* **2011**, *18*, 53–78. [CrossRef]
139. Döhler, M.; Mevel, L. Modular subspace-based system identification from multi-setup measurements. *IEEE Trans. Automat. Contr.* **2012**, *57*, 2951–2956. [CrossRef]
140. Döhler, M.; Lam, X.-B.; Mevel, L. Uncertainty quantification for modal parameters from stochastic subspace identification on multi-setup measurements. *Mech. Syst. Signal Process.* **2013**, *36*, 562–581. [CrossRef]
141. Loh, C.H.; Chen, M.C. Modeling of environmental effects for vibration-based shm using recursive stochastic subspace identification analysis. In *Key Engineering Materials*; Trans Tech Publications Ltd.: Stafa-Zurich, Switzerland, 2013; Volume 558, pp. 52–64.
142. Brehm, M.; Zabel, V.; Bucher, C. Optimal reference sensor positions using output-only vibration test data. *Mech. Syst. Signal Process.* **2013**, *41*, 196–225. [CrossRef]
143. Cara, F.J.; Juan, J.; Alarcón, E.; Reynders, E.; De Roeck, G. Modal contribution and state space order selection in operational modal analysis. *Mech. Syst. Signal Process.* **2013**, *38*, 276–298. [CrossRef]
144. Ashari, A.E.; Mevel, L. Auxiliary input design for stochastic subspace-based structural damage detection. *Mech. Syst. Signal Process.* **2013**, *34*, 241–258. [CrossRef]
145. Tondreau, G.; Deraemaeker, A. Numerical and experimental analysis of uncertainty on modal parameters estimated with the stochastic subspace method. *J. Sound Vib.* **2014**, *333*, 4376–4401. [CrossRef]
146. Rainieri, C.; Fabbrocino, G. Influence of model order and number of block rows on accuracy and precision of modal parameter estimates in stochastic subspace identification. *Int. J. Lifecycle Perform Eng.* 10 **2014**, *1*, 317–334. [CrossRef]

147. Cho, S.; Park, J.-W.; Sim, S.-H. Decentralized system identification using stochastic subspace identification for wireless sensor networks. *Sensors* **2015**, *15*, 8131–8145. [CrossRef]
148. Markovsky, I. The most powerful unfalsified model for data with missing values. *Syst. Control Lett.* **2016**, *95*, 53–61. [CrossRef]
149. Banfi, L.; Carassale, L. Uncertainties in an Application of Operational Modal Analysis. In *Model Validation and Uncertainty Quantification*; Springer: Cham, Switzerland, 2016; Volume 3, pp. 107–115.
150. Spiridonakos, M.D.; Chatzi, E.N.; Sudret, B. Polynomial Chaos Expansion Models for the Monitoring of Structures under Operational Variability. *ASCE-ASME J. Risk Uncertain. Eng. Syst. Part A Civ. Eng.* **2016**, *2*, B4016003. [CrossRef]
151. Huynh, T.C.; Park, J.H.; Kim, J.T. Structural identification of cable-stayed bridge under back-to-back typhoons by wireless vibration monitoring. *Measurement* **2016**, *88*, 385–401. [CrossRef]
152. Li, D.; Ren, W.-X.; Hu, Y.-D.; Yang, D. Operational modal analysis of structures by stochastic subspace identification with a delay index. *Struct. Eng. Mech.* **2016**, *59*, 187–207. [CrossRef]
153. Pepe, M.; Costantino, D.; Restuccia Garofalo, A. An Efficient Pipeline to Obtain 3D Model for HBIM and Structural Analysis Purposes from 3D Point Clouds. *Appl. Sci.* **2020**, *10*, 1235. [CrossRef]
154. Kovačević, S.M.; Bačić, M.; Stipanović, I.; Gavin, K. Categorization of the Condition of Railway Embankments Using a Multi-Attribute Utility Theory. *Appl. Sci.* **2019**, *9*, 5089. [CrossRef]
155. Yang, K.; Ding, Y.; Sun, P.; Zhao, H.; Geng, F. Modeling of Temperature Time-Lag Effect for Concrete Box-Girder Bridges. *Appl. Sci.* **2019**, *9*, 3255. [CrossRef]
156. Vanlanduit, S.; Parloo, E.; Cauberghe, B.; Guillaume, P.; Verboven, P. A robust singular value decomposition for damage detection under changing operating conditions and structural uncertainties. *J. Sound Vib.* **2005**, *284*, 1033–1050. [CrossRef]
157. Naseralavi, S.S.; Salajegheh, E.; Fadaee, M.J.; Salajegheh, J. A novel sensitivity-based method for damage detection of structures under unknown periodic excitations. *J. Sound Vib.* **2014**, *333*, 2776–2803. [CrossRef]
158. Döhler, M.; Hille, F. Subspace-based damage detection on steel frame structure under changing excitation. In *Structural Health Monitoring*; Springer: Cham, Switzerland, 2014; Volume 5, pp. 167–174.
159. Banan, M.R.; Mehdi-Pour, Y. Detection and assessment of damage in 2D structures using measured modal response. *J. Sound Vib.* **2007**, *306*, 803–817. [CrossRef]
160. Siegert, D.; Döhler, M.; Mekki OBen Mevel, L.; Goursat, M.; Toutlemonde, F. Vibration monitoring of a small span composite bridge. In *Structural Dynamics*; Springer: New York, NY, USA, 2011; Volume 3, pp. 53–61.
161. Huang, M.C.; Wang, Y.P.; Chang, M.L. Damage Detection of Structures Identified with Deterministic-Stochastic Models Using Seismic Data. *Sci. World J.* **2014**, *2014*. [CrossRef]
162. Kim, J.; Lynch, J.P. Subspace system identification of support excited structures-part II: Gray-box interpretations and damage detection. *Earthq. Eng. Struct. Dyn.* **2012**, *41*, 2253–2271. [CrossRef]
163. Wang, Y.P.; Lin, Y.T.; Huang, G. Damage Localization of Output-Only Frame Systems Using Stochastic Subspace Identification. Adv. *Mater. Res.* **2012**, *3*, 1352–1359. [CrossRef]
164. Moaveni, B.; Stavridis, A.; Shing, P.B. System identification of a three-story infilled RC frame tested on the UCSD-NEES shake table. In *Dynamics of Civil Structures*; Springer: New York, NY, USA, 2011; Volume 4, pp. 135–143.
165. Weng, J.H.; Loh, C.H.; Yang, J.N. Experimental Study of Damage Detection by Data-Driven Subspace Identification and Finite-Element Model Updating. *J. Struct. Eng.* **2009**, *135*, 1533–1544. [CrossRef]
166. Karami, K.; Akbarabadi, S. Developing a Smart Structure Using Integrated Subspace-Based Damage Detection and Semi-Active Control. *Comput. Civ. Infrastruct. Eng.* **2016**, *31*, 887–903. [CrossRef]
167. Belleri, A.; Moaveni, B.; Restrepo, J.I. Damage assessment through structural identification of a three-story large-scale precast concrete structure. *Earthq. Eng. Struct. Dyn.* **2014**, *43*, 61–76. [CrossRef]
168. Shinagawa, Y.; Mita, A. Verification of structural health assessment method using full-scale collapse test of four-story steel building. In *Key Engineering Materials*; Trans Tech Publications Ltd.: Stäfa, Switzerland, 2013; pp. 174–183. [CrossRef]
169. Zhou, W.; Li, H.; Mevel, L.; Döhler, M.; Lam, X.B.; Mao, C.; Ou, J. Seismic Damage Assessment for a Residential Masonry Building Using Aftershock Monitoring of Wenchuan Earthquake. In Proceedings of the 24th International Conference on Noise and Vibration Engineering (ISMA2010), Leuven, Belgium, 20 September 2010; pp. 773–782.

170. Yoshimoto, R.; Mita, A.; Okada, K. Damage detection of base-isolated buildings using multi-input multi-output subspace identification. *Earthq. Eng. Struct. Dyn.* **2005**, *34*, 307–324. [CrossRef]
171. Allahdadian, S.; Ventura, C.E.; Andersen, P.; Mevel, L.; Dohler, M. Sensitivity Evaluation of Subspace-Based Damage Detection Method to Different Types of Damage. In *Structural Health Monitoring and Damage Detection*; Springer: Cham, Switzerland, 2015; pp. 11–18. [CrossRef]
172. Dohler, M.; Hille, F.; Mevel, L.; Rucker, W. Structural health monitoring with statistical methods during progressive damage test of S101 Bridge. *Eng. Struct.* **2014**, *69*, 183–193. [CrossRef]
173. Deraemaeker, A.; Reynders, E.; De Roeck, G.; Kullaa, J. Vibration-based structural health monitoring using output-only measurements under changing environment. *Mech. Syst. Signal Process.* **2008**, *22*, 34–56. [CrossRef]
174. Hu, W.H.; Moutinho, C.; Caetano, E.; Magalhaes, F.; Cunha, A. Continuous dynamic monitoring of a lively footbridge for serviceability assessment and damage detection. *Mech. Syst. Signal Process.* **2012**, *33*, 38–55. [CrossRef]
175. Loh, C.-H.H.; Chao, S.-H.H. Centralized vs. Pattern-level Feature Extraction for Structural Damage Detection. *Theor. Appl. Mech.* **2014**, *79*, 479–489. [CrossRef]
176. Mevel, L.; Goursat, M. Stochastic subspace-based structural identification and damage detection and localisation - Application to the Z24 bridge benchmark. *Mech. Syst. Signal Process.* **2003**, *17*, 143–151. [CrossRef]
177. Kullaa, J. Damage detection of the Z24 bridge using control charts. *Mech. Syst. Signal Process.* **2003**, *17*, 163–170. [CrossRef]
178. Reynders, E.; Wursten, G.; De Roeck, G. Output-only structural health monitoring in changing environmental conditions by means of nonlinear system identification. *Struct. Heal. Monit. Int. J.* **2014**, *13*, 82–93. [CrossRef]
179. Nguyen, V.H.; Mahowald, J.; Maas, S.; Golinval, J.-C. Use of time-and frequency-domain approaches for damage detection in civil engineering structures. *Shock Vib.* **2014**, *2014*. [CrossRef]
180. Cruz, P.J.S.; Salgado, R. Performance of Vibration-Based Damage Detection Methods in Bridges. *Comput. Civ. Infrastruct. Eng.* **2009**, *24*, 62–79. [CrossRef]
181. Nigro, M.B.; Pakzad, S.N.; Dorvash, S. Localized structural damage detection: A change point analysis. *Comput. Civ. Infrastruct. Eng.* **2014**, *29*, 416–432. [CrossRef]
182. Döhler, M.; Andersen, P.; Mevel, L. Variance computation of modal parameter estimates from UPC subspace identification. *Irnia* **2017**, *16*, 416–432.
183. Zhou, Y.; Prader, J.; Weidner, J.; Moon, F.; Aktan, A.E.; Zhang, J.; Yi, W.J. Structural Identification Study of a Steel Multi-Girder Bridge Based on Multiple Reference Impact Test. *Int. Symp. Innov. Sustain. Struct. Civ. Eng.* **2013**, *12*, 315–356.
184. Hu, W.-H.; Cunha, Á.; Caetano, E.; Magalhães, F.; Moutinho, C. LabVIEW toolkits for output-only modal identification and long-term dynamic structural monitoring. *Struct. Infrastruct. Eng.* **2010**, *6*, 557–574. [CrossRef]
185. Chang, M.; Pakzad, S.N.; Leonard, R. Modal identification using smit. In *Topics on the Dynamics of Civil Structures*; Springer: New York, NY, USA, 2012; Volume 1, pp. 221–228.
186. Goursat, M.; Mevel, L. COSMAD: Identification and diagnosis for mechanical structures with Scilab. In Proceedings of the 2008 IEEE International Conference on Computer-Aided Control Systems, San Antonio, TX, USA, 3–5 September 2008; pp. 353–358.
187. Hoofar, S.; Hooman, S.; Norhisham, B.; Heidarrezaei, M.; Koloor, S.S.R.; Petru, M. Vehicle-assisted techniques for health monitoring of bridges. *Sensors (Basel)* **2020**. (Under review).
188. Shokravi, H.; Shokravi, H.; Bakhary, N.; Heidarrezaei, M.; Koloor, S.R.K.; Petru, M. A review on vehicle classification methods and the potential of using smart-vehicle-assisted techniques. *Sensors (Basel)* **2020**. (under review).

© 2020 by the authors. Licensee MDPI, Basel, Switzerland. This article is an open access article distributed under the terms and conditions of the Creative Commons Attribution (CC BY) license (http://creativecommons.org/licenses/by/4.0/).

MDPI
St. Alban-Anlage 66
4052 Basel
Switzerland
Tel. +41 61 683 77 34
Fax +41 61 302 89 18
www.mdpi.com

Applied Sciences Editorial Office
E-mail: applsci@mdpi.com
www.mdpi.com/journal/applsci